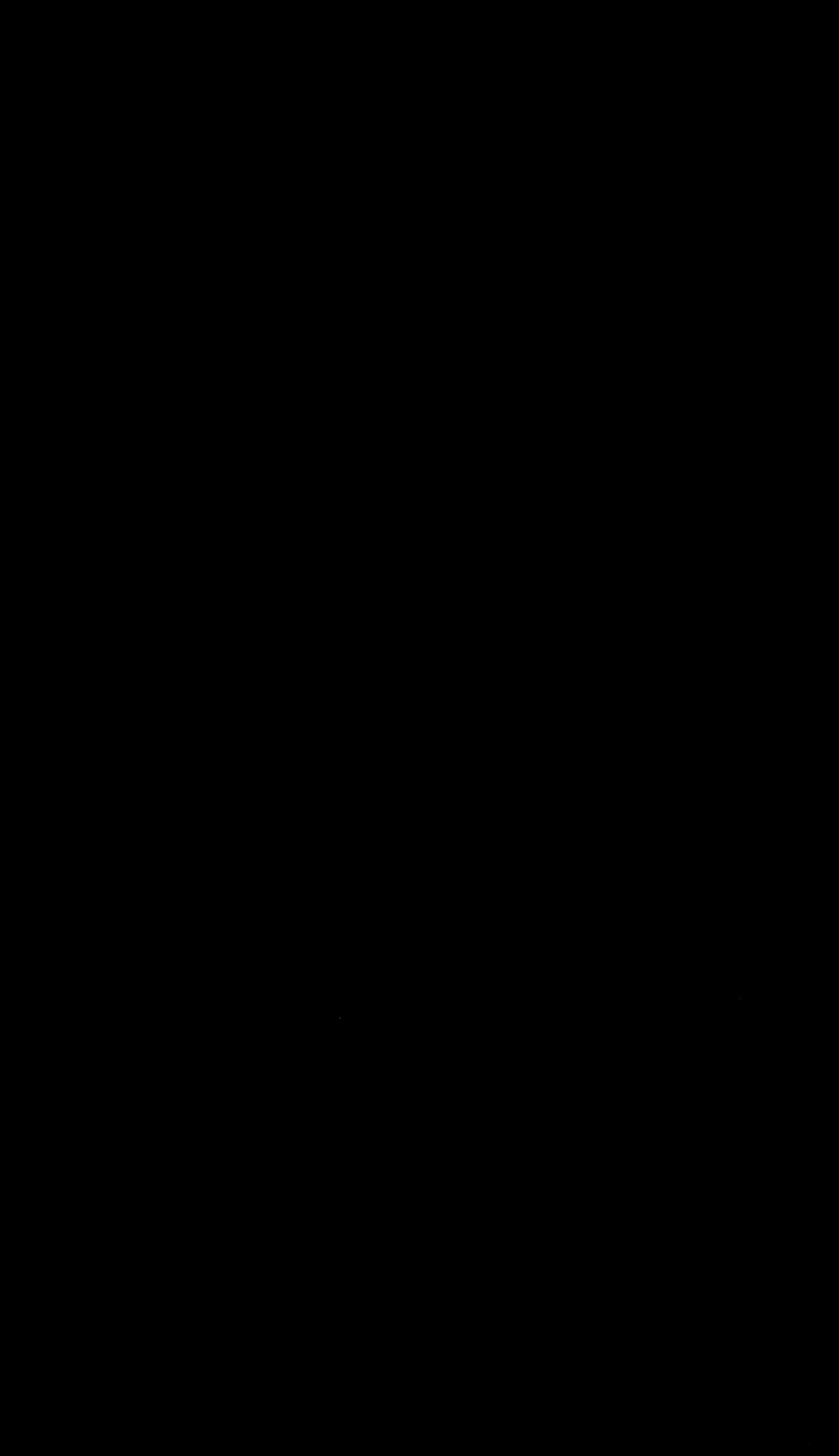

Antivivisection and

Richard D. French

Medical Science

Princeton University Press

in Victorian Society

Published by Princeton University Press, Princeton and London

Library of Congress Cataloging in Publication Data will
be found on the last printed page of this book

This book has been composed in Linotype Caledonia

Printed in the United States of America
by Princeton University Press, Princeton, New Jersey

Acknowledgments

Any historian is indebted to the work of other scholars. In my case, the inspiration and encouragement of Brian Harrison of Oxford, Roy MacLeod of Sussex, Lloyd Stevenson of Johns Hopkins, and Robert Young of Cambridge were indispensable, and I am deeply grateful. Others to whom I owe thanks include Dennis Chitty and William Gibson of the University of British Columbia, who first interested me in the history of science and medicine and who helped me to get an institutional foothold in the field; Alistair Crombie of Oxford, who, as my supervisor, has been extraordinarily indulgent of my intellectual enthusiasms; Charles Webster of Oxford, whose encouragement and criticism were most valuable on many occasions; Sir Edgar Williams, Warden of Rhodes House, Oxford, for his support and especially for his dazzling demonstration of the old boy network in action, in obtaining for me access to restricted government documents; Jeanne Peterson of Indiana University, whose knowledge of the London medical profession helped me greatly; and Dr. R. J. Lebowich, of New York, London, and Athens, for his kindness and unfailing support.

A number of my colleagues were good enough to permit me to read and quote from as yet unpublished studies, and I thank Gerald Geison ("Michael Foster and the Rise of the Cambridge School of Physiology, 1870–1900," Yale University Ph.D. thesis, 1970), Frank Turner ("Between Science and Religion: The Reaction to Scientific Naturalism in Late Victorian England," Yale University Ph.D. thesis, 1971), Jeanne Peterson ("Kinship, Status, and Social Mobility in the Mid-Victorian Medical Profession," University of California at Berkeley Ph.D. thesis, 1972), Paul Cranefield (*The Way In and the Way Out. François Magendie, Charles Bell and the Roots of the Spinal Nerves,* New York, 1974), Robert Young ("Natural Theology, Victorian Periodicals, and the Fragmentation of the Common Context," *Victorian Studies,* in press), and Brian Harrison ("Animals and the State in Nineteenth-Century England," *English Historical Review,* lxxxviii (1973), 786–820).

This book has benefited greatly from the criticism and suggestions of those who have read all or part of the manuscript upon which it is based, among whom are Sydney Eisen, Gerald Geison, Brian Harrison, Jack Lesch, Jerrold Seigel, Tom Settle, Lloyd Stevenson, James Turner, Steven Turner, Robert Young, and an anonymous referee for the Princeton University Press. For their valuable comments and their forebearance, I thank them. The disparity between their ideals and the reality of this volume is of course my own responsibility.

Finally, to my colleague at Princeton, Gerry Geison, I must extend special gratitude, not only for his intellectual aid, but most importantly, for his friendship and encouragement.

The staffs of many libraries have assisted in the research upon which this book is based. Eric Gaskell and Eric Freeman of the Wellcome Institute of the History of Medicine were particularly kind to a fledgling scholar who combined inexperience with impatience. Klaus Beltzner and Leo Fahey of the Science Council of Canada helped in the processing and presentation of the statistical data in the final chapter.

My research was carried out with the financial support of a Rhodes Scholarship, a Canada Council Pre-Doctoral Fellowship, and a Canada Council Social Sciences and Humanities Grant.

I am grateful to the following institutions for permission to quote from unpublished manuscript material held in their archives: the British Museum, the Imperial College of Science and Technology, University College London, the Royal Botanic Gardens, the Wellcome Institute of the History of Medicine, the Bodleian Library of Oxford University, the Woodward Library of the University of British Columbia, the New York Academy of Medicine, and the Henry Huntington Library of San Marino, California. Transcripts of Crown-copyright records in the Public Record Office appear by permission of the Controller of H.M. Stationery Office. I also wish to thank the following individuals for permission to quote from the letters of their ancestors: the Earl of Derby, the Earl of Carnarvon, Sir Philip Christison, Dr. Noble Frankland, and Sir Robert Foster. For their assistance and their permission to reproduce many of the illustrations included here, I am grateful to the Wellcome Institute of the History of Medicine, the British Museum Newspaper Library, and the Tate Gallery.

Contents

List of Figures and Tables

Abbreviations

A.A.M.R.	Association for the Advancement of Medicine by Research
B.M.A.	British Medical Association
B.M.J.	*British Medical Journal*
G.M.C.	General Medical Council
H.O.	Home Office
I.C.S.T.	Imperial College of Science and Technology
P.R.O.	Public Record Office
R.S.P.C.A.	Royal Society for the Prevention of Cruelty to Animals
S.P.C.A.	Society for the Prevention of Cruelty to Animals
Y.M.C.A.	Young Men's Christian Association

Antivivisection and Medical Science in Victorian Society

1. Introduction

For some few years now it has been evident that the democratic societies, and especially their younger members, have become increasingly ambivalent, suspicious, indeed outrightly hostile, toward science. Science has been portrayed as a Pandora's box from which all sorts of material blight have been thoughtlessly sprung; as an established institution so calcified and self-absorbed that it is incapable of responding effectively to urgent social needs; and as the epitome of a hyper-rational world view that suppresses such essential elements of the human experience as spontaneity, creativity, and mysticism. Indictments of science have spawned considerable hand-wringing among its proponents: articles are written, conferences are called, new ways of presenting science to a supposedly disillusioned public are sought.

The discussion, however, is not a debate, for none of the parties can enter sufficiently into the points of view of the others to controvert them. In fact, attempts to engage on the accuracy or justice of the attacks on science are searches for simple answers that do not exist. The issues are never as straightforward as they are perceived to be, and both polemic and carefully marshalled evidence miss the mark when one is faced with a collision of values in a milieu of rapid social and intellectual change.

Concurrently with swings in popular attitudes toward science have come problems of funding and increasing focus upon political representations by scientific communities. The transition from the sixties to the seventies has brought curtailment of hitherto rapidly expanding science budgets. As the exponential curves representing increases in research funding inevitably round off into logistic ones, and as governments move increasingly to support research on a mission-oriented basis, scientists agonize over what they view as their failure to communicate effectively with politicians and bureaucrats. Once again, the situation is hardly as black and white as it is sometimes portrayed, for the developments in question are less a sudden and inexplicable turnabout from previous practice than symbols of the ultimate obsolescence of the political styles and traditions that have served the scientific community for more than a century. The roots of that political style are deep, and attention to its obsolescence is, I would argue, at least as important a priority for the community as agonizing over the fickle tastes of undergraduates. There is a linkage between the two, to be sure. But it is the historical, sociological, and political analysis of the whole phenomenon that must be pursued.

This book is not an attempt to resolve any of the outstanding issues between science and its critics or between science and its patrons. It is, however, written on the premise that one way to approach these issues is to try to understand some of their previous manifestations. While scientists are clearly somewhat taken aback by what they view as a sudden fall from grace in both the popular mind and the political process, historians are well aware that science has survived and even thrived while the subject of cultural and political tensions. For various reasons—the reluctance of historians to deal with science, the interest of historians of science in the internal intellectual development of various disciplines—there have been relatively few systematic studies of scientists under cultural and political pressure. This is especially true for the last century and a half, during which science professionalized and expanded rapidly, such that it evolved into the corporate entity that we call the scientific community. The growing body of literature on scientific communities operating in hostile environments has, however, delineated one or two basic points, which this book attempts to exemplify and amplify. These basic points are worth some examination, because it seems they are often overlooked in contemporary discussion of popular antiscientism and of the political relations of the scientific community.

In the first place, this and other studies indicate that monocausal explanations for public attitudes toward science are inevitably inadequate. There are no simple explanations because the scientific community is interconnected in a multitude of ways to the society in which it exists; hence, popular attitudes and behavior toward it are embedded deeply in the social, intellectual, cultural, and political history of the period. Public opinion of science is always the product of a complex interaction of forces for which alleged causal factors, such as the atom bomb, the theory of evolution, sputnik, or the shortcomings of general scientific education, must be regarded as at best no more than convenient shorthands. One objective of this book, then, is to explore as thoroughly as possible the Victorian historical context within which the antivivisection movement developed its critique of medical science. The essential simplicity of the image of antivivisection—a group of animal lovers rising in defense of their pets—is seen to be no more than a single aspect of the movement. By analyzing the multiple influences that conjoined to produce the agitation that threatened experimental medicine, the intention here is to demonstrate the degree to which understanding of the phenomena in question may be impoverished by a superficial approach.

A second point that has emerged from recent studies speaks to the problems of political representation by the scientific community, to which attention has been drawn by recent reductions in funding. The point is that scientists and scientific communities have a characteristic style and strategy in their dealings with governments and public bodies. That style revolves around certain goals: the attempt to maximize the freedom, access to resources, and prestige of scientists, while using the claims of expertise to minimize the degree of accountability deemed to be due to the society that supplies the means to these goals. Such goals are scarcely novel; they are shared by all professions. In general, however, the scientific community has been uniquely successful in achieving its objectives, and it has done so largely through conciliatory, quietist, informal, and nonpublic political relations. Personal contact, minimally structured channels of communication, vague delineation of formal responsibility in the public eye, little or no debate in the press, emphasis on general utilitarian goals, coupled with insistence on autonomy for the community and the investigator, have typified the political style of science. This style owes its success to the very strong socialization involved in apprenticeship to the community and to the fact that the scientific community does not operate on a fee-for-service basis with the public, but rather serves the public on a basis of collective production of knowledge. Only the deepest and most divisive issues are likely to be argued before the lay public, and such cases represent a rare failure of the community to preserve a united, politically neutral, and noncontroversial front or to police the practices of its members. Thus governments and major public patrons, eager to reap what they believe to be the practical benefits of science, have ceded very substantial resources and autonomy to the community, and particularly to the leadership emerging from the political infrastructure of the community (its universities and scientific associations).

Present discussion of the political problems of science has been largely uninformed by knowledge of the realities of the formulation of public policy and of the historical relationship of scientists to that process. Thus it has not broached the possibility that the political style and strategy that have served scientists so well for more than one hundred years may, in an age of increasing public scrutiny and runaway costs of research, be obsolete. In any case, this book aims to put flesh and blood upon the abstraction of this point about the traditional political style of the scientific community, by exploring its success for the nascent profession of medical science in Victorian England. In the analysis of the emerging profession's debt to the power and political skills of its more robust elder brother, the medical profession, and of its successful defense of its prerogatives against the antivivisection movement and the legislation that the movement initiated, the political style of scientists is clearly outlined. Examining the development and operation of that style may instigate an examination of its suitability to the conditions of the present day.

The link between current attacks upon science and simultaneous limitations on growth of research funding undoubtedly exists, but it has yet to be analyzed and clarified. One way of viewing this relationship might be to see the frequent, and by now rather trite, pleas for scientists to exercise something called "social responsibility" as the popular public counterpart to the governmental demand for an increasingly rigorous kind of accountability from a scientific community absorbing a significant proportion of public money. In the case study to be presented here, the political problems of experimental medicine—the threat to its development and autonomy—were certainly a rather direct translation of public sentiment into legislation. It seems unlikely that significant infringement upon the perquisites of science by government could occur without the appropriate climate of public opinion, or that popular hostility would not be reflected in political and administrative decisions. This aspect of the contemporary scene must await some future historian; insofar as Victorian antivivisection is concerned, however, the clarity and directness of the relationship and the typicality of the scientists' political response represent a classic case for such phenomena.

That case ought to be a salutary reminder that the present tensions are scarcely as unprecedented as some believe. How far have we come from Frances Cobbe's charging science with "overestimate of Knowledge as compared to Love" to Theodore Roszak's indictment of the modern scientific world view? How far has the "social responsibility of science" debate progressed since Auberon Herbert and George Gore staked out the two extreme positions? To the *Times*, Herbert wrote of scientists:

> They have seen the moral chaos which in a great measure, their own work was producing, and old landmarks and rules of life disappearing. They have heard on all sides of them the cry of bewilderment from men and women, and they have scarcely lifted a finger to lighten the burden or given an hour's thought to the question how men should face the old fact called life in the new world which had sprung into existence. . . . with rare exceptions, science has practically said, "Adjust your burdens for yourselves; leave us in peace in our laboratories!" . . .[1]

Such sentiments left most of the scientific community unmoved. George Gore pointed out:

> It requires very little skill to ask complex questions; it is also far more easy to challenge experimentalists to prove their statements than to acquire the ability to understand their proofs. It is not a duty of scientific men to answer every question, because, however much may be explained, there always remain unexplored portions of science for anti-scientific persons to cavil about. Most scientific investigators consider that by making costly and difficult experiments and giving the results of them to the public, they have thereby more than fulfilled their duty in the matter.[2]

If, then, one can understand previous experience and behavior surrounding these expressions of opinion, one is perhaps closer to an accurate perception of the contemporary situation and of the essentially unresolved issues that underly it. What are the multifaceted social and cultural changes that interact to form lay attitudes toward science? How can the lay community, which funds science, insure that public values and priorities are scrupulously respected without so encumbering research as to vitiate its value? How can the scientific community respond to public aspirations and preoccupations while maintaining its own integrity? Is its past strategy in this regard still viable? The case study presented here is intended not so much to

[1] *Times* (London) 17 January 1876. See also S. H. Harris, *Auberon Herbert, Crusader for Liberty* (London, 1943), pp. 169–172.

[2] *The Utility and Morality of Vivisection* (London, 1884), p. 14.

answer these questions as to highlight their pertinence and complexity.

There is a second theme in this book. It speaks to another area of contemporary interest where the bulk of professional opinion and practice currently encounters significant criticism: highly technological, science-intensive, research-oriented, modern medicine. For a vocal group of commentators, some of them members of the profession, medicine has lost sight of prevention of disease and care of individual patients in the headlong pursuit of machine-based therapies, radical surgical procedures, and research strategies that suppress clinical objectives in favor of scientific ones. Huge expense and tremendous effort are, we are told, wasted in prolonging the lives of the very ill, while relatively little systematic attention is paid to the basic medical needs of large underprivileged sections of the populace. The race for professional prestige and the idealization of laboratory research have, it is said, distorted the goals of medicine until, despite the efforts of its practitioners, there is little of substance left to its humane ethic.

There are, of course, at least two strains of critical thought at work here, and it is necessary to emphasize the value of a historical perspective upon both. Physicians have always been liable to attack for quackery, ineffectuality, greed, and mutually protective irresponsibility. Such negative comment is difficult for any major social institution, such as a profession, to avoid. It is, in contrast, only within rather less than the last one hundred years that systematic research in science has become the underpinning of medical practice. While there have been minority countermovements within the profession, the public has been impressed by the triumphs of immunology, anesthetics, antibiotics, and the like into a clear endorsement of the invasion of medicine by scientific research and expensive technology. Recent lay criticism of the hitherto seemingly unquestioned assumptions of modern medicine is not, however, as novel as it may appear.

The original attempts to recast the foundations of medicine on the results of laboratory research began in earnest during the last few decades of the nineteenth century. These attempts involved two fundamental changes: the development of a new profession of practitioners of experimental medicine, men who were perhaps more natural scientists than healers per se, and the systematic adjustment of clinical procedures as specified by research results. Both changes met with violent opposition from within the medical profession proper and from segments of the public. The most effective and articulate popular resistance to such changes came from the antivivisection and antivaccination movements, and these movements were strongest in Victorian England. The movements also partook in healthy measure of the more general hostility toward the profession, which merely awaited an appropriate focus to elicit its expression.

The book that follows, then, explores public perceptions of a really crucial turning point in the history of medicine. There is insufficient space even to begin a discussion of the ramifications of the reconstitution of medicine upon the basis of science. Rather, what is attempted is a delineation of popular sentiment on this subject and of the various means employed to attempt to affect the attitudes and behavior of the profession. It is perhaps not unreasonable to suggest the relevance of these phenomena to the contemporary situation, in which the medical profession sees itself embattled as never before. For not only did the nineteenth-century debate contain the seeds of today's, it also involved systematic practical attempts to modify the activities of offending institutions and individuals within the medical profession.

The persisting appeal of the alternative future for medicine, which the antivivisectionists and others tried to conceptualize and promote, is quite astonishing. One should not fail to note, however, that the degree to which these alternative futures could be impressed upon or realized within the profession differs significantly between the latter part of the nineteenth century and the present time. The account that follows shows that the political power of the medical profession, in terms of public pressure, was enormous during the late Victorian period. (This power provided the base for the exercise of a very different kind of political strategy by the embryonic profession of medical scientists.) In fact, British medicine had just emerged from the intensive political activity of the mid-century period with substantial statutory autonomy, one of the classic hallmarks of a profession. Thus here, and on the whole elsewhere in the West, the introduction of science into medicine took place during the flush of professional independence and deference from the politicians. The initiation of this transition, where the new experimental approaches of physiology, pharmacology, and bacteriology were sanctified under the mantle of ultimate therapeutic utility, was by no means unopposed. It was, however, spearheaded by the medical scientists and, most importantly, by many of the members of the medical elite formally empowered by the recent developments in the machinery of professional self-government. In short, at a time when the medical leadership enjoyed unprecedented influence upon both the rank and file of the profession and upon the politicians, experimental medicine had the majority of these men as its advocates. What, in contrast, is the contemporary situation? The medical profession, although still very powerful, is nevertheless highly dependent upon, and interpenetrated by, the health bureaucracy. The rising costs and political significance of medical practice and research have forced medicine into the administrative apparatus of the welfare state. The consequent erosion of professional autonomy may loom very large indeed if the present critiques of medicine continue, for the health bureaucracy is politically accountable in the partisan arena in a way that the profession's own governing bodies have never been. It is the intention of this book not only to demonstrate the previous manifestations and flavor of critiques of certain prevailing assumptions and practices of modern medicine, but also, by examining the conditions under which such pressures were historically withstood, to inspire reflection about similar phenomena today.

The attempt to place Victorian antivivisection in its historical context has demanded the use of a variety of historiographic approaches. At various times, this book utilizes the literature and analytical tools of the history of science, political and administrative history, social history, cultural, intellectual, and religious history, psychohistory, and so forth. Inevitably, I have been far less proficient in any of these historiographies than I could wish. I hope that the reader will indulge me in these flaws, in the interest of a more comprehensive view of the subject.

There are, of course, limits to the substantive coverage of the book. The principal among these is simply that this is a history of antivivisection, not of vivisection. It is not to any extent directly concerned with the debates within the scientific and medical communities over the validity or ultimate medical utility of vivisection as a research method, although antivivisectionist exploitation of these debates is discussed. Nor is it a study of the development of technique in experimental medicine. Rather, it focuses upon public attitudes toward science and medicine, their background, and their consequences. The subject of the book, then, is the antivivisection movement in England from about 1870 to the mid-nineties. This movement was essentially the first of its kind, the intellectual and institutional godfather of counterpart agitations in other countries. It was, furthermore, the greatest threat to the actual existence of experimental medicine in any major country. The chronological purview of the study corresponds to the tenure of the first—and most intellectually and socially distinguished—generation of movement leadership. Twentieth century antivivisectionist literature shows a very heavy dependence upon arguments and appeals originally developed by the Victorians. Thus, a study of nineteenth-century English antivivisection is a study of the movement at its high point of cultural and political consequence.

A final word of guidance to the reader. I have been greatly struck by the consistency with which colleagues and reviewers who have read the thesis upon which this book is based find themselves preferring one or other of what they regard as two halves of the study. I am thus anxious to provide a basis for the reader to select, from what is a rather lengthy book, those parts that are most germane to his interests. Chapters 2 to 7 are a narrative and an analysis of the formal institutional consequences of the development of experimental medicine and of opposition to it by the antivivisection movement. Exclusive of the prehistory in Chapter 2, the chronological period covered intensively in these chapters is from about 1870 to the early 1880s. The approach here is primarily that of institutional, political, and administrative history, with a leavening of history of experimental medicine. This contrasts with Chapters 8 to 11, which are organized for a thematic treatment of antivivisection as a social, intellectual, religious, and cultural phenomenon, and which attempt to bring appropriate historiography to bear. The period covered is from the later seventies to the mid-nineties, but no narrative is involved. There are a number of reasons for the book being organized in this way. The principal one is that the institutional consequences of antivivisection developed largely from the events of 1874 to 1876, while the flavor and texture of the movement cannot be portrayed clearly without reference to material it produced over a much longer period of time, beginning in the late seventies and continuing for nearly two decades. Thus it seemed to me that complete integration of these two broad areas of analysis would not only involve tremendous problems of organization, but would also tend to misrepresent the degree to which the characteristics of the movement had emerged to public view by that point in the mid-seventies when it enjoyed its greatest political influence. In any case, I would urge readers who are primarily concerned with, for example, Victorian attitudes toward science, or tactics of popular agitations, to begin by reading the chapter or chapters that seem most relevant, rather than to read the chapters seriatim. I am convinced that for many readers, who will be interested in the behavior and mentality of the movement itself rather than in its political and administrative implications, a start in Chapters 8, 9, 10, and 11 would be most appropriate. The book has been designed with this in mind.

In the course of this study, I have been enormously impressed with the amount of research that could usefully be carried out in and around the subject area. If there is one thing that modern scholarship has conclusively shown, it is that studies beget more studies. More to the point, however, the philosphy behind the book is not transmittal of yet more factual information, nor the arbitration of academic debates, nor the resolution of the problems of science and society. It is rather the attempt to aid the reader to transcend contemporary perspectives through the more synoptic viewpoints that can be recreated from history. Surely, one function of that discipline is to explore the multitude and complexity of causal factors of historical phenomena, in order to heighten perception and analysis of the present and the future.

2. Animal Experiment and Humanitarian Sentiment before 1870

Green Teddington's serene retreat
For Philosophic Studies meet,
Where the good Pastor Stephen Hales
Weighed moisture in a pair of scales,
To lingering death put Mares and Dogs,
And stripped the Skins from living Frogs.
Nature he loved, her works intent,
To search or sometimes to torment.

—Thomas Twining

Man's desire to learn more about the structure and function of living organisms lead him early to experiment upon them; vivisection[1] only *seemed* novel by the time it became a matter of public concern in the late nineteenth century. It was practiced in the ancient world, notably by Herophilus, Erisistratus, and Galen. The latter's legacy was developed by the anatomist-physiologists of the sixteenth and seventeenth centuries until, slowly, "vivisection became a routine procedure."[2] William Harvey's brilliant demonstration of the circulation of the blood by means

[1] I will use "vivisection" interchangeably with "experiment upon living animals," to include such cases as the administration of drugs or bacteria where its literal implication of surgical incision is not present. This is the general, though not invariable, controversial usage.

[2] J. Schiller, "Claude Bernard and Vivisection," *J. Hist. Med.* xxii (1967), 246–260. Schiller's useful summary of the early development of the practice of vivisection is on pp. 246–250; the quotation is from p. 247.

of vivisection stimulated a spate of animal experiment following his *De Motu Cordis* (1628).[3]

By the eighteenth century, the experiments of the Rev. Stephen Hales (1677–1761) and lesser practitioners occasioned disapproving comment in literary circles. Of Hales, Alexander Pope opined ". . . he is a very good man, only I'm sorry he has his hands so much imbued in blood," and ". . . he commits most of these barbarities with the thought of being of use to man. But how do we know that we have a right to kill creatures that we are so little above as dogs, for our curiosity?"[4] Joseph Addison noted at about the same time—early in the century—that "there are . . . innumerable retainers to physic, who, for want of other patients, amuse themselves with the stifling of cats in an air pump, cutting up dogs alive, or impaling insects upon the point of a needle for microscopical observations. . . ."[5] Samuel Johnson foreshadowed much of the rhetoric of the antivivisection movement when in 1758 he descanted with magnificent scorn:

> Among the inferior professors of medical knowledge is a race of wretches, whose lives are only varied by varieties of cruelty. . . . What is alleged in defense of these hateful practices everyone knows, but the truth is, that by knives, fire, and poison knowledge is not always sought and is very seldom attained. The experiments that have

[3] Shugg has carefully examined the writings of Robert Boyle, Robert Hooke, John Evelyn, Richard Lower, and others in the Harveian tradition of the 1660s and 1670s. He detects among them a genuine concern for animal suffering overcome only by their desire to advance medicine and science. Boyle, Hooke, and the rest are acquitted by Shugg of Harwood's accusation that they were "fiendishly and morbidly cruel." W. Shugg, "Humanitarian Attitudes in the Early Animal Experiments of the Royal Society," *Annals of Science* xxiv (1968), 227–238. Cf. D. Harwood, *Love for Animals and How It Developed in Great Britain* (New York, 1928), pp. 78–81, 98–114.

[4] A. E. Clark Kennedy, *Stephen Hales, D.D., F.R.S.: an Eighteenth Century Biography* (Cambridge, 1929), pp. 55–57.

[5] J. Addison in *The Spectator* xxi, quoted in Harwood, *Love for Animals*, p. 113.

> been tried are tried again . . . I know not that by living dissections any discovery has been made by which a single malady is more easily cured. And if knowledge of physiology has been somewhat increased, he surely buys knowledge dear, who learns the use of the lacteals at the expense of his humanity. It is time that universal resentment should arise against these horrid operations, which tend to harden the heart, extinguish those sensations which give man confidence in man, and make physicians more dreadful than gout or stone.[6]

By the mid-eighteenth century, then, explicit discussion of the justification for animal experiment had begun, though the question was far from a public issue, no "universal resentment" having arisen. Further, Johnson's initial phrase indicates that only a minority of medical men experimented upon living animals. The evidence on the extent of vivisection in eighteenth-century Britain is sparse, but it would be difficult to demonstrate that the practice was anything but rare to nonexistent among the vast majority of doctors, despite Addison's "innumerable retainers to physic." Most were presumably interested in the practical benefits of "any discovery . . . by which a single malady is more easily cured," but few indeed pursued physiological knowledge pure and simple, which Johnson implied to be the somewhat incidental and virtually useless product of idle curiosity, a dilletantish luxury.

The quotation from Johnson captures the flavor of mid-eighteenth-century indignation and skepticism toward animal experiment. For a century to come such opinion was to be largely confined to scattered literati and the occasional humanitarian pamphleteer. Its widespread expression awaited the development and institutionalization of experimental medicine and the rise of an organized movement to protect animals. The former occurred initially in France and Germany, the latter in Britain.

[6] S. Johnson in *The Idler* xvii (5 August 1758), quoted in "Samuel Johnson," *Zoophilist* iii (1884), 288, or Harwood, *Love for Animals,* pp. 296–299.

MEDICINE AND VIVISECTION IN BRITAIN, 1800–1860

A survey of British experimental physiologists during the first half of the nineteenth century reveals that the issue of the moral justification or scientific validity of animal experiments was likely to become an issue for virtually any man who performed them. A. P. Wilson Philip (1770–1847) seems to have been charged with gratuitous cruelty by enemies within the Royal Society of London, demonstrating that the issue of animal experiment was a lively enough question of morals and taste to be used as a political weapon.[7] James Blundell (1790–1878) found it necessary to "defy the penny drivellers of the press, the declamatory and spurious orators of the day" and "defend the sacrifice of animals insofar as it is calculated to contribute to the improvement of science."[8] Sir Charles Bell (1774–1842) was less certain of the ethical status of animal experiment, despite some degree of success with the method: ". . . I cannot perfectly convince myself that I am authorized in nature or religion, to do these cruelties—for what?—for anything else than a little egotism or self-aggrandizement; and yet, what are my experiments in comparison with those which are daily done? And done daily for nothing." According to Bell, his means of discovery was deduction from anatomy, not experiment. The latter was a persuasive artifice intended "not to form my own opinions, but to impress them upon others." He went on, "It must be my apology, that my utmost efforts of persuasion were lost, while I urged my statements on the grounds of anatomy alone."[9]

Bell's views on animal experiment contrast markedly with

[7] W. H. McMenemey, "Alexander Philip Wilson Philip (1770–1847). Physiologist and Physician," *J. Hist. Med.* xiii (1958), 289–328, esp. 303–304.

[8] J. H. Young, "James Blundell (1790–1878). Experimental Physiologist and Obstetrician," *Medical History* viii (1964), 164.

[9] G. Gordon-Taylor and E. W. Walls, *Sir Charles Bell. His Life and Times* (Edinburgh and London, 1958), pp. 111, 128–129. See also A. Pichot, *The Life and Labours of Sir Charles Bell* (London, 1860), pp. 127–128.

the attitude of his French rival François Magendie (1783–1855), whose experimental demonstration in 1822 of the sensory and motor functions of the dorsal and ventral spinal nerve roots respectively gave rise to the bitter priority dispute with regard to what became known as the Bell-Magendie law. Cranefield's study of the Bell-Magendie controversy brings out clearly the essence of the vivisectional method for the solution of questions of physiology and the enormous complications and difficulties inherent in its development.[10] Bell's antipathy to vivisection and his attachment to deductive inference from anatomy characterized the early nineteenth-century British tradition as surely as Magendie's relentless animal experimentation, empiricism, and hostility to deduction and system-building presaged the future of French physiology. Magendie's insistence upon the inseparability of mind and hand in experimentation upon living animals not only brought scientific results, but also crucially influenced Claude Bernard's classic, *Introduction to the Study of Experimental Medicine*.[11] Publication of Bernard's book during the 1860s had a significant impact on the then emerging school of British experimental physiologists, but as regards the first half of the nineteenth

[10] P. F. Cranefield, *The Way In and the Way Out. François Magendie, Charles Bell, and the Roots of the Spinal Nerves* (New York, 1974). Cranefield shows that Bell and his allies took advantage of antivivisectionist sentiment to stigmatize Magendie and belittle his crucial contributions. "It is quite incorrect to say that Bell was not an experimentalist. It is true that he was primarily an anatomist and that he experimented reluctantly and badly, but he was definitely an experimentalist. . . . Only after it became clear that Mayo and Magendie had obtained correct results by means of experiments similar to those that had led him to incorrect results did Bell begin to denounce experimentation as a source of knowledge about the nervous system." For an illuminating discussion of the background to Bell's methodological convictions, see L. G. Stevenson, "Anatomical Reasoning in Physiological Thought," in C. McC. Brooks and P. F. Cranefield (eds.), *The Historical Development of Physiological Thought* (New York, 1959), pp. 27–38.

[11] See J. Schiller, "The Genesis and Structure of Claude Bernard's Experimental Method," in *The Foundations of Scientific Method: The Nineteenth Century* (Bloomington, Ind., 1973), pp. 139–144.

century, the Bell-Magendie comparison neatly symbolizes the contrasting approaches of British and continental medical thought.

There can be little doubt that in the days before anesthetics, Magendie's numerous experiments and radical surgical procedures must have caused animals a great deal of suffering.

Magendie's public lecture-demonstrations during a visit to London in 1824 provoked considerable outcry.[12] In a leading article on legislation for the study of anatomy in 1829, the editor of the *London Medical Gazette* noted:

> We recollect, some few years ago [probably 1824], a violent clamour was raised against the practice of experimenting upon living animals; indeed we believe the ferment has not yet subsided. Certain lecturers were represented in the most odious light as unnecessarily torturing and sacrificing the lives of rabbits, frogs, dogs, and cats. The attention of Parliament was called to the subject; the infliction of pains and penalties was threatened; and conviction, under a special statute, was with difficulty evaded. The appalling experiments of Magendie were the topic of the day; and the correspondence of Mr. Abernethy, Sir Everard Home, and others, with various members of Parliament, excited a strong sensation.[13]

The *London Medical Gazette*, while defending the principle of animal experiment, continued to hammer Magendie for his profligacy and cruelty in repeating allegedly fruitless experiments.[14] James Macaulay, an antivivisectionist doctor, recalled his impression of Magendie's Paris lectures in a letter to a friend:

[12] J. M. D. Olmsted, *François Magendie. Pioneer in Experimental Physiology and Scientific Medicine in XIX Century France* (New York, 1944), pp. 138–143.

[13] "Dissection of the Living," *Lond. Med. Gaz.* iii (1829), 644–645. Referred to are the surgeons John Abernethy (1764–1831) and Sir Everard Home (1756–1832).

[14] "On Experiments on Living Animals," *Lond. Med. Gaz.* xx (1837), 804–808; "Experiments on Living Animals," *Lond. Med. Gaz.* xxiv (1839), 212–215.

> In 1837 I attended, along with Edward Forbes and others known to you, the class of Magendie; at least we went to some of his lectures. The whole scene was revolting, not the cruelty only, but the "tiger-monkey" spirit visible in the demoralized students. We left in disgust, and felt thankful such scenes would not be tolerated in England by public opinion.[15]

In such an atmosphere, few Britons cared to pursue the new experimental physiology. One of the few was Marshall Hall, (1790–1857), who probably bore the greatest burden of antivivisectionist opprobrium. His researches into the phenomena of reflex action led to repeated accusations of cruelty in the medical press. In 1831 and again in 1847 he carefully outlined the circumstances and conditions under which he regarded experiments upon living animals to be justified: such experiments must have a useful end in view, be necessary to attain that end, be original, be as free of pain as possible, and be competently witnessed. "In order to fully accomplish these objects," he continued, "it would be desirable to form a SOCIETY for physiological research. Each member should engage to assist the others." In such a society, experiments could be proposed, discussed, executed, attested, and published by the members in a format that would insure against endless repetition of vivisections resulting from points of contention.[16] Hall was the first of a long line of medical men and physiologists who attempted to set guidelines for animal experiment and called for formal organization to foster experimental medicine. He admitted that unnecessary, inept, and cruel experiments sometimes took place and felt that control and prevention of these would remove the stigma attaching to vivisection.

[15] Macaulay to Playfair, 13 May 1875. Imperial College of Science and Technology, Playfair Papers #469. Edward Forbes (1815–1854) was a British naturalist.

[16] M. Hall, *A Critical and Experimental Essay on the Circulation of the Blood* (Philadelphia, 1835; British ed., 1831), pp. xvii–xxiii, or "Experiments in Physiology as a Question of Medical Ethics," *Lancet* i (1847) 58–60, 135, 161.

Among Hall's critics were practitioners familiar with the medical literature of the period. Throughout the nineteenth century, there is abundant evidence of doubt within the profession as to the methodology of animal experiment[17] and as to its ethical standing. The *London Medical Gazette* provided a forum for R. M. Hull and George Macilwain to attack painful experiments as immoral and unenlightening, and to challenge their professional brethren to demonstrate otherwise.[18] Macilwain went so far as to publish a pamphlet attacking vivisection with that blend of natural theology and inductive philosophy quite characteristic of English scientific circles in the early part of the century. Hull and Macilwain seem to have stimulated little in the way of overt reaction, though Macilwain claimed in one of his letters to the *Gazette* in 1843 that ". . . I have received communications by which I find that the profession take more interest in the question than I had anticipated, and that Dr. Hull and myself are by no means so singular in our views as I had feared."[19]

In the mid-nineteenth century, the attitude of the medical

[17] For an early nineteenth-century example of criticism of the method, see J. Barclay, *The Muscular Motions of the Human Body* (Edinburgh, 1808), pp. 270–273.

[18] R. M. Hull, "On Vivisection," *Lond. Med. Gaz.* xxxii (1842–1843), 862–864 and xxxiii (1843–1844), 55–57, 219–220. G. Macilwain, "On Vivisection," *Lond. Med. Gaz.* xxxii (1842–1843), 897–898, and xxxiii (1843–1844), 30, 188–189. See also "S.," Experiments on Living Animals," *Lond. Med. Gaz.* iii (1829), 670–671, and R. M. Glover, "Vivisection," xxxiii (1843–1844), 81–82.

[19] G. Macilwain, *Remarks on Vivisection and on Certain Allegations as to Its Utility, and Necessity, in the Study and Application of Physiology* (London, 1847). Macilwain was a Fellow of the Royal College of Surgeons, and for a long time the most visible medical antivivisectionist. He testified before the Royal Commission of 1876, and published a spirited commentary upon it, but seems to have had little to do with the specialized antivivisection societies of the later period. See G. Macilwain, "Vivisection," *Lond. Med. Gaz.* xxxiii (1843–1844), 30; *Medicine and Surgery One Inductive Science* (London, 1838), esp. pp. 48–50, 54; *Vivisection. Being Short Comments on Certain Parts of the Evidence Given Before the Royal Commission* (London, 1877).

profession toward criticisms of vivisection emanating from within its own ranks was generally indifference or at least public silence. In contrast, protests by laymen, especially when arising from such institutional sources as the Royal Society for the Prevention of Cruelty to Animals (R.S.P.C.A.) or its offshoot the Animals' Friend Society aroused rather more concern. Macilwain thought it advantageous to note on the title page of his supposedly scientific discussion of the question in 1847 that it was "Published by the Desire of some Medical and other Supporters of the Royal Society for the Prevention of Cruelty to Animals." It is to this more or less organized movement that we must now turn.

THE RISE OF ANIMAL PROTECTION

Dorothy George notes "a growing feeling against cruelty to animals" in mid-eighteenth-century London, citing Hogarth's *Four Stages of Cruelty* (1751) and quoting his "hope of, in some degree, correcting that barbarous treatment of animals, the very sight of which renders the streets of our metropolis so distressing to every feeling mind. . . ."[20] Shugg, Harwood, De Levie, and others have traced the rise of sentimental feeling toward animals to roughly the same period when evangelicalism, humanitarianism, and romantic poetry were beginning to alter public manners and mores fundamentally.[21] The latter part

[20] D. George, *London Life in the Eighteenth Century* (London, 1925), p. 17.

[21] W. Shugg, "Humanitarian Attitudes in the Early Animal Experiments of the Royal Society," *Annals of Science* xxiv (1968), 227n 3; D. Harwood, *Love for Animals* (New York, 1928), chaps. 3–5; D. DeLevie, *The Modern Idea of the Prevention of Cruelty to Animals and its Reflection in English Poetry* (New York, 1947). On eighteenth-century humanitarianism in general, see R. S. Crane, "Suggestions Toward a Genealogy of the 'Man of Feeling,'" *Engl. Lit. Hist.* i. (1934), 205–230. For an entertaining and erudite exploration of the religious and intellectual origins of animal protection, see L. G. Stevenson, "Religious Elements in the Background of the British Anti-Vivisection Movement," *Yale J. Biol. and Med.* xxix (1956), 125–157.

of the century saw the first few trickles of the meliorist literature of animal protection, the first few articles and tracts that heralded the nineteenth-century flood.[22] A pamphlet by Henry Crowe, M. A., Vicar of Buckingham, was perhaps typical of the state of mind and art that this genre had reached by 1819. Crowe's *Zoophilos, or, Considerations on the Moral Treatment of Inferior Animals* included a substantial chapter "On Cruelty in philosophical researches" in which he made the first of many subsequent comparisons between vivisectional cruelties and those of the Inquisition. Further, said Crowe, "Such researches are begun with the best intentions, both to advance science and benefit mankind. But the zeal of elucidating favourite theories, and pursuing discoveries, misleads the philosopher, till he unfortunately loses sight of the original end, and imperceptibly falls into . . . extravagencies, to call them by no harsher name."[23] Experiments upon animals joined bull-baiting, cock-fighting, abuses in animal husbandry, and casual cruelty to beasts of burden as evils to be stamped out.

A good example of the particularly strong impact that vivisection could make upon the literate public is charmingly embodied in a letter from the young Quaker W. E. Forster to his friend Barclay Fox in September, 1836:

> I find my mother, supposing thee to have influence with scientific men, has been writing thee an epistle on cruelty. Don't let it bother thee; but if thou shouldst have a good and easy opportunity to preach to anybody upon these abominable living experiments, and let her know thereof, she will never be tired of holding thee up to the admiration of all the lads and lasses within hearing, and it will be a great kindness to her, at any rate, for she has been reading those dreadful things about galvanized frogs and impaled dogs, etc., till she is the same

[22] For a bibliography of some of this literature, mostly of the eighteenth century, see H. S. Salt, *Animals' Rights Considered in Relation to Social Progress* (New York and London, 1894), pp. 105–132.

[23] H. Crowe, *Zoophilos, or, Considerations on the Moral Treatment of Inferior Animals* (London, 1819), p. 71. See pp. 54–73.

herself as if she had a continual shock of galvanism about her.[24]

In 1846 the Rev. David Davis, dissenting minister, presented Queen Victoria with a petition drawing her attention to the fact that living animals are "dissected" in England. He included a copy of his petition of the same year to the King of France in which he referred to the dissection of living horses at the veterinary college at Alfort and asked the King to have legislation passed against such practices. Davis did this "Having . . . in me . . . a thorough conviction that *science* should give way to *humanity* (if the interest of either be incompatible with that of the other)."[25] As David Mushet wrote, in *The Wrongs of the Animal World,* "The labourer's heavy anger—the indulgence of the heedless parent—the sportsman's habitual wont of thought, bring each their suffering, and foster evil habits in the heart; but what are these, and all the cruelties of avarice, or custom, or wanton gluttony, compared with acts committed coolly, for abstract speculation and for purposes of *fame?*"[26]

Meanwhile, the animal-protection movement gradually institutionalized, providing a common focus for hitherto disparate and intermittent individual efforts. The younger Sir William Pulteney had failed to pass his Bill outlawing bull-baiting in 1800, and in 1809 a similar fate had met Lord Erskine's Bill to prevent malicious wounding or wanton cruelty to certain animals. In 1822, the combined efforts of Richard "Humanity Dick" Martin (1754–1834) in the Commons and Erskine (1750–1823) in the Lords succeeded in passing Martin's Act, which outlawed cruelty to the larger domestic animals such as horses and cattle. The Act did not apply to cats and dogs. Martin attempted

[24] T. W. Reid, *Life of the Right Honourable William Edward Forster* (4th ed., London, 1888), i. 84.

[25] P.R.O., Home Office Papers, 45/O.S. 1550. Italics in the original.

[26] D. Mushet, *The Wrongs of the Animal World* (London, 1839), p. 192. See pp. 189–249. Italics in the original.

to improve the measure by a series of unsuccessful bills until he left the House in 1826.[27]

In 1825, Martin introduced the question of experiment upon animals into the House of Commons for the first time with an attack upon the cruelty of Magendie's experiments, concluding:

> He had heard that this fellow [Magendie] was again coming to this country to repeat his experiments. He therefore had mentioned it to the House in the hope that it would give publicity and excite against the perpetrator of such unnecessary cruelty, the odium he merited. He trusted that when it was known, the fellow would not find persons to attend his lectures and would thus be compelled to wing his way back to his own country.

Magendie was defended by Robert Peel and Sir James Mackintosh. Martin himself refused to condemn all animal experiment out of hand, or at least he refused to do so in Parliament.[28]

Early and abortive attempts to found societies to promote animal welfare were made in Liverpool as early as 1809 and in London in 1822. By 1824 the London endeavors finally took root. The "Society instituted for the purpose of preventing cruelty to Animals" inaugurated 16 June 1824 at a coffee house meeting. It is no part of my intention here to recount the rise of the society from early struggles, when its debts sent its secretary, the Rev. Arthur Broome, to prison, to its heyday in the late nineteenth century, when it was a wealthy and complex philanthropic organiza-

[27] A. W. Moss, *Valiant Crusade. The History of the R.S.P.C.A.* (London, 1961), pp. 14–19.

[28] *Records of Proceedings in Parliament, Letters and Articles in the "Times" and other Publications, and of the general Progress of Public Opinion, with reference to the Prevention of Cruelty to Animals and the Promotion of their proper Care and Treatment. 1800–1895.* Royal Society for the Prevention of Cruelty to Animals, London. (Hereafter cited as *R.S.P.C.A. Records*), ii. 207–228. These Parliamentary discussions of vivisection, prompted by Martin, took place on 24 February and 11 March 1825. They are the only such discussions I have found prior to 1875.

tion engaged in a massive program of public education and enforcement. The story has been enthusiastically narrated elsewhere.[29] It is germane to note, however, that the Royal Society for the Prevention of Cruelty to Animals (as it fairly soon became) was committed to a two-pronged attack on animal cruelty: on the one hand, wide publicity to the principle of kindness to animals represented as congruent with and, indeed, entailed by prevailing social and religious values; and, on the other hand, enforcement of existing laws and passage of new ones, through inspection and lobbying, respectively. As the society increased its membership and developed its ties with the aristocracy, its power and prestige in English life became such that no group whose interests it threatened could afford to ignore it.[30] During the first thirty years of the Society's operations the issue of animal experiment slowly emerged from a position of concern but little urgency to a relatively significant priority.

ANIMAL PROTECTION AND VIVISECTION BEFORE 1870

The numerous and accessible cruelties to animals in streets, marketplaces, cockpits, and the like, often involving members of the working class, were the primary focus of S.P.C.A. activity in the nineteenth century. Vivisection was simply one among many problem areas and, involving as it did educated individuals of recognized status in the community, a particularly delicate one.

The first prospectus of the S.P.C.A. contained a passage that, although critical of cruel experiments and contending

[29] See E. G. Fairholme and W. Pain, *A Century of Work for Animals. The History of the R.S.P.C.A., 1824–1934* (2nd. ed., London, 1934), and A. W. Moss, *Valiant Crusade. The History of the R.S.P.C.A.* (London, 1961), esp. pp. 20–64. The best analytical work is B. Harrison, "Animals and the State in Nineteenth-Century England," *English Historical Review*, lxxxviii (1973) 786–820. Further insight may be expected from the research of James Turner of Harvard University.

[30] For perspective on the class relations of the R.S.P.C.A., see B. H. Harrison, "Religion and Recreation in Nineteenth Century England," *Past and Present* xxxviii (1967), 98–125, esp. 116–119.

"that Providence cannot intend that the secrets of Nature should be discovered by means of cruelty," nevertheless allowed that certain necessary experiments "under the control of a benevolent mind" might be justifiable.[31] In the very early days of the society, its first secretary, the Rev. Arthur Broome, attempted a quite sophisticated approach to the problem through sympathetic medical allies. In 1827 the S.P.C.A. published a pamphlet on vivisection that included undertakings by several dozen physicians and surgeons "to prevent and discourage this 'physiological butchery.' "[32] Broome took advantage of public controversy over Magendie's visit to England to elicit support from distinguished surgeons for "a board composed of the most humane and eminent members of the profession, to whom all proposed experiments must be submitted with their object and the ultimate views of the proposers concerning them."[33] This proposal, similar to Marshall Hall's of 1831 in its attempt to set up appropriate machinery within the profession, represents the most sophisticated policy on the part of the S.P.C.A. until the 1860s. Broome resigned as secretary in 1828 and his program for attacking vivisectional abuses through the medical profession seems to have left with him.

A splinter group from the S.P.C.A., the Animals' Friend Society, attempted a simple extension of the strategy of inspection—which was proving so successful in obtaining convictions under Martin's Act for offenses by the working class—to the problem of animal experiment. In the *Morning Herald* newspaper for 15 November 1838 it advertised:

> SURGICAL EXPERIMENTS ON LIVING ANIMALS—It being a common practice of many Surgeons and Students to cut up Dumb Animals Alive, and inflict the most revolting and protracted torture on them to please their own vanity, and for the interests of their own species, the ANIMALS' FRIEND SOCIETY will liberally REWARD any person who

[31] E. G. Fairholme and W. Pain, *A Century of Work for Animals*, p. 191.

[32] *R.S.P.C.A. Records* iii (1827–1830), 6.

[33] D. Mushet, *The Wrongs of the Animal World*, p. 209.

> will enable it to CONVICT such delinquents before a Magistrate; and will be glad of any particulars of such experiments, with the names of the operators. . . .[34]

This strategy failed. To the best of my knowledge, no prosecution for cruelty in experiments upon living animals was instituted under Martin's Act (extended in 1835 to cover all domestic animals) under 1874.

Through the mid-century period, the S.P.C.A. grew from strength to strength, gaining experience, prestige, financial power, and political influence. There were murmurings about vivisection, occasional insistence that it was a subject for legislative action, but, in sum, little more than talk and a £100 prize essay on the subject.[35] S.P.C.A. success in other spheres insured that the discussion of possible legislation in 1837, 1843, and 1847 would not go unanswered, either in the correspondence columns of the *Times* or in the medical press.[36] In any case, S.P.C.A. policy

[34] Quoted in J. Chippendale, "Experiments on Animals," *Lancet* i (1838–1839), 357–358. The society repeatedly attacked vivisection in the nine volumes of its journal, eagerly controverting medical men who defended the practice. See *Animals' Friend* i–ix (1833–1841), *passim,* and *R.S.P.C.A. Records* v (1830–1846), 88, 93, 98–99.

[35] The strongest expression, from the 1843 presidential address: "I do not wish to harrow up your feelings, or to work on them more than is necessary to impress you with the necessity of ultimately looking to legislative remedies for the correction of these monstrous abuses. Nor is it only for the sake of the victims that the legislature should interfere; they are bound to rescue the youth of England, who are intended for the profession of surgery, from such contaminating scenes; so horrible that no religious parent could bear to see his child exposed to them." Quoted in R. M. Hull, "On Vivisection," *Lond. Med. Gaz.* xxxii (1842–1843), 863–864. See also "Experiments on Living Animals," *Lond. Med. Gaz.* xxiv (1839), 160–163, and Fairholme and Pain, *A Century of Work for Animals,* pp. 192–193.

[36] *R.S.P.C.A. Records* vi (1842–1849), 36–39. See also "Bibliographical Notices," *Lancet* i (1844), 283–284, a review of R. Jameson, *Remarks on the Use of Vivisection as a Means of Scientific Research, in a letter addressed to the Earl of Carnarvon, President of the Society for Preventing Cruelty to Animals* (London, 1844). I have not seen the Jameson pamphlet, nor G. F. Etherington, *Vivisection Investigated and Vindicated* (Edinburgh, 1842).

on vivisection seems to have been of little consequence until the late 1850s. It was the decade beginning in 1857 that saw the (by then) *Royal* Society for the Prevention of Cruelty to Animals begin active involvement in the practical problems of the control of vivisection.

French veterinary schools at Alfort, Lyon, Toulouse, and Paris employed a system of student surgical training on living but aged and valueless horses. The system, begun, according to the *British Medical Journal,* in 1761,[37] involved a large number of standard veterinary operations carried out on the same horse, which was under restraint and without benefit of anesthetics (these had been widely used for medical purposes since the mid-nineteenth century).[38] The operations were intended to develop the manual dexterity necessary for the practice of veterinary surgery and not to elucidate points of physiology. News of these cruelties, of mutilated and suffering horses left overnight to die, reached the R.S.P.C.A. in 1857. The Society, faced with an abuse strongly criticized even by the British medical press[39] and directed toward the animal most beloved by the British gentry,[40] organized a delegation to the Emperor Napoleon.

At the delegation's audience with the Emperor in April 1861, prohibition of vivisection at the veterinary schools was strongly urged.[41] So began more than two years of twists and turns in French policy toward the practices in question, during which they were now forbidden, then freely pursued, then restricted by ministerial decree. The question seems to have been dropped after 1864, but flared

[37] "The Discussion on Vivisections," *B.M.J.* ii (1863), 323–325.

[38] For one of many descriptions of the practice, see A. Perry, "Vivisection Cruelties," *Lancet* ii (1860), 517.

[39] The *B.M.J.*, the *Lancet,* and the *Medical Times and Gazette* condemned the practice: see, e.g. *B.M.J.* i (1861), 502–503; *Lancet* ii (1860), 143–144 or ii (1863), 224–225; *Medical Times and Gazette* ii (1860), 383.

[40] A series of letters and articles by the Paris correspondent of the *Times* kept the topic before the public.

[41] See "Vivisection," *Lancet* i (1861), 402, and *Times* (London) 20 April 1861.

up again in 1867. The ultimate fate of vivisection in certain French schools is not of primary concern here.[42] What is clear is that the R.S.P.C.A.'s initiative stimulated wide discussion of experiments on living animals on both sides of the channel, much of it colored by the undoubted brutality of the French veterinary schools.

In Britain, the R.S.P.C.A. quickly turned to the question of possible cruelties in the home country itself.[43] Initial enthusiasm for the society's French foray on the part of the British medical press soon faded into a wary ambivalence. Unsupervised French veterinary students were one thing, eminent British medical practitioners quite another, as the *Lancet* and *B.M.J.* hastened to point out. Everyone agreed that the Alfort system was cruel. While British veterinary students did not vivisect, in the interests of medical science a few British medical men reluctantly, but necessarily and justifiably, did.[44] R.S.P.C.A. initiatives—such as Sir John Scott Lillie's request to the British medical

[42] In France, a commission composed of Cruveilhier, Cloquet, Claude Bernard, Robin, Moquin-Tandon, and Le Blanc reported to the Academy of Medicine in 1863 that vivisection was necessary for both training in veterinary surgery and scientific advance, but that it should be carried out by or under the direction of professors, with great care to minimize pain. The Academy of Medicine rejected the report of the commission, allegedly because of the report's indirect admission of possible abuses connected with vivisection. Instead, it approved resolutions labelling the R.S.P.C.A.'s charges "unfounded" and leaving the practice of vivisection to the discretion of men of science. According to W. O. Markham, editor of the *Lancet* at the time, the French were offended by the English interference and by the indignant and sentimental language in an English pamphlet, *Observations on the Dissection of Living Animals,* which the Academy erroneously took to be an R.S.P.C.A. publication. On the other hand, a few French medical journals were very critical of vivisectional abuse. The story can be traced in *B.M.J., Lancet, Times* (London), *L'Abeille Médicale,* and *L'Union Médicale,* esp. 1860–1864 and 1867.

[43] See letter by John Colam, Secretary of the R.S.P.C.A., in the *Times* (London) 11 August 1863, quoted in *R.S.P.C.A. Records* viii (1858–1865), 123–126.

[44] See, e.g., "Vivisection at Alfort," *Lancet* ii (1860), 395–396; *B.M.J.* ii (1861), 509–510; W. Sharpey, "Address in Physiology," *B.M.J.* ii (1862), 163–164.

profession for advice as to the utility of vivisection,[45] inquiries as to vivisection in British institutions,[46] an international congress on the subject of vivisection held at the Crystal Palace in August 1862,[47] and a prize essay contest in 1864[48]—stimulated the development of a specific policy toward vivisection by the society.

Discussion at the Crystal Palace, which included George Macilwain among other medical men, and at later annual meetings, seems to have originated the R.S.P.C.A.'s policy against all *painful* vivisection—that is, any potentially painful operation or experimental procedure carried out without anesthesia. The Earl of Harrowby (1831–1900), president of the society 1861–1877, and sometime physiologist Sir Benjamin Ward Richardson (1828–1896) argued in 1863 that vivisection under anesthesia was permissible, since without pain there could be no cruelty.[49] While various members of the society held widely divergent views on the issue of vivisection,[50] which has always been a divisive

[45] J. S. Lillie, "Vivisectional Cruelties," *Lancet* i (1861), 44–45.

[46] *Medical Times and Gazette* i (1861), 49; "Report of the Royal Commission on the Practice of Subjecting Live Animals to Experiments for Scientific Purposes," *Parl. Papers 1876* xli (C.–1397), Q.1568.

[47] See "Vivisections," *Lancet* i (1862), 642; "Vivisection," *Lancet* ii (1862), 221; "Vivisection and Cruelty to Animals," *B.M.J.* ii (1862), 216. The date of the congress is erroneously given as 1863 in Fairholme and Pain, *A Century of Work for Animals*, p. 193. They are also in error in implying the R.S.P.C.A.'s complete success in its campaign against veterinary vivisection in France.

[48] A. W. Moss, *Valiant Crusade*, p. 156.

[49] E. G. Fairholme and W. Pain, *A Century of Work for Animals*, pp. 193–194. On Richardson's attitudes, see L. G. Stevenson, "Science Down the Drain," *Bull. Hist. Med.* xxix (1955), 1–26. Richardson became progressively less enthusiastic about experiments upon living animals as he aged. Antivivisectionists late in the century, who viewed Richardson as sympathetic to their cause, were shocked to discover his earlier activity in experimental medicine. See the handbill *Sir Benjamin Ward Richardson as a Vivisector*, (no place, name or date, c. 1885), which challenged Richardson to repudiate his vivisections.

[50] E. G. Fairholme and W. Pain, *A Century of Work for Animals*, pp. 195–196, and *Vivisection, The Royal Society for the Prevention*

and contentious point for the animal protection movement,[51] the position of 1863 remained the most important determinant of R.S.P.C.A. policy for the remainder of the century.[52] This policy was to play a crucial role in the disputes of the 1870s, for its recognition that vivisection was permissible under certain circumstances, such as anesthesia, was to prove too moderate for many antivivisectionists who were initially part of the R.S.P.C.A.'s constituency.

For the medical press in the 1860s, it was essential to distinguish vivisection in the pursuit of scientific research from that for instructional purposes. The journals unanimously condemned vivisection for the achievement of surgical dexterity and were highly suspicious or outrightly critical of vivisection for lecture demonstration. In rare cases, however, the practice of even painful vivisection was justified for research purposes. Editorial opinion was against parliamentary legislation and called for the profession itself to specify the circumstances and conditions under which experiments on living animals were to be allowed.[53] Although there were occasional attacks on the R.S.P.C.A. for its meddling, hypocrisy, and ignorance with regard to vivisection,[54] the medical press remained remarkably sympathetic to the society's general mission. The *Lancet* in particular noted its successes from time to time and in 1871 called for a medical practitioner to be struck off the rolls when he was convicted of (nonvivisectional) cruelty following an R.S.P.C.A. prosecution.[55] Antivivisec

---

of Cruelty to Animals and the Royal Commission (London, 1876), pp. i–ii.

[51] C. D. Niven, *History of the Humane Movement* (New York, 1967), pp. 79–96.

[52] Fairholme and Pain, op. cit., p. 200.

[53] See "Vivisection at Alfort," *Lancet* ii (1860), 395–396. Also see *Lancet* ii (1863), 224–225; *B.M.J.* i (1861), 502–503, ii (1861), 509–510, "Vivisections," ii (1862), 260, ii (1863), 268–270, "Justifiable Vivisection," i (1864), 71; *Medical Times and Gazette* ii (1860) 383, and i (1861), 227.

[54] See *Medical Press and Circular* ii (1866), 256, and "Vivisection and Torture Traps," *B.M.J.* i (1864), 643–644.

[55] "Cruelty to Animals," *Lancet* i (1871), 358.

tionist opinion within the medical profession certainly remained and occasionally made itself heard.[56]

It is clear that public consciousness of the existence of vivisection and of opposition to the practice was periodically stimulated by these sporadic outbursts of controversy. The *Times* and the *Daily Telegraph* at this point enthusiastically supported the antivivisection efforts of the R.S.P.C.A., the former declaring in an 1867 editorial, "We must all respect, though we cannot worship science; but what is science that it should trample, like the Moloch of old, or the goddess Reason of a later day, on the better feelings of human nature?"[57] Charles Dickens reviewed animal experiments for "the improvement of the resuscitation of the apparently drowned," sponsored by the Royal Humane Society, asking "Will the Society for the Prevention of Cruelty to Animals be good enough to look after the Royal Inhumane Society?"[58] The experimental approach had its proponents, some of them prepared to countenance legislation rendering vivisection "amenable to the supervision of the inspector of anatomy."[59] Another defender of the practice felt that "If there is any one thing which the mass of educated, and even half-educated, Englishmen undoubtedly believe, it is the connection of their material prosperity with the progress of science . . . they would, we believe, be very slow to put any new obstacle in the way even of . . . unpopular branches of knowledge."[60]

Despite these discussions, however, no mass coverage,

[56] See Henry MacCormac's challenge to the claims made for vivisection in research: H. MacCormac, "On Vivisection," *Medical Press and Circular* i (1868), 280–282.

[57] *Times* (London) 12 January 1867, quoted in *R.S.P.C.A. Records* ix (1865–1867), 112–117.

[58] [C. Dickens], "Inhumane Humanity," *All the Year Round* xv (1866), 238–240.

[59] W. Bowman, *The Address in Surgery, Read at Chester, August 9, 1866, at the Twenty-fourth Annual Meeting of the British Medical Association* (London, 1882), pp. 35–37. Dr. Scoffern, "Vivisection," *Belgravia* ii (1867), 101–109, 216–222. The quote is from 221.

[60] "Physiological Experiments: Vivisection," *Westminster Review* lxxxv (1866), 132–155. The quote is from 151.

and, therefore, no mass awareness of the issue could emerge until it became clear that experiments on animals in Britain paralleled those on the continent in either extent or potential for abuse. Until such time the few British medical scientists could feel, like the *Westminster* reviewer, relatively safe from the threat of active interference, if only because of the popularity and prestige of science. We have seen that the medical press was at pains to differentiate British practices from those on the continent. This kind of distinction was made early in the century and was maintained, by most parties to the discussion, well into the 1860s.[61] The distinction was indeed valid through the mid-century period, but it was to break down after 1870, thereby allowing antivivisectionists to argue with some plausibility that British experimental medicine in the seventies was neither more modest in its demands nor more ethical in its conduct than the French or German varieties.

[61] See, e.g., F. P. Cobbe, "The Rights of Man and the Claims of Brutes," *Fraser's Magazine* lxviii (1863), 594–596, and "Vivisection," *Lancet* ii (1863), 252.

3. Experimental Medicine in Britain

It is clear that through the mid-nineteenth century experimental medicine in Britain was greatly underdeveloped, both in technical sophistication and degree of professionalization, in comparison to developments in Germany and France.[1] One major factor was a highly centralized university system, dominated by Oxford and Cambridge, which had little incentive or inclination to innovate. In these universities, science meant natural theology, a set of beliefs that emphasized the Creator's design of the universe and thus viewed science as primarily a minute observation of structure, from which function could be inferred. In the case of medicine, these attitudes clearly underlay the reliance on anatomical inference as the basis of physiology, which dominated British medical thought through most of the century.[2] Combined with widespread antipathy to experiments upon living animals, such attitudes kept Britain in substantial isolation from the laboratories of Germany and France, an isolation that became most acute and most evident during the 1850s and 1860s.

The lengthy if tenuous tradition of vivisection in Britain, from Harvey, through Hales, John Hunter (1728–1793),

[1] Much of the following discussion is based upon G. L. Geison, "Social and Institutional Factors in the Stagnancy of English Physiology, 1840–1870," *Bull. Hist. Med.* xlvi (1972), 30–58, and R. D. French, "Some Problems and Sources in the Foundations of Modern Physiology in Great Britain," *History of Science* x (1971), 28–55.

[2] See, e.g., G. Macilwain, *Medicine and Surgery One Inductive Science* (London, 1838). For an excellent historical treatment, see L. G. Stevenson, "Anatomical Reasoning in Physiological Thought," in C. McC. Brooks and P. F. Cranefield (eds.), *The Historical Development of Physiological Thought* (New York, 1959), pp. 27–38.

Benjamin Brodie (1783–1862), Philip, Blundell, Hall, Augustus Waller (1816–1870), and a few others, does not belie the abundant evidence that the extent of vivisection in Britain was extremely slight until the late nineteenth century. The *Lancet* was almost certainly accurate in making the following complacent response to a letter of 1836 protesting the cruelty and immorality of experiments on animals:

> This expostulation is partly directed against a "practice" which does not exist. Our correspondent is not justified by facts in connecting this word with "experiments on living animals". British professors of medicine *very rarely* prove by such experiments that they feel themselves warranted in testing theories which hold forth a promise of ulterior benefit to man.[3]

Jameson's estimation in 1844 of the number of vivisections per year in Britain at about a thousand was wildly high, as the *Lancet* pointed out.[4] In 1862 Sir Benjamin Ward Richardson, both physiologist and physician, told the R.S.P.C.A., "I know of no institution in the kingdom where vivisection is carried on publicly, or even privately, in a systematic manner."[5] In 1863, the *Lancet* averred that

> . . . as a general rule, neither our students nor teachers are wont to carry on experiments upon living animals even in a private way. The utmost that can be said is, that perhaps some two or three, or at most six, scientific men in London are known to be pursuing certain lines of investigation which require them occasionally during the course of a year to employ living animals for the purpose of their inquiries.[6]

[3] J. D. P., "Experiments on Living Animals," *Lancet* ii (1835–1836), 390–391.

[4] "Bibliographical Notices," *Lancet* i (1844), 283–284. The number of vivisections performed under license in Britain did not reach one thousand until 1887, even though by the mid-eighties the inhibiting effect of the Act of 1876 upon experiments had been almost completely circumvented.

[5] "The Vivisection Act," *The Animal World* vii (1876), 133.

[6] *Lancet* ii (1863), 252–253. See also *Medical Times and Gazette* xvii (1858), 249.

As an anonymous British reviewer of some of Claude Bernard's work put it in 1858:

> In this country, experimental physiology labours under peculiar disadvantages. There still exists some of that feeling of repugnance to vivisections which prevented Sir Charles Bell from discovering those functions of the spinal cord which M. Brown-Séquard is now elucidating. Those tender feelings touching the sufferings of animals, embodied in acts of parliament and incorporated in societies for the prevention of cruelty to animals, not to speak of our innate sympathies, stand in the way of our prosecuting inquiries which demand vivisections for their elucidation. We allow ourselves a certain latitude respecting cold-blooded animals, and we may gratify ourselves with a private battue of "grenouilles" and "tortues"; but as to practicing physiological experiments upon such universal favourites as "chats", "lapins", and "chiens", not to speak of any individual of the "equine" or "bovine" species, is a thing itself not to be mentioned. Practically and legally, then, we may consider ourselves as quite *hors de combat* respecting these physiological "nouveautés" which we see issuing from time to time from the Parisian press.[7]

If an aversion to vivisection was partly responsible for technical backwardness in physiology, Britain's low level

[7] *The Salivary Glands and Pancreas, Their Physiological Actions and Uses in Digestion: Being a Review of the Doctrines Taught by M. Claude Bernard* (Glasgow, 1858), pp. 1–2, from *Glasgow Medical Journal* xxi (1858). I am indebted to G. L. Geison and F. L. Holmes for this reference. The same journal had this to say in 1860, ". . . in reference to all such experiments by vivisection, . . . the repugnance felt by physiologists in this country to their performance, has been one of the chief causes why the investigation of many physiological questions has been materially interfered with, and many important discoveries, which could not be arrived at in any other way, have not been made. . . . In France there appears to be less scruple in this matter than has hitherto prevailed in this country; and we could almost go the length of saying, that we should feel content that the status quo should remain as it is, even though our continental brethren should continue in advance of us in matters pertaining to experimental physiology," ibid. (1860), 244–245. I am grateful to Sandra Black for this reference.

of professional opportunity in the discipline had roots in a related phenomenon: the perceptions of British medical men as to what constituted potentially fruitful areas of research. From the earliest part of the century, an important segment of medical opinion refused to accept laboratory research as as important as the traditional avenues of clinical casework and human anatomy. There was a profound tension within the profession between those who saw the future of medicine as inextricably linked with basic scientific research, involving vivisection, and those who preferred to see the practice of medicine as an art to which any research method overlooking the patient must be inconsequential. As Stevenson has clearly pointed out, the sanitary tradition in British medical thought frequently entailed a "sanitarian syndrome" of "active religiosity, anti-vivisectionist sentiment, and disbelief in the value of immunization"; one could be progressive and yet uncommitted to animal experiment.[8]

While English medical journals, not without a certain pride, were asserting English freedom from the stigma of vivisection, workers in dozens of laboratories in Germany, France, Austria, Switzerland, and Italy were establishing the tenets of the new experimental medicine.[9] Continental researchers had few qualms about animal experimentation. The new developments were decisively influenced by systematic recourse to experiments upon living animals, in the attempt to confirm or deny hypotheses about physiological functions. Vivisection was not an end in itself but part of an activist, interventionist, empiricist, approach to

[8] L. G. Stevenson, "Science Down the Drain," *Bull. Hist. Med.* xxix (1955), 1–26.

[9] The following discussion is indebted to J. Schiller, "Claude Bernard and Vivisection," *J. Hist. Med.* xxii (1967), 246–260; "Physiology's Struggle for Independence in the First Half of the Nineteenth Century," *History of Science* vii (1968), 64–89; "The Genesis and Structure of Claude Bernard's Experimental Method," in *The Foundations of Scientific Method: The Nineteenth Century* (Bloomington, Ind., 1973), pp. 133–160. See also J. M. D. Olmsted and E. H. Olmsted, *Claude Bernard and the Experimental Method in Medicine* (New York, 1952), esp. pp. 41–43.

problems of bodily function and disfunction. This approach sought to supplement the observations and diagnoses of the clinical school and the deductions of function from structure made by students of morbid anatomy. Experimental medicine rejected the system-building of the clinical and anatomical approaches, emphasizing the need for the painstaking accumulation of facts based upon specified experimental procedures and reproducible phenomena. It conceived a function of the body in a mechanistic hypothesis and tested that hypothesis in an experiment designed to isolate the function in question so as to yield a clear positive or negative result. Physical and chemical techniques were utilized for experimental purposes whenever possible, and slowly the physiological laboratory developed a unique technology of its own. The introduction of anesthesia in 1847, in the form of ether, greatly aided control of extraneous variables and precise observation by eliminating animal pain in those experimental procedures where it could be utilized. According to Cranefield, "the discovery of ether made modern experimental physiology possible" and "was at least as important to experimental medicine as it was to clinical surgery." Anesthesia made practical many experiments hitherto invalidated by the surgical conditions and physiological consequences associated with pain. Furthermore, it overcame the repugnance of many scientists to experiments upon living animals: "there can be no doubt that people who might otherwise have shunned experimental physiology (in favor, say, of anatomy, animal chemistry, embryology, or some other field) were led to take it up because it no longer required the use of surgical procedures in the absence of anesthesia."[10] Thus the early groping steps of the likes of Harvey, Haller, and Magendie evolved into a methodology systematically employed by a community of practitioners, engaged in checking and developing one another's contributions to the field. By the 1860s, experimental medicine in France and Germany had

[10] P. F. Cranefield, *The Way In and the Way Out. François Magendie, Charles Bell, and the Roots of the Spinal Nerves* (New York, 1974), pp. 53–54, and personal communication, 23 July 1973.

achieved legitimacy within key universities, recognition by the medical profession, and substantial intellectual stature, reflected in a burgeoning literature.

Of all this, England knew little although a few frustrated enthusiasts were trying to promote experimental medicine in the face of indifference and hostility. They did not fail to draw attention to England's inability to emulate the spectacular advances of continental physiology.[11] Slowly, by the late 1860s, the complex of intellectual, educational, and social factors that maintained the medical profession's extreme conservatism were beginning to break down under the duress of legislation dating from the late 1850s. Foreign advances involving sophisticated physiological concepts, methods, and instruments began to make part-time professorships in anatomy *and* physiology, held by busy medical practitioners, look increasingly inadequate to the task of maintaining the supremacy of British medicine. The emergence of institutional niches for would-be physiologists, relieving them of the burden of maintaining a medical practice,[12] went hand in hand with the reform of medical curricula and examination requirements for professional certification. This process was a gradual one, which extended in its initial phase into the 1890s,[13] but if a single year be chosen as a turning point, that year must be 1870.

[11] "New Editions of Physiological Works," *British and Foreign Medico-Chirurgical Review* xxxvi (1865), 51–55, and "British and Foreign Science," *Reader* (15 July 1865), 61–62. For further discussion and bibliography, see the articles cited in note 1.

[12] Charlotte Hall, in her biography of her husband Marshall Hall, noted "the supposed general rule, that physiologists do not succeed in practice" and referred to the pecuniary sacrifice that the study of physiology entailed for the physician. C. Hall, *Memoirs of Marshall Hall, M.D., F.R.S.* (London, 1861), pp. 121–123. The more apposite formulation, for our purposes, would have been "the supposed general rule, that practitioners do not succeed in physiology."

[13] The resistance to the development of the study of physiology as a specialization was exemplified by Andrew Clark, doyen of fashionable London physicians, when he spoke against the bill to regulate animal experiment by licensure, in 1876: "This Bill, as it stood, if it were to pass (even if it were modified, as some of the leaders amongst the physiologists would have it), would be a grievous thing,

In 1870 a small group of experimentally inclined British physiologists, many of them with continental educations, accepted recently developed institutional positions from which they and their allies and proteges were to dominate the science. With the model of French and German physiology clearly in mind, they transformed education and research in the subject in Britain from a poor and rather suspect tributary of gross and microscopic anatomy to a thriving, prestigious, and independent discipline whose members were clearly in a position of world leadership by the turn of the century. In 1870, J. S. Burdon Sanderson (1828–1905) was appointed Professor of Practical Physiology and Histology at University College London, and in the following year became first Superintendent of the Brown Institution. In the same year, E. A. Schäfer (1850–1935) became Assistant Professor of Physiology under Burdon Sanderson, and Michael Foster (1836–1907) was selected as first Praelector of Physiology at Trinity College, Cambridge (see Figure 1).[14] The immediate impact of these men and a few colleagues on their respective institutions is revealed in the contrast between the evidence given above as to the extent of vivisection in the sixties and that gathered by the Royal Commission in the mid-seventies.

Also in the year 1870, the Royal College of Surgeons began the first of a series of reforms in their examining procedures, the effect of which was to require more and more physiological knowledge on the part of would-be practitioners. Inevitably a further effect was to place progressively stronger pressure on teaching institutions, such

---

as it would tend to establish a class distinction between professional physiologists and general practitioners. Such a result would be most deplorable; he [Clark] would have men physiologists and general practitioners as well." *B.M.J.* ii (1876), 154. Cf. Edwards Crisp, ibid., 155.

[14] For short biographical sketches, see E. A. Sharpey-Schafer, *History of the Physiological Society during its First Fifty Years 1876–1926* (London, supplement to the *Journal of Physiology,* 1927).

Figure 1. John Burdon Sanderson (1828–1905), left, and Michael Foster (1836–1907), pioneers of professional physiology in Great Britain. *Courtesy of the Wellcome Trustees.*

as the London hospital medical schools, to create separate teaching posts in the subject and to find the most competent physiologists to fill them.[15] As mentioned above, this process was not completed until the nineties, but the turning point was undoubtedly the Royal College of Surgeons' realization of the need for specialist training in physiology. Comparative figures on the success in the Royal College examinations of men from the various institutions must have been a powerful spur to reform, especially in the highly competitive atmosphere of medical education in the metropolis.

Medical witnesses before the Royal Commission in 1876 were unanimous in testifying that there was much more vivisection performed on the continent than in Britain. More to the immediate point, however, many of the witnesses also noted a recent increase in the amount of vivisection in Britain. A survey carried out by the Royal Commission showed that by 1875 medical schools in Great Britain carried out a significant number of experiments on living animals for purposes of instruction and research. Among the institutions particularly prominent were University College London, the Brown Institution, Cambridge University, and Edinburgh University, in all of which changes or additions of teaching personnel since 1870 had undoubtedly greatly stimulated such activity.[16] In the light of these

[15] Z. Cope, *The Royal College of Surgeons of England. A History* (London, 1959), p. 143. In testimony before the Royal Commission of 1876, William Sharpey, for many years Professor of General Anatomy and Physiology at University College London, readily agreed with the suggestion of his colleague, commissioner J. E. Erichsen, that, "physiological laboratories have been established recently in a great measure, have they not, under the direction of the examining authorities and bodies of this country, such as the College of Surgeons?" "Report of the Royal Commission on the Practice of Subjecting Live Animals to Experiments for Scientific Purposes," *Parl. Papers 1876* xli (C.–1397), Q.484.

[16] "Report of the Royal Commission on the Practice of Subjecting Live Animals to Experiments for Scientific Purposes; with Minutes of Evidence and Appendix," *Parl. Papers 1876* xli (C.–1397), 277–733; "General Analytical Index," *Parl. Papers 1877* xxvii, 663–686; *Digest of Evidence Taken Before the Royal Commission on the Practice of Subjecting Live Animals to Experiments for Scientific Purposes;*

events, it is interesting that in 1870 the General Committee of the British Association for the Advancement of Science resolved to form a committee to enunciate guidelines for physiological experiment and

> . . . . to consider from time to time whether any steps can be taken by them or by the Association, which will tend to reduce to its minimum the suffering entailed by legitimate physiological inquiries; or any which will have the effect of employing the influence of this Association in the discouragement of experiments which are not clearly legitimate on live animals.

This committee, consisting of ten anatomists and physiologists (Burdon Sanderson and Foster were among them), produced a report that was submitted to the Association at its 1871 meeting. The report amounted to just such a code to govern experiment as the medical press had called for in the sixties. It made four main points. First, every experiment that could utilize anesthesia ought to do so. Second, teaching demonstrations on living animals ought to be painless or to utilize anesthesia. Third, painful experiments for the purposes of research ought to be performed only by skilled persons with appropriate instruments and facilities in a laboratory, "under proper regulations." Finally, vivisection ought not to be performed in veterinary education for the purpose of obtaining manual dexterity.[17]

---

with an Alphabetical List of Witnesses (London, H.M.S.O., 1876).

Appendix iii of the Report of the Royal Commission shows the results of a survey of the medical and veterinary schools of Great Britian and Ireland with regard to frequency and type of animal experiments performed. The survey's categories and the responses to them do not appear to me to be sufficiently consistent to abstract any meaningful statistical data. For example, whether operations on pithed frogs or indeed the act of pithing itself are to be considered as experiments on living animals is unspecified in the survey.

[17] *Report of the British Association for the Advancement of Science* (41st meeting, Edinburgh, 1871; London, 1872), p. 144. For a copy of the report, see Appendix I. The report was signed by only seven of the ten committee members. I have found no evidence, however, that it was controversial. See also "Vivisection," *B.M.J.* ii (1871), 185.

The inclusion of this last point shows clearly the impact of the revelations of French practices, for British veterinarians never used vivisection to any extent and had, in fact, been in the forefront of protest against Alfort and the other French schools. Nevertheless, the committee apparently felt it necessary to use the imprimatur of the British Association to reassure the public on the point. The physiologists knew they were treating a subject of considerable sensitivity, but they still trusted that the statement of careful guidelines and the avowal of good intentions might suffice.

Events showed differently. The ambitions of British physiologists inevitably collided with an antivivisection tradition that had deep, if sporadic and disparate, roots. British physiologists were neither circumspect nor sophisticated enough to avoid the collision, and they barely succeeded in minimizing the damage resulting. They faced an organization that was wealthy, prestigious, politically influential, adept in courts of law, publicity conscious, and, most important, vigilant. The R.S.P.C.A. was ready to test the applicability of Martin's Act to the practice of vivisection. British experimental medicine was not slow to provide it with an opportunity. "*Lex*," a correspondent of the R.S.P.C.A. magazine, *Animal World*, wrote in 1870, "My opinion is, that if this matter were agitated, it might lead to the whole subject being regulated by legislative enactment. . . ."[18] Quite so.

PRELUDE TO CONTROVERSY

The R.S.P.C.A. was devoting quite minimal attention and resources to the problem of vivisection during the period from 1870 to 1874, while British physiology was establishing itself at Cambridge and University College London. "The entire suppression of experiments upon animals made for discoveries in science when conducted with torture" was only one of a number of specific goals listed by *Animal World* when the society's periodical was launched in 1869.[19]

[18] *Animal World* ii (1870–1871), 46.

[19] "Our Object," *Animal World* i (1869–1870), 8.

There were occasional skirmishes in the press,[20] but no sustained discussion arose. An advertisement in the *Scotsman* seeking cats and dogs to be used in the University of Edinburgh's physiological laboratory angered the *B.M.J.* by its lack of tact. It editorialized in June 1873:

> It is evident that, unless the doings within the walls of the Edinburgh University Physiological Laboratory are kept publicly quiet, it is very probable that the whole question of experiments on living animals will be once more brought before the public in the usual sensational style, to the injury of harmless physiological research.[21]

The inclusion in John Burdon Sanderson's 1873 lectures at University College London, of a section "On the propriety of using the lower animals for the purpose of experimentation"[22] indicated at least some sense by physiologists of popular concern about the subject. That that sense was woefully deficient was revealed in the same year with the publication of a two-volume work, *Handbook for the Physiological Laboratory*, edited by Burdon Sanderson.[23] The other contributors were Emanuel Klein, histologist and bacteriologist of the Brown Institution,[24] Michael Foster, physiologist of Cambridge,[25] and T. Lauder Brunton, pharmacologist of St. Bartholomew's Hospital, London.[26] The

[20] See *Animal World* i (1869–1870), 43, 58–59, 80, 125–126; ii (1870–1871), 46, 138; iii (1871–1872), 176; and *Pall Mall Gazette* (1871), passim; *Nature* ix (1873), 144–145, 242–243.

[21] Edinburgh University was the only British university with pretentions to experimental medicine before 1870. "Physiological Research on the Lower Animals," *B.M.J.* i (1873), 662.

[22] H. Cushing, *The Life of Sir William Osler* (London, 1940), p. 99. I am told that such lectures are still given in the course on general physiology at University College.

[23] J. Burdon Sanderson (ed.), *Handbook for the Physiological Laboratory* (London, 1873) 2 vols. On Burdon Sanderson see G. Burdon Sanderson, *Sir John Burdon Sanderson—A Memoir* (Oxford, 1911).

[24] W. Bulloch, "Emanuel Klein," *J. of Path. and Bac.* xxviii (1925), 684–697.

[25] G. L. Geison, "Michael Foster and the Rise of the Cambridge School of Physiology, 1870–1900" (Yale Univ. Ph.D. thesis 1970).

[26] W. F. Bynum, "Thomas Lauder Brunton," *Dictionary of Scientific Biography* (New York, 1970), ii. 547–548.

Handbook was the first manual for the physiological laboratory published in the English language. One volume contained precise step-by-step instructions for the repetition of dozens of classical experiments in physiology—successful solutions to problems in the subject. The second volume contained 123 plates illustrating the experiments. Burdon Sanderson, in the preface, wrote, "*This book is intended for beginners in physiological work.* It is a book of methods, not a compendium of the science of physiology, and consequently claims a place rather in the laboratory than in the study."[27]

The importance of the publication of the *Handbook* at this time can scarcely be overestimated. It embodied the vivisectional methodology of what was to Britain a new physiology, and thus disclosed to concerned lay readers the continental models adopted by the country's rising young physiologists. Including among its authors, as it did, men of key institutional positions, it was taken to presage widespread vivisection for instructional purposes in medical education. Failing to specify the use of anesthetics, as it largely did, it gave no assurance as to the painless nature of such vivisection. Prefaced for "beginners," as it was, it seemed to set no limits to callow youth being indoctrinated in animal experiment. The impact of the *Handbook* among laymen was not sudden and widespread, it was progressive and limited to those primarily interested in the problem of experiments on living animals. It is not clear how many readers became antivivisectionists from a perusal of the book, but innumerable references in the pamphlet literature of the movement make it clear that appeals to the *Handbook* as proof of the nature and extent of vivisection *in Britain* exerted a very powerful cumulative influence.[28] At last British antivivisectionists had a clear mandate to scrutinize the institutions and literature of British physiology for cruelty in the new and growing practice

[27] *Handbook,* vol. i, p. vii. Italics are my own.

[28] See, for example, *Prof. Burdon Sanderson's Portion of the "Handbook of the Physiological Laboratory": Extracts therefrom, with Explanations* (London, c. 1884).

of animal experiment. Witnesses before the Royal Commission were often asked their opinions of the *Handbook*,[29] and Burdon Sanderson was cross-examined at length on it.[30] He admitted that the preface was misleading, "beginners" being meant to apply to the small class of persons who wished to do actual research in physiology and not to all medical students, much less to any others who might study the subject. Further, he testified,

> May I make a general observation in reference to this book, namely, that we had not in view the criticisms of people who did not belong to our craft in writing it, and that we did not guard against all possible misunderstandings of that sort. It is generally understood that we use anaesthetics whenever we possibly can, and consequently that is a thing taken for granted. That ought to have been stated much more distinctly at the beginning in a general way; but it was not stated for the reason I have given.[31]

Unfortunately, the damage had already been done by 1875, when Burdon Sanderson spoke these words. Both the Report of the Royal Commission[32] and British teachers of physiology (including Foster and Burdon Sanderson)[33] agreed in attributing public feeling on the question to the

[29] "Report of the Royal Commission on the Practice of Subjecting Live Animals to Experiments for Scientific Purposes," *Parl. Papers 1876* xli (C.–1397), Q. 205–207, 354–356, 767, 970–971, 999, 1110–1111, 1270, 1351, 1623–1631, 1939–1942, 2170–2176, 2411, 2415, 3533, 3537, 5226, 5776–5790. Hereafter cited as "Report of the Royal Commission."

[30] "Report of the Royal Commission," Q. 2209–2284, 2746–2749.

[31] "Report of the Royal Commission," Q. 2265. Burdon Sanderson's 1873 lecture had declared painful demonstrations to be unjustified.

[32] "Report of the Royal Commission," p. viii. See also the R.S.P.C.A. submission on the *Handbook*, Appendix iv, pp. 658–663.

[33] *Memorandum of Facts and Considerations Relating to the "Cruelty to Animals Bill," by Teachers of Physiology in England, Scotland and Ireland* (London, 1876), p. 1. Cf. F. P. Cobbe, *Life of Frances Power Cobbe* (Boston and New York, 1894), ii. 567–568.

publication of the *Handbook* in 1873. As the *Saturday Review* wrote in 1876:

> It may be thought . . . surprising that the proceedings of so few persons—some of them persons whose names are scarcely known to the general public—should excite throughout the country a feeling of indignation so universal that the most remote country villages and the most obscure religious communities should express abhorrence of their acts by petitioning the legislature against them. The truth is, we imagine, that the present strong feeling on the subject is mainly due to the publication of the notorious *Handbook for the Physiological Laboratory*, the tendency of which was supposed to be to import into this country not only the practices, but the principles, which are believed to prevail in France and Germany with respect to the relation of man to animals. The appearance of this book seemed ominous. It was regarded as the introduction into England of a new moral contagion . . .[34]

The *Handbook* provided ample cause for vigilance on the part of the animal protection movement. The movement still, however, lacked a cause celebre.

1874: A CONTROVERSY SUSTAINED

In late 1873 a surge of agitation against the experiments of Moritz Schiff, Professor of Physiology at the Royal Superior Institute in Florence, proved the starting point of a lengthy discussion of the rights and wrongs of vivisection in British periodicals. This discussion directed increased attention to the *Handbook for the Physiological Laboratory* and to the issue of animal experiment in Britain. An article by the *Times's* Rome correspondent,[35] very favorable to

[34] *Saturday Review* xli (1876), 773.

[35] *Times* (London), 24 December 1873. The original agitation against Schiff had been initiated in 1863 by the English colony in Florence. See F. P. Cobbe, *Life of Frances Power Cobbe* (London, 1894), ii. 561–565. For a synopsis of an 1863 pamphlet written by Schiff in his own defense, see "Schiff on Vivisection," *British and Foreign Medico-Chirurgical Review* lvii (1876), 134–136.

efforts against Schiff's experiments, was followed by a number of similar articles in the *Morning Post* and other newspapers. George Macilwain wrote to the *Times* and he was joined in the *Times* and the *Spectator* by a second medical critic of vivisection, A. De Noë Walker. Walker condemned the practice in strong terms and called for its legal restriction, though not its abolition.[36] The first public defender of the practice was not a medical man but a zoologist, Edwin Ray Lankester (1847–1929), a student of T. H. Huxley. Lankester, who was no man to pour oil on troubled waters, enthusiastically vindicated Schiff and vivisection in general, helpfully adding that experiments on living animals took place in Britain as well as on the continent (despite the *Times* article's report to the contrary). In a second letter, Lankester outdid himself with a prescient but impolitic dictum, which was to be cited again and again by antivivisectionists: "If you allow experiment at all, you must admit the more of it the better, since it is very certain that for many years to come the problems of physiology demanding experimental solution will increase in something like geometrical ratio, instead of decreasing."[37]

Talk like this was quite enough for Richard Holt Hutton, editor of the *Spectator*, who declared that there could be no justification for painful experiments on living animals and that physiologists could not be trusted to limit themselves (an obvious reference to "geometrical" rates of increase). Perhaps, wrote Hutton, scientific advances had been made through vivisection, but one could not approve any painful vivisection unless one were willing to approve all that any qualified scientist wishes to perform. It was only too clear where such a policy would lead.[38]

It did not take the medical press long to realize that the discussions in the *Times* and the *Spectator* were doing the cause of experimental medicine more harm than good.

[36] *Spectator* xlvii (1874), 47, 110–111. Walker was a former military surgeon who had been a student in physiological laboratories on the continent. "Report of the Royal Commission," Q. 1703, 1707.

[37] *Spectator* xlvii (1874), 13–14, 46–47.

[38] *Spectator* xlvii (1874), 44–45, 176–178, 591–592.

Only the *Lancet* ventured to "hope the discussion will not close until a solution be arrived at in accordance with the dictates of humanity and the interests of science."[39] The *Medical Times and Gazette* sniffed at the public ignorance displayed and regretted the unproductive nature of the discussion.[40] Where earlier medical men attacking vivisection had been generally ignored, now the *Medical Times and Gazette*,[41] the *British Medical Journal* and the *Practitioner* closed ranks in explicitly criticizing Macilwain and Walker in order to exert professional pressure upon them. The *B.M.J.* referred to Macilwain's nom de plume ("Author of *Medicine and Surgery One Inductive Science*") and dismissed him: ". . . persons . . . are often found of eccentric character but high pretensions, who are ignorant of the fundamental facts of research with which they profess to be conversant."[42] The *Practitioner* was more patronizing: "Mr. de Noë Walker is presumably a young man and the temptation of occupying a large space in a public sensational discussion has probably been too much for his common sense as well as for his sense of justice and propriety."[43] Later medical men who opposed vivisection found themselves subjected to similar criticism, labeled by antivivisectionists as "trades-union" measures.

The conversion of Richard Hutton (1826–1897) to the cause of antivivisection was highly significant. Mathematician, theologian, editor, Hutton was a brilliant and powerful figure whose opinions clearly mattered in the intellectual life of London. He did not wait long to turn his newly enunciated principles into action. As a member of the Senate of the University of London, Hutton launched a concerted campaign to end experiment upon animals at the Brown Institution, a university affiliate for research on the diseases of animals. Hutton's lengthy and unsuccess-

[39] *Lancet* i (1874), 22–23.
[40] *Medical Times and Gazette* i (1874), 10, 70–71.
[41] Ibid.
[42] *B.M.J.* i (1874), 18–19.
[43] *Practitioner* xii (1874), 38–39.

ful battles, which ultimately led to his tendering his resignation to the Senate, were widely reported.[44]

Hutton's efforts kept the issue before the public when the controversy initiated by the agitation against Schiff began to die down. The *Medical Times and Gazette* was sharply critical of Hutton,[45] but the *Lancet*, while disagreeing with him, was pleased at his directing further attention to the question, for it felt that enlightened public opinion rather than legislation must be the pathway to a solution.[46] A relatively uninfluential medical monthly, the *Doctor*, condemned the Senate for rejecting Hutton's resolution, and unqualifiedly contradicted John Burdon Sanderson's defense of the Institution.[47] Medical opinion was as divided as public opinion, if the medical press can be taken as any indication.

The discussion surrounding the Schiff and Brown Institution affairs was undoubtedly followed closely in the offices

[44] See R. H. Hutton, *The University of London and Vivisection* (London, 1876), and R. H. Hutton to J. S. Burdon Sanderson, 29 May 1875, The Library, University College, London. Burdon Sanderson Papers, Ms. Add. 179/2 f.21. For any amount of detail on subsequent attacks on the Brown Institution, see G. R. Jesse's two pamphlets, *The Animal Hospital Founded by Thomas Brown* (London, 1876) and *Man's Injustice to Animals. The Brown Animal Sanatory Institution* (2nd ed., London, 1888). See also "The Brown Institution," *Verulam Review* i (1888–1889), 133–144, and *Lancet* i (1883), 795.

[45] *Medical Times and Gazette* i (1874), 534.

[46] *Lancet* i (1874), 733–734; see also "Debate on Vivisection at the Meeting of the Convocation of the University of London," *Lancet* i (1874), 709–710.

[47] *Doctor* iv (1874), 101, 140. The *Doctor* was published in London, 1871–1878. In 1875, it declared vivisection to be justifiable for purposes of original research but not for demonstration or "sensation." Further, ". . . the cruelties to which we have adverted are regularly practised at the Brown Institution, and that for the purpose of teaching. In fact, that institution may be looked upon as the centre of vivisection, whence its pupils go forth to practise elsewhere. What would the benevolent Mr. Brown have thought had he foreseen to what his legacy for a hospital for animals would lead! We commend the case to the Royal Society for the Prevention of Cruelty to Animals". See *Doctor* v (1875), 21–22.

of the R.S.P.C.A. The issue was dealt with at length at the Sixth International Congress of Societies for the Prevention of Cruelty to Animals, held in London in June 1874. After the reading of a letter from Queen Victoria, alluding to "sufferings . . . from experiments in the pursuit of science,"[48] the congress heard a number of speakers on the subject of vivisection. De Noë Walker renewed his attack on the practice and called for its regulation by law, adducing a number of vivid examples in support of his views. Richard Hutton and John Colam, secretary of the R.S.P.C.A., spoke in a similar vein. According to Benjamin Ward Richardson, however, stopping vivisection would cause more suffering than it would prevent. Richardson rejected statements that experiments on living animals are useless and argued ably for their indispensability to medicine. He opposed suggested restrictions and inspection of experiments: "If . . . you strive to suppress experiment, by making it impossible in open day, you will only relegate it to dark places in which it should never be concealed." Colam challenged the existing secrecy with regard to experiments on living animals and, in the end, the congress passed Colam's resolution "that painful experiments on living animals, if not already illegal, should be forbidden by law except under licence and precautions for publicity, and that no experiments on living animals should be permitted except under the same precautions."[49] There could be no doubt that a strong antivivisection policy was now an idea whose time had come. The problem of vivisection moved rapidly higher in the priorities of the R.S.P.C.A., though its annual report in August 1874 revealed no specific

[48] "Report of the Royal Commission," Appendix i.

[49] See *Animal World* v (1874), 134; "Congress of the Society for the Prevention of Cruelty to Animals," *B.M.J.* i (1874), 847; "The Societies for the Prevention of Cruelty to Animals and Vivisection," *Medical Times and Gazette* i (1874), 696–697; B. W. Richardson, "On Experimentation on Animals for the Knowledge and Cure of Disease," *Lance*t ii (1874), 5–7. Richardson's attitude toward animal experiment became progressively less favorable as he aged. See L. G. Stevenson, "Science Down the Drain," *Bull. Hist. Med.* xxix (1955), 1–26.

proposals for legislation. Instead, Colam pledged to the membership continued publicity to abuses connected with experiments on living animals, stating that discussions had already "tended to check the practices of experimental physiologists." The executive of the society, Colam said, would remain vigilant for any opportunity to act.[50]

R. H. Hutton is one figure crucial for an understanding of how antivivisection developed from disorganized and impotent strands of opinion into an agitation with national and international repercussions. John Colam, secretary of the R.S.P.C.A. from 1861 to 1905, is a second. Colam's indefatigability, ingenuity, and tact dominated the R.S.P.C.A. during its greatest days in the latter part of the nineteenth century. It was during that time that the society first achieved broad scale enforcement of animal protection laws, greatly extended such legislation, and undertook systematic education in humanitarian values. Colam's R.S.P.C.A. of the mid-seventies was highly organized (instigating hundreds of prosecutions a year), wealthy (receiving thousands of pounds in subscriptions and legacies annually), prestigious (with highly cultivated and very public aristocratic connections), and powerful (with influential members in both Houses of Parliament). It had its own periodical and virtually unlimited access to others.[51] Into the hands of this humanitarian monolith the British Medical Association (B.M.A.) marched, obliging with a particularly ill-chosen experimental demonstration at its annual meeting at Norwich in August 1874. In the aftermath, the train of events toward a national agitation could not have been reversed.

At that meeting, a French experimentalist, Eugene Magnan, was invited to lecture on the physiological effects of alcohol. After his lecture, a separate experimental demonstration of the induction of epilepsy in the dog by intravenous injection of absinthe was to be carried out for those

[50] *Animal World* v (1874), 118.

[51] My discussion of the R.S.P.C.A. here and elsewhere in this study owes much to a lecture by B. H. Harrison at Corpus Christi College, Oxford on 12 May 1970.

medical men who were interested. At the appointed time and place, with two canine subjects appropriately bound on an experimental bench in front of him, Magnan spoke briefly and commenced his injection into a vein in the thigh of the first dog. What happened thereafter is not completely clear, but something of a scene took place. Protests at the cruel and unnecessary nature of the experiments were voiced by certain of those present, notably Samuel Haughton (1821–1897) of Trinity College, Dublin; T. Jolliffe Tufnell (1819–1885), President of the Royal College of Surgeons of Ireland; and a layman who happened to be present quite by accident, one Knight Bruce. During the spirited dispute that followed, a vote was taken which resulted in a considerable majority of those present in favor of the experiments proceeding. At some point Jolliffe Tufnell cut the restraints of one of the experimental animals, leaving the staggering beast to be apprehended by Magnan. When Tufnell returned with two county magistrates, the gathering broke up in confusion.

The vigilant Colam now had his opportunity. Informed of the B.M.A. debacle—perhaps by a Mr. Smith of the Norwich R.S.P.C.A., who had attended the sessions of the congress—he dispatched one of the society's inspectors, Richard Roard, to the scene to make inquiries. In October, Colam formally proceeded against Magnan and three Norwich doctors responsible for arranging the demonstration, charging them with wanton cruelty to a dog under Martin's Act (which had been amended to include all domestic animals within its purview). He prosecuted the case in person at Norwich Petty Sessions, 9 December 1874. With Magnan in France, only the three Norwich medical men appeared as defendants.

Richard Roard testified as to arrangements for the demonstration and as to purchase of the dogs, which had been carried out by the defendants. Knight Bruce, Jolliffe Tufnell, and Samuel Haughton described the events in question with vehemence and indignation. The last two, as expert medical witnesses, testified as to the worthless and unnecessary nature of the experiments. Vivisection in general was condemned as a demoralizing and dangerous practice. Tufnell's and Haughton's opinions were augmented by the testimony of the president of the B.M.A. for 1875, Sir William Fergusson (1808–1877). Fergusson was sergeant-surgeon to Queen Victoria, and among the most influential surgeons in the country. Although Fergusson admitted he was not au courant with current research on the physiological effects of alcohol, and had not attended the demonstration, he did not hesitate to stigmatize Magnan's experiments in terms similar to those of the previous witnesses.

Defense contentions that the prosecution had failed to demonstrate the involvement of the present defendants in the operation, which was supposed to have constituted the alleged cruelty, proved successful when the nine-man Board of Magistrates handed down such a judgment. The Board further held, by a five to four majority, that the R.S.P.C.A. was justified in instituting the proceedings, and denied a defense motion for costs.[52]

Press reaction to the various events of 1874 shows steadily increasing public interest in the issue of experimentation on living animals, culminating in national coverage of the Norwich trial. Editorial opinion during 1874 was extremely suspicious of the practice of vivisection. Both *Punch* and *Saturday Review*, ultimately to be bitterly attacked by anti-vivisectionists for their support of medical interests, were in 1874 highly critical of vivisection, though they would not see it completely prohibited.[53] Experimental medicine seems to have been in retreat before an increasingly nega-

[52] For accounts of the demonstration and the trial see the *Norwich Chronicle* (12 December 1874); *R.S.P.C.A. Records* xiv (1873–1875), 83–89; "Prosecution at Norwich. Experiments on Animals," *B.M.J.* ii (1874), 751–754; "Vivisection at Norwich," *Medical Times and Gazette* ii (1874), 691; "Vivisection Prosecution at Norwich," *Lancet* ii (1874), 851–852; "Vivisection at Norwich," *Lancet* ii (1874), 348; *Vivisection Prosecution. Report of a Prosecution of Physiologists by the R.S.P.C.A., at the Town Hall, Norwich, for alleged cruelty to two dogs* (London, 1875). It seems very likely from its content that this last pamphlet was published by the R.S.P.C.A. itself.

[53] "Vivisection," *Saturday Review* xxxvii (1874), 46–47; "Cruelty to Animals," *Saturday Review* xxxvii (1874), 807–808; "Vivisection and Cheek," *Punch* lxvi (1874), 28; and "Vivisection and Science," *Punch* lxvii (1874), 257.

tive public opinion. Word of the Norwich dispute reached the public in garbled forms even before the legal proceedings,[54] and the trial itself undoubtedly confirmed suspicions and generated new ones throughout the country. The reaction of the *Liverpool Daily Post* was probably not untypical: "We are not ourselves disposed to go to the length of saying that vivisection ought in no case to be allowed. But we emphatically declare that we do not trust the physiological conscience in this matter." Clearly, the R.S.P.C.A.'s legal defeat at Norwich was a moral and propaganda victory. If the applicability of Martin's Act to scientific experiments on living animals was left in doubt, the cruelty of physiologists was not. The advice of the Liverpool editor, therefore, was "The Society for the Prevention of Cruelty to Animals are bound, if they wish to continue in existence, to persevere till they have tested the applicability of the law, in its present state, to vivisection."[55] The implication was that should Martin's Act prove insufficient, special legislation would be necessary. December 1874 saw the first two antivivisection pamphlets, *Need of a Bill* and *Reasons for Interference*, appear in London.[56]

The medical press remained greatly divided in this atmosphere of increasing urgency. The *B.M.J.* adopted the hard line that was to characterize it throughout the controversy. It justified Magnan's experiments and gave a blistering criticism of Tufnell and Sir William Fergusson: "The quality of mind of gentlemen who can go into court as accusers of their brethren on a matter as to which, according to their own statements, they are profoundly ignorant, is certainly somewhat beyond ordinary appreciation."[57] Despite Fergusson's eminence in the medical world and in

[54] *Lancet* ii (1874), 348.

[55] For cross-sections of press response to the Norwich trial, see "The Vivisection Trial at Norwich," *Public Opinion* xxvi (1874), 763–764, and "Prosecutions against Vivisections," *Animal World* vi (1875), 2–3. *The Liverpool Daily Post* is quoted in the former.

[56] I have been unable to locate these two pamphlets by Frances Power Cobbe. See *Dates of the Principal Events connected with the Anti-Vivisection Movement* (London, 1897), p. 1.

[57] "M. Magnan's Experiments," *B.M.J.* ii (1874), 828–829.

the B.M.A. (which published the *Journal*), editor Ernest Hart (1835–1898) maintained strong pressure upon him in subsequent issues.[58] Moreover, according to Hart, "We do not hesitate to stigmatise the prosecution as a grave misuse of the forms of justice. A more lamentable display of fanatical ignorance on the part of the prosecution and a board of local magistrates has never been witnessed in a court of justice."[59] The *Medical Times and Gazette* and the *Lancet,* on the other hand, found the prosecution quite understandable and reasonable. The *Medical Times and Gazette* felt "that the manner, time and place of performing the experiments were ill-chosen and injudicious" and was "not at all satisfied that such a demonstration . . . was either necessary or useful. . . . the whole proceeding was an unhappy mistake."[60] The *Times and Gazette* was also critical of Fergusson.[61] The *Lancet* justified Magnan's experiments, and, while not condemning the R.S.P.C.A., pointed out that ". . . the question cannot be properly discussed except by those who possess the physiological knowledge which shall enable them to distinguish between appearances and realities." To keep the important decisions from resting with such ignorant bodies as the Norwich Bench of Magistrates, the *Lancet* was willing to countenance some formal arrangement. It rejected arguments that it was impossible to draw a firm line between justified and unjustified experiments with the contention that it was at least possible to ensure that experimenters were skilled persons with adequate scientific objectives. "We believe . . . that an attempt might be made to institute something in the way of regulation and supervision. . . . Someone, in whose

[58] "The Recent Vivisection Trial at Norwich," *B.M.J.* i (1875), 25–26 and "Sir William Fergusson and the Norwich Prosecution," ibid., 63. For an antivivisectionist defense of Fergusson, see "Vivisection at Norwich," *Medical Press and Circular* i (1875), 149–150.

[59] See note 57. This was surely hard on the magistrates, who betrayed themselves by no more than the occasional ejaculation at the more sensational points of evidence, and by denial of costs.

[60] "Vivisection at Norwich," *Medical Times and Gazette* ii (1874), 691.

[61] "Vivisection Again!" i (1875), 116–117.

knowledge, judgement, and character the profession and the public had confidence, might be invested with the requisite authority to inquire into and report upon the aims and methods of such experimenters, whenever and wherever any investigation was necessary."[62] It was clear that such arrangements were unlikely to evolve in the absence of legislation.

A favorable public opinion and a divided and confused medical press, making frantic efforts to force recalcitrant members to toe the professional line, signalled a golden opportunity for organized animal protection to step in with a coordinated campaign. The momentum of Norwich was indeed to invigorate the antivivisection movement, putting experimental medicine on the defensive for nearly a decade and threatening its vulnerable foothold in British institutions.

[62] *Lancet* i (1875), 19–21.

4. The Politics of Experimental Medicine

The publication of the *Handbook for the Physiological Laboratory*, the enthusiastic predictions by Ray Lankester in the *Spectator*, the campaign by physiologists and their medical allies to have experimental medicine on the continental model entrenched in British medical schools, all seemed part of an ominous pattern of events to antivivisectionists. Until Norwich, however, there had been few avenues of practical action open to them. The R.S.P.C.A. was, by the seventies, a large, multifaceted organization unprepared to move too far ahead of public opinion and thus jeopardize any part of its very substantial constituency. Its prosecution of what it took to be obvious abuse at Norwich was to initiate a train of events beyond its control. The publicity surrounding the Norwich affair permitted the evolution of an explicitly antivivisectionist movement, a movement whose independent power derived from a somewhat different public than the R.S.P.C.A.'s and whose relations with the larger society were ambivalent to say the least.

Slowly the leading actors in the Victorian vivisection controversy emerged. Michael Foster, Cambridge physiologist, wrote a defense of vivisection for *Macmillan's Magazine*.[1] Richard Hutton used the *Spectator* and his prominence in London life to draw public attention to the issue. John Colam, secretary of the R.S.P.C.A., capably pursued the Norwich affair and remained the most important influence within the movement for animal protection. It was, however, the somewhat improbable figure of Frances Power Cobbe who grasped the initiative during the crucial months

[1] "Vivisection," *Macmillan's Magazine* xxix (1874), 367–376.

in the aftermath of Norwich. By doing so, she was to dominate the new movement until she became in the public mind the personification of antivivisectionism (see Figure 2).

FIRST STEPS: STRATEGIES FOR RESTRICTION

Cobbe (1822–1904) was an indomitable Anglo-Irish spinster; as a journalist she wrote a great deal on feminist, religious, and philanthropic topics.[2] She was well known in London literary circles, writing editorials for the *Echo* newspaper from 1868 to 1875. Cobbe had been sensitized to vivisection as an issue by the R.S.P.C.A.'s campaign in France during the sixties. In 1863, she led the English community in Florence against the experiments of Moritz Schiff, Professor of Physiology at the Royal Superior Institute. Her wide reputation and extensive personal acquaintance with the likes of Benjamin Jowett, J. A. Froude, W. E. H. Lecky, James Martineau, Bishop Colenso, and Herbert Spencer were to stand her in good stead in her attempts to legitimize the antivivisection movement. She brought other assets to the movement as well, as the following brilliant, if hostile, capsule description indicates:

> For superabounding energy, for absolute self-conviction, for magnificent unscrupulousness of assertion, for entire imperviousness alike to reason and ridicule, and for inextinguishable eloquence, especially in the direction of vituperation, she is, in these latter days, almost, if not quite, unsurpassed. And for a public agitation of any kind, these are the gifts required; the indispensable, if not the only, requirements for practical success.[3]

As Cobbe reflected on the Norwich trial, it became evident to her that Martin's Act was insufficiently strong to deal with abuses of vivisection. In particular, she felt that

[2] See her autobiography, *Life of Frances Power Cobbe* (Boston and New York, 1894), 2 vols. Discussion of Cobbe's activities and of events with which she was concerned is based upon her autobiography, especially vol. ii. 556–634, unless otherwise noted.

[3] C. Adams, "The Anti-Vivisection Movement and Miss Cobbe," *Verulam Review* iii (1892–93), 201.

Figure 2. Frances Power Cobbe (1822–1904), philanthropist, journalist, feminist, founder and doyenne of Victorian antivivisectionism. *Courtesy of the Wellcome Trustees.*

the Act would require medical testimony that given operations were not only painful, but scientifically useless, in order to obtain convictions. Clearly, such testimony would rarely be forthcoming, for it was to just such a purpose that Tufnell, Fergusson, and others had been belabored by the medical press. Precisely as the *Lancet* had wished for formal procedures that removed decisions about animal experiments from the power of lay ignorance, Cobbe strove for special legislation removing such decisions from the sphere of medical and scientific self-interest.

It was clear to Cobbe that the body with the wealth, organization, and power to effect the needed legislation was the R.S.P.C.A. In January 1875, a few weeks after the Norwich trial, she drew up a memorial to the committee of the R.S.P.C.A. urging the organization to mobilize to restrict the practice of vivisection. She circulated the memorial together with her two pamphlets, *Reasons for Interference* and *Need of a Bill,* to a large number of influential people, succeeding in obtaining some six hundred signatures, "every one of which represented a man or woman of some social importance."[4] The R.S.P.C.A. was aware of Cobbe's activity and amenable to receiving her memorial, but the memorial was not, as the *B.M.J.* believed, "issued . . . in blank from the offices of the Society."[5] Indeed, though there can be no doubt where Colam's sympathies lay—and he was the most powerful single individual in the society—the Norwich prosecution had aroused a significant medical and scientific interest group within the R.S.P.C.A. Even before presentation of the memorial by an influential deputation on 25 January 1875, the *Animal World* was warning its readers, "If the committee [of the society] should appear to be less energetic than the deputation, it must be remembered that they are equally opposed to the practices complained against, and their reticence is the natural consequence of responsibility and personal contact with various opposing forces."[6]

Nevertheless, the impact of the memorial, both on the

[4] *Life of Frances Power Cobbe,* ii. 570.

[5] "Experiments on Animals," *B.M.J.* i (1875), 159.

[6] *Animal World* vi (1875), 19.

members of the General (i.e., gentlemen's) and Ladies' Committees of the society, to whom it was directly addressed, and on the general public, must have been considerable. The memorial was signed by eminent members of the church, the aristocracy, the military, and the House of Commons, with a smattering of literary figures, educators, and medical men. Signatories included the Archbishop of York, the Archbishop of Westminster (Manning), the Lord Chief Justice, John Bright, Tennyson, Browning, Carlyle, Martineau, Ruskin, Lecky, Jowett, and F. W. Newman. The memorial-cum-deputation ploy was by this time a tried and true means of obtaining for an issue the requisite public and institutional visibility, but no one could deny that Cobbe had executed it uncommonly well.

After alluding to the recent spread of the practice of experiments on living animals ("A recent correspondence in *The Spectator* shows that many English physiologists contemplate the indefinite multiplication of such vivisections"), the memorial instanced a number of abuses of the practice, which it "characterized as gratifications of the 'dilettantism of discovery.'" If the events at Norwich exemplified what might take place in public, the memorialists asked, how much more is there to fear what "inconsiderate young students" and others might do in private laboratories. The memorial suggested that the R.S.P.C.A. form a subcommittee to look into practical steps to restrict vivisection and that it instruct Colam to further test the applicability of Martin's Act by prosecutions. Should this latter step prove inadequate, the R.S.P.C.A. might consider special legislation to prohibit painful experiments except by registered persons in authorized laboratories, to prohibit painful experiments in illustration of lectures, and to render liable publishers of periodicals printing accounts of cruel experiments. The last provision was recommended since ". . . the ambition for scientific notoriety . . . may be deemed a not insignificant motive for the performance of many of these experiments."[7]

[7] For an account of the deputation, the text of the memorial, and a list of the more eminent signatories, see "Memorial against Vivisection," *Animal World* vi (1875), 38.

The response of the R.S.P.C.A. to the proposals of the memorial was crucial. Really decisive action by the society at this point, of the kind Cobbe envisioned it taking, might have maintained for it an exclusive franchise on effective action from the humanitarian side and an undivided constituency in antivivisectionist opinion. Instead, the society would not or could not react with more than a restrained fact-finding enterprise, as the notice in the *Animal World*, probably by Colam, had foreseen. The organization was undoubtedly hamstrung by a great variety of opinion on the subject in question within its membership, and was forced to begin a lengthy campaign of justification for its caution with regard to the issue.[8] Richard Hutton, in the *Spectator*, charged the society with intentionally dragging its feet in the face of "great professional influences raised against them" and foresaw "that if they do not show a little more courage and a little more zeal, some other Society will grow up in their place which, by boldly doing the work from which they shrink, will succeed in their popularity and influence."[9] Cobbe was understandably disgusted at the outcome of her efforts:

> . . . I was requested to attend (for the occasion only) the first meeting of the sub-committee for vivisection of the R.S.P.C.A. On entering the room my spirits sank, for I saw round the table a number of worthy gentlemen, mostly elderly, but not one of the more distinguished members of their committee or (I think) a single Peer or Member of Parliament. In short, they were not the men to take the lead in such a movement and make a bold stand against the claims of science. After a few minutes the chairman himself asked me: "Whether *I* could not undertake to get a bill into Parliament for the object we desired?" As if all my labor with the memorial had not been spent to make *them* do this very thing! It was

[8] See J. Colam, *Vivisection. The Royal Society for the Prevention of Cruelty to Animals and the Royal Commission* (London, 1876), esp. pp. i–ii.

[9] "The Moral Consequence of Vivisection," *Spectator* xlviii (1875), 369–370. See also 144.

> obviously felt by others present that this suggestion was out of place, and I soon retired, leaving the sub-committee to send Mr. Colam round to make enquiries among the physiologists, a mission which might, perhaps, be represented as a friendly request to be told frankly "whether they were really cruel." I understand, later, that he was shown a painless vivisection on a cat and offered a glass of sherry; and there (so far as I know or ever heard) the labors of the subcommittee ended.[10]

The R.S.P.C.A., represented by Colam, was later to make voluminous and important submissions to the Royal Commission, but in February 1875, Cobbe had not the patience to bear with the moderation of the eminent gentlemen who ran the society. Colam's requests that medical societies permit him and two other R.S.P.C.A. representatives to attend the next vivisection performed by the society were treated just as cavalierly by the profession as they were by Cobbe. The *Lancet* asserted, ". . . we suspect that the real object of the circular is to get as many refusals as possible to the absurd request, and then to print them and circulate them as proofs that vivisection is extensively practised, and that the operators preserve the utmost secrecy."[11] As Cobbe knew only too well, the R.S.P.C.A.'s endeavors were nowhere so Machiavellian. How far the R.S.P.C.A. was from militant antivivisectionism was emphasized when its president, Lord Harrowby, wrote anxiously to the *Times* to disavow on behalf of the society any responsibility for the statements of Cobbe's memorial, or for its promotion.[12]

Discussion of vivisection continued apace in both the medical and daily press. The leading medical journals and the *Times* called for a dispassionate commission of inquiry, having the benefit of scientific expertise.[13] Clearly there was a fear that the public sentiment displayed in sensational correspondence in the likes of the *Echo, Morning*

[10] *Life of Frances Power Cobbe* ii. 577–578.
[11] *Lancet* i (1875), 246–247.
[12] *B.M.J.* i (1875), 186.
[13] *B.M.J.* i (1875), 159; Lancet i (1875), 173–174, 204.

Post, Morning Advertiser, and *Daily News* might be channelled into support for legislation similar to that proposed in the memorial.

It was precisely such a letter, published in the *Morning Post* (2 February 1875), that so further aroused public opinion that Hutton, Cobbe, and other activist antivivisectionists no longer needed the wealth and power of the R.S.P.C.A. to gain access to Parliament. Of all medical testimony against vivisection, before or since, George Hoggan's letter had the greatest impact. Hoggan was a pensioned officer of Her Majesty's Navy, who had recently taken his M. B. at Edinburgh University as "a sort of pastime," going on to spend four months in the laboratory of Claude Bernard in Paris.[14] His letter was a feeling and eloquent protest against experiments on animals, based upon his Paris experiences. Some familiar features of later antivivisectionist literature were blended by Hoggan with surpassing skill, in this his first attempt at the genre: pathetic and appealing experimental subjects, cynical and indifferent physiologists, ghastly laboratory detail, inefficacious anaesthetics.[15] This and later Hoggan efforts of 1875 were a major force in creating and maintaining antivivisectionist public opinion;[16] his status as a "medical man" was an important factor. Hoggan's assertion that "I am inclined to look upon anaesthetics as the greatest curse to vivisectible animals" made a tremendous impression.[17]

[14] "Report of the Royal Commission," Q. 3433, 3496.

[15] *Morning Post* (2 February 1875), reprinted in *Spectator* xlviii (1875), 177–178; *Life of Frances Power Cobbe,* ii. 579–580; G. Hoggan, *Vivisection* (London, 1876); and Appendix II.

[16] See G. Hoggan, "Vivisection" *Fraser's Magazine* n.s. xi (1875), 521–528; *Anaesthetics and the Lower Animals* (London, 1875); *Urari or Curare* (London, 1875); *Paul Bert's Observations on a Curarized Dog* (London, 1875).

[17] The assertion, from *Anaesthetics and the Lower Animals,* was in fact Hoggan's principal legacy to the movement. His appearance before the Royal Commission was unimpressive and he was dominated by Cobbe in later institutional efforts. Hoggan's statements as to the extreme difficulty of anesthetizing lower animals were strongly disputed. See "Vivisection and Anaesthetics," *B.M.J.* i (1875), 749, 829–830.

Cobbe contacted Hoggan soon after his first letter appeared (they had originally met at the home of feminist Barbara Bodichon) and "thenceforth took counsel frequently as to the policy to be pursued in opposing vivisection."[18] Her effective freedom from the R.S.P.C.A. owed to a state of public opinion in which the *Morning Advertiser* could refer to "the demoniacal practice of vivisection" and "the cynical coxcombry of science"[19] and in which a society calling for the total abolition, rather than the mere restriction, of vivisection could achieve sufficient support for an expensive series of advertisements in the *Times* and the *Morning Post.* Without Cobbe, Hutton, or Hoggan as yet prepared to go as far as calling for total abolition of vivisection, their demands for legislative restriction looked like a moderate and reasonable response to public concern.

Cobbe seems to have had little difficulty in developing a legislative proposal for the restriction of vivisection. Making use of her connections, notably with the Conservative whip, Sir William Hart Dyke, she arranged to have a bill for "Regulating the Practice of Vivisection" drafted by Sir Frederick Elliot and approved by a number of other parliamentarians, notably Lord Chief Justice Coleridge and Robert Lowe, M. P. for the University of London. The bill was presented in the House of Lords by Lord Henniker (Lord Hartismere) on 4 May 1875.[20]

Henniker's bill provided that vivisection be confined to premises annually registered with the Home Secretary for such purposes and subject to inspection. Anesthetics were to be used in all experiments, except under a six-month personal licence costing £10 and issued at the discretion of the Secretary. The influence of Hoggan's expose of curare—revealed by Bernard to affect only nerves of motion

[18] *Life of Frances Power Cobbe,* ii. 580.

[19] *Morning Advertiser,* 13 May 1875, quoted in "Sir Henry Thompson on Vivisection," *Public Opinion* xxvii (1875), 608–609. Most public expression was more measured but no less adamant. See, for example, "Vivisection" *Leisure Hour* (1875), 515–518, 620–621, and cf. T. Watson, "Vivisection," *Contemporary Review* xxv (1875), 867–870.

[20] *Life of Frances Power Cobbe,* ii. 581.

and not nerves of sensation—was evident in the provision that curare not be deemed an anesthetic for the purpose of the bill. Penalties under the bill were to be no more than £20. Vivisection was defined as "the cutting or wounding or treating with galvanism or other appliances, any living vertebrate animal for purposes of physiological research or demonstration, also the artificial production in any living vertebrate animal of painful disease for purposes of physiological research or demonstration."[21]

Cobbe's strenuous attempts to crystallize antivivisectionist sentiment into active steps to counter the abuse did not go unnoticed by the handful of physiologists, biologists, and medical men whose scientific interests were at stake, although the *Lancet* taxed the scientific interest for its seeming indifference:

> On the other side, the question has been conducted as if it concerned physiologists alone, who were to be a law unto themselves, and each to do what seemed right in his own eyes; that the matter was one into which outsiders had no right whatever to intrude; in fact, "whatever is, is right," and so unquestionably right as to stand in no need of investigation or restriction. We have from the first striven to take a middle course . . .[22]

A highly influential group had in fact determined in early 1875 to prepare and present a bill that would appease popular feeling and forestall the danger that humanitarian legislation, by the R.S.P.C.A. or some other body, might pose to scientific research.[23] Charles Darwin, who had re-

[21] For the text of the bill, see "Report of the Royal Commission," Appendix iii, section 6.

[22] *Lancet* i (1875), 204. The *Lancet* was always somewhat more moderate toward antivivisection than the hard-line *B.M.J.*

[23] My description of this activity is based upon F. Darwin (ed.), *Life and Letters of Charles Darwin* (London, 1887), iii. 201–204; F. Darwin and A. C. Seward (eds.), *More Letters of Charles Darwin* (London, 1903), ii. 435–436; L. Huxley, *Life and Letters of Thomas Henry Huxley* (London, 1900), i. 437–438; W. Reid, *Memoirs and Correspondence of Lyon Playfair* (London, 1899), p. 220; Imperial College of Science and Technology, Playfair Papers, #120–123, 176,

fused to sign Cobbe's memorial when circularized,[24] wrote shortly thereafter, in January 1875, to T. H. Huxley proposing the preparation of a legislative measure to take the wind out of the enemy's sails.[25] Neither Darwin (1809–1882), by 1875 a famous and popular figure, confined by illness largely to his home in Kent, nor Huxley (1825–1895), then the most powerful man in British science, was a physiologist and neither performed experiments on living animals. But their strong sympathy with British physiology, at that time struggling to emerge from the German and French shadow, stimulated them to initiate a petition and a bill designed to protect the interests of experimental medicine. Through February and March, they conferred closely with John Burdon Sanderson, Professor of Physiology at University College, London. Burdon Sanderson took responsibility for coordinating consultations with physiologists, anatomists, and medical men in London, Oxford, Cambridge, and Edinburgh. Among those contacted were Sir John Simon (1816–1904), then in his last year as medical officer of the Local Government Board; William Sharpey (1802–1880), Professor of Physiology and Anatomy at University College London, 1836–1874; Sir Robert Christison (1797–1882), Professor of Toxicology at Edinburgh, 1822–1877; Sir Henry Acland (1815–1900), Regius Professor of Medicine at Oxford, 1858–1894; George Rolleston (1829–1881), Professor of Anatomy and Physiology at Oxford, 1860–1881; Sir William Gull (1816–1890) an eminent London physician; and Michael Foster (1836–1907), Praelector and Professor of Physiology at Cambridge, 1870–1903.

From this group, a scientists' lobby to protect experimental medicine was to emerge. Even within such a limited

---

181, 182, 206, 207, 280, 366, 455, 469; I.C.S.T., Huxley Papers, 5.316, 5.322, 15.135; The Library, University College London, Burdon Sanderson Papers, MS Add. 179/2 ff. 1–23, 70, 71, MS Add. 179/3 f.26, MS Add. 179/61; Woodward Library, University of British Columbia, Sinclair Collection, Darwin-Burdon Sanderson Correspondence, 1875.

[24] F. Darwin, *Life and Letters of Charles Darwin*, iii. 203.

[25] L. Huxley, *Life and Letters of Thomas Henry Huxley*, i. 437.

cross-section of the elite, however, very considerable disagreements arose over the extent of commitment to vivisection and over the best tactics to protect scientific research. The weakness of experimental medicine is illustrated by the fact that of the group, only Burdon Sanderson, Christison, and Foster regularly used living animals in their research. Tensions arose within the group over distinctions between the profession of medicine and the would-be profession of experimental medicine, and over differences in institutional interests and in perceptions of antivivisection. Acland and Rolleston, for example, refused to sign a petition originally drawn up by Huxley and circulated by Burdon Sanderson on the grounds that its claims for vivisection were too strong. Acland denied what he took part of the petition to imply, namely that vivisection was the chief way of advancing therapeutic medicine.[26] The reception of the proposed bill, drawn up by R. B. Litchfield, Darwin's son-in-law, in consultation with Darwin, Huxley, Burdon Sanderson, Simon, Foster, and Sir James Paget was no more unanimous; Christison, Acland, Gull, and Sharpey were extremely reluctant to see the issue as a matter for legislation at all.[27] Nevertheless, the promoters of the bill, when it became obvious that their activity was becoming public knowledge, took steps to ensure its presentation in Parliament with their scientific imprimatur and prestigious

[26] A copy of the petition, which, to my knowledge, was never presented in any formal way, is in the Burdon Sanderson Papers MS Add. 179/61. For the refusals see Rolleston to Burdon Sanderson, 14 April 1875, MS Add. 179/2 f.5 and Acland to Burdon Sanderson, 1 May 1875, MS Add. 179/2 f.13.

[27] See Christison to Burdon Sanderson, 13 April 1875, Burdon Sanderson Papers, MS Add. 179/2 f.3; Christison to Playfair, 12 May 1875, Playfair Papers, #176; for Acland see note 26; for Gull, see Gull to Burdon Sanderson, 6 May 1875, Burdon Sanderson Papers, MS Add. 179/2 f.15; for Sharpey, see Sharpey to Burdon Sanderson, 19 May 1875, Playfair Papers, #623. Rutherford, Professor of Physiology at Edinburgh, was likewise antipathetic to legislation. See *Lancet* ii (1875), 285–286. At least one antivivisectionist medical man was most unwilling to see legislation on the subject. See J. Macaulay's letter, *B.M.J.* ii (1875), 569–570, where social sanctions against men performing cruel experiments were advocated.

backing in both houses.[28] Darwin and Burdon Sanderson approached Lyon Playfair (1818–1898), chemist and Liberal M. P. for the University of Edinburgh, who had expressed his interest in the matter. Playfair, a frequent Parliamentary spokesman for the medical profession, agreed to undertake charge of the bill in the Commons. He succeeded in inducing Lord Cardwell (1813–1886), an army reformer and one-time Peelite, to support it in the Lords. The Earl of Shaftesbury (1801–1885), the eminent philanthropist who was to play an important role in the antivivisection movement, also agreed to support the measure. Joining Playfair in sponsoring the bill in the Commons were Shaftesbury's son, Evelyn Ashley (1836–1907), and Spencer Walpole (1806–1898).[29]

After failing to persuade Henniker to withdraw his bill, presented on 4 May, Playfair presented the Darwin-Burdon Sanderson bill to the Commons on 12 May, trusting that it would at the very least prevent the Henniker measure from passing.[30] Earlier, steps had been taken to obtain assurance from Richard Cross, Home Secretary of the Conservative government under Disraeli, that the government would not commit itself until it had seen the scientists' bill.[31]

The principle of this bill was the regulation of *painful* experiments on living animals. Painless experiments, such as those under anesthesia, that were "for the purpose of scientific discovery, but for no other purpose," were legal-

[28] See esp. L. Huxley, *Life of Thomas Henry Huxley,* i. 437 and F. Darwin, *Life and Letters of Charles Darwin,* iii. 204.

[29] See Burdon Sanderson to Darwin, 30 April 1875, 6 May 1875, Sinclair Collection, and Burdon Sanderson to Playfair, 27 April 1875, Playfair Papers #120.

[30] See Burdon Sanderson to Darwin, n.d. (c. 7 May 1875), Sinclair Collection, and Burdon Sanderson to Acland, 6 May 1875, Burdon Sanderson Papers, MS Add. 179/2 f.71.

[31] See Burdon Sanderson to Darwin, 12 April 1875, Sinclair Collection; Derby to Cross, 17 April 1876, Brit. Mus., Add. MSS. 51266 (Cross Papers), f.24: "I enclose a letter just received from Darwin—the Darwin"; and F. Darwin, *Life and Letters of Charles Darwin,* iii. 204.

Figure 3. Lyon Playfair (1818–1898), parliamentary spokesman for the medical and scientific interest. *Courtesy of the Wellcome Trustees.*

ized and clearly removed from both state regulation and the purview of Martin's Act. Individuals could apply for a five-year license to perform potentially painful experiments after obtaining the signatures of a president of one of the medical or scientific bodies and of a professor of physiology, medicine, or anatomy. Such licences were to be granted at the discretion of the Home Secretary, experiments without anesthetics being permissible under it, (i) "for the purpose of new scientific discovery and for no other purpose," (ii) when anesthetics or insensibility would frustrate the purposes of the experiment, (iii) when suffering was kept to an absolute minimum. A licensee was to maintain a register of all such experiments he performed. Penalties for violations under the bill were to be no greater than £50 or three months imprisonment.[32]

At first blush, both the Henniker and Playfair bills seem to indicate a surprising coextension of purpose between the two sponsoring groups. The very significant differences revealed by closer examination of the bills should not obscure the fact that the two parties were very close. The differences in question were more the product of two independent attacks on the same complex question, without benefit of precedent or any extensive experience with similar legislative problems, than they were of deep seated irreconcilable viewpoints on what was or was not permissible in the way of experiments upon living animals. After all, both bills allowed for painful experiments under appropriate conditions. Colam and the R.S.P.C.A. were slowly and deliberately developing their own legislation and had been in contact with the physiologists,[33] but would never

[32] For the text of the bill, see "Report of the Royal Commission," Appendix iii, section 7.

[33] According to Burdon Sanderson, the R.S.P.C.A. scheme in April 1875 was "certainly not more restrictive" than the scientists' bill. This is hard to reconcile with the R.S.P.C.A.'s stated policy against painful experiments and was certainly not embodied in the bill that the R.S.P.C.A. recommended to the Royal Commission. See below, and Burdon Sanderson to Darwin, 30 April 1875, Sinclair Collection, and *B.M.J.* i (1875), 614, which was less sanguine about the R.S.P.C.A measure.

have admitted the justifiability of painful experiments. They refused to support either of the Playfair or Henniker bills.[34]

Although a detailed consideration of the bills shows a number of ambiguities or difficulties in their practical applicability, one issue on which the bills differed may be discussed to illustrate the legislative pitfalls that might be encountered. This was the question of vivisection for demonstration or teaching purposes. Paradoxically, the Henniker bill, which made no distinction between vivisection "for purposes of physiological research or demonstration," could be interpreted as sanctioning even painful vivisection for demonstration, while the Playfair bill, allowing vivisection "for the purpose of new scientific discovery, but for no other purpose,"[35] unquestionably outlawed demonstrational vivisection, *even under anesthesia.* When this aspect of Playfair's bill became clear, the members of the informal scientists' lobby complained bitterly to Burdon Sanderson and Playfair. Burdon Sanderson was caught in the middle between Playfair, who claimed he was no expert in the subject matter and had merely followed Burdon Sanderson's recommendations, and Darwin and Huxley, who denied acquiescing to this aspect of the bill and renounced responsibility for it. Although it is not entirely clear whether Playfair was forced to compromise on the issue in order to gain the support of Cardwell and Shaftesbury, as Burdon Sanderson at one point suggested, or whether Burdon Sanderson had simply overlooked the consequences of the measure as he presented it to Playfair, the incident

[34] For R.S.P.C.A. comment on the bills, see "Vivisection Legislation," *Animal World* vi (1875), 84 and vii (1876), 66, 82. For the R.S.P.C.A. bill as finally presented to the Royal Commission, see "Report of the Royal Commission," Appendix iii, section 5.

[35] The question of just what constituted a "new scientific discovery" would of course have been a pregnant source of confusion. As William Sharpey pointed out, "there is no more fruitful source of dispute than the question of originality in scientific discovery." Sharpey to Burdon Sanderson, Playfair Papers, #623; "Report of the Royal Commission," Q. 2314.

reveals a chink of naivete in the apparent political sophistication of the scientists' lobby.[36]

The most important difference between the two bills related to the question of enforcement, and here, despite the Playfair bill's stiffer penalties for infractions, substantive differences seem to arise. Just before presentation, the Playfair bill had provided that prosecutions under it had to be approved by the Home Secretary,[37] and the bill as presented did not provide for specific individuals responsible for enforcement. On the other hand, the Henniker bill utilized the Inspectors of Anatomy appointed under the Anatomy Act passed earlier in the century; furthermore, the Playfair bill regulated only painful experiments, while the Henniker bill required all experiments on living animals to take place in registered premises that were to be open to inspection. To R. H. Hutton and F. P. Cobbe, this difference was crucial: all that would be required to avoid regulation by the Playfair bill would be the *avowal* that one always used anesthetics, and hence this measure would be quite useless in practical terms.[38] The absence of inspection in the Playfair measure was less a devious plot to take in the humanitarian party, however, than simply a social distaste for the idea: "It would be . . . absurd to appoint one man to inspect half a dozen others of at least equal status with himself," wrote Burdon Sanderson. In

[36] On this incident, see L. Huxley, *Life and Letters of Thomas Henry Huxley*, i. 438; F. Darwin and A. C. Seward (eds.), *More Letters of Charles Darwin*, ii. 435–436; "Report of the Royal Commission," p. xx and Q. 4665. Litchfield to Burdon Sanderson, 22 May 1875, Playfair Papers #455; Burdon Sanderson, Darwin, and Huxley to Playfair, 24 May–2 June 1875, Playfair Papers #122, 206, 207, 366; Darwin to Huxley, 21 May 1875, Huxley Papers 5.136; Burdon Sanderson to Darwin, 23 May 1875, Sinclair Collection. The difference between the bills in this respect became widely known when a letter from Evelyn Ashley, cosponsor of the Playfair bill, in the *Spectator* xlviii (1875), 660, claimed it as an advantage for his measure.

[37] See *Zoophilist* i (1881), 33–34.

[38] See [R. H. Hutton], "The Vivisection-Restriction Bills," *Spectator* xlviii (1875), 623–625 and *Life of Frances Power Cobbe*, ii. 584.

addition, inspection was scarcely justified for such a rare phenomenon: "There is no comparison between experimentation and anatomical dissection. Every student must dissect, but not one in a hundred need take part in any vivisection."[39]

Public reaction to the proposed legislation was mixed. The *Medical Press and Circular,* the *Lancet,* and *Nature* criticized the Henniker bill and preferred Playfair's, as did the *Saturday Review.*[40] The latter's earlier antipathy to experiments had become by 1875 less marked than its concern over the passion, exaggeration, and personalities indulged in by extreme antivivisection opinion.[41] This latter segment of opinion had little use for either bill. R. H. Hutton, for all his suspicion of Playfair's bill, pleaded with extremist humanitarians not to reject the principle of restriction as unworthy compromise: "What we now most fear is that the scruples of the humanitarians may prevent us from gaining even the advantage of this [scientific] admission on behalf of the poor creatures. . . ."[42] The *B.M.J.,* for its part, also declared both bills unsatisfactory: Henniker's for obvious reasons, Playfair's because it yielded to an unjust and ignorant clamor.[43]

Competing and controversial bills, introduced late in the session by private members, stood little chance of success,

[39] Burdon Sanderson to Darwin, 14 April 1875, Sinclair Collection; see also Acland to Burdon Sanderson, 5 May 1875, Burdon Sanderson Papers, MS Add. 179/3 f.26.

[40] *Medical Press and Circular* i (1875), 473; *Nature* xii (1875), 21, 52; *Lancet* i (1875), 697–698; "The Vivisection Bills," *Saturday Review* xxxix (1875), 647–648. *Nature,* as the organ of scientific opinion, criticized registration of premises as a threat to the research of nonmedical biologists who might work in their own homes, rather than in officially recognized institutions.

[41] See, for example, *Saturday Review* xxix (1875), 140–141.

[42] *Spectator* xlviii (1875), 624, 646. In Hutton's view, public opinion simply would not support the abolition of all painful experiments. Baroness Burdett Coutts, a prominent philanthropist, had written to the *Echo* rejecting the bills before Parliament because they sanctioned vivisection under certain conditions. See *Animal World* vi (1875), 91.

[43] *B.M.J.* i (1875), 716.

as the scientists' lobby had realized when they presented their bill. A considerable body of editorial opinion, including the *Lancet* and the *Times,* had called again for a commission of inquiry, and there was little dissent, except from Henniker and Hutton,[44] when the Home Secretary, Richard Cross, announced on 24 May 1875 a Royal Commission "on the practice of subjecting live animals to experiments for scientific purposes."[45] Although most people, and especially the scientists and medical men, agreed with the sporting magazine, *Field,* when it said "we must not allow ourselves to be carried away by our feelings on the subject, and thereby led to support any hastily devised scheme of legislation . . . ,"[46] Henniker hung on until 3 August, when he finally joined Playfair in withdrawing his bill.[47]

The general welcome accorded the Royal Commission signaled a considerable reduction in the level of public controversy over the issue of experiments on living animals, though important activity took place in the remainder of 1875, both in and out of the hearings of the Royal Commission. The spring of 1875 had seen the issue of vivisection achieve wide publicity. The first practical steps toward a response to public concern were taken by slowly emerging pressure groups. The alignments of certain individuals within the contending groups were, as events over the next twelve months showed, highly fluid, because of the novelty of the issue and the complexity of its regulation. The involvement of Robert Lowe in drawing up the Henniker bill contrasted to his bitter opposition in 1876 to any legislation whatsoever on the subject. The support of the Earl

[44] Hutton's declaration that he agreed with Lord Henniker that the Royal Commission was not really necessary was ironic in the light of his subsequent acceptance of a position as a member of the Commission. *Spectator* xlviii (1875), 711.

[45] *Hansard,* 3rd series, 1875, ccxxiv, 794. Queen Victoria was very sympathetic to antivivisection. She had been pressuring Prime Minister Disraeli for a Royal Commission. See Brian Harrison, "Animals and the State in Nineteenth-Century England," *English Historical Review,* lxxxviii (1973) 786–820.

[46] Quoted in *Public Opinion* xxvii (1875), 667.

[47] *Hansard,* 3rd series, 1875, ccxxvi, 431.

of Shaftesbury and Evelyn Ashley for the scientists' bill contrasted with their later institutional roles in the antivivisection movement. These changes of alliance exemplify the tentative, exploratory nature of the 1875 commitments.

Outside the R.S.P.C.A., both sides in the dispute had shown a certain amount of initiative and political sophistication in their lobbying activities, together with a final failure to come to grips successfully with the problem in their legislative proposals. In a way, however, it is a pity that the announcement of the Royal Commission precluded some sort of legislative compromise between the Henniker and Playfair bills. Never again were the differences between the two parties to appear so eminently negotiable; never again would contacts between them be so readily exploitable. The events of the next year served only to increase contentious priorities and harden positions on either side.

THE R.S.P.C.A. AND ANTIVIVISECTION SOCIETIES

During the sitting of the Royal Commission and the political activity following its report in early 1876, the first formal antivivisection societies emerged. The development of these societies was largely the result of the R.S.P.C.A.'s very cautious approach to the issue of experiments on living animals. Before sketching the founding of the major antivivisection organizations, it may be worthwhile to analyze why the R.S.P.C.A. failed to embrace militant antivivisectionism.

The vivisection issue had plagued the R.S.P.C.A. since its origins, but intervals of interest did not crystallize into serious practical activity until the decade from 1857 to 1867, when the society mounted its campaign against abuse in the French veterinary schools. Even during this period, a host of more common and more accessible evils dominated the society's attention and resources. After the French campaign, a watching brief on developments in Britain was instituted. The institutionalization of experimental physiology in England during the early seventies alerted the R.S.P.C.A.'s secretary, John Colam, culminating in the Norwich prosecution. Meanwhile, the society had adopted its policy against all *painful* vivisection, while recognizing that painless experiments on living animals, such as those under anesthetics, were permissible for appropriate purposes under appropriate safeguards.

Without the R.S.P.C.A., there would have been no Norwich trial, no concrete abuse exposed to generate national publicity, and, therefore, perhaps nothing but a lengthy series of impotent protests in correspondence columns, to show for antivivisection sentiment in Great Britain. Because it had the resources and the organization to undertake the Norwich prosecution, the society created the possibility for an effective agitation. By temporizing and factfinding in the face of the memorialists' requests for action, however, it lost the initiative. The R.S.P.C.A. remained a highly influential factor in the antivivisection controversy, but it no longer made events. It simply reacted to them.

The R.S.P.C.A.'s caution on vivisection in 1875 was entirely typical of its program for the remainder of the century. The society's moderate position may be explained by reference to its general modus operandi and to the particular nature of antivivisection as an issue.

By the seventies, the R.S.P.C.A. was a large, wealthy, powerful, and prestigious organization. The society had not arrived at this point by vigorous crusading alone. It had carefully cultivated the aristocracy. It had nurtured a growing relationship of mutual trust with government, slowly building an impressive body of legislation for animal protection. It had embarked on a broad program of public education. Most importantly, it had never moved too far ahead of public opinion and, thus, had maintained as broad a constituency as possible. As Harrison has pointed out:

> Prudence was . . . a major element in the Society's successful strategy. The R.S.P.C.A., as the organization for the general defence of animals, was never a single-issue group; it therefore always found itself in a political situation where opposition to the powerful on one issue had to be moderated in the hope of attracting support from the powerful on another.

But zeal could be harnessed without jeopardizing general public sympathy:

> One way of following this strategy without alienating the enthusiasts was to encourage them to form distinct organizations for their narrower purposes . . . for successful reform involves buttressing a movement with all the extremists' energy, imagination, and resourcefulness without encumbering it with their tactical errors and political naivety.[48]

Thus Colam and the R.S.P.C.A. refused to respond to Cobbe's initiative with more than investigation and a call for judicious inquiry. The years 1875 and 1876 saw the beginnings of the hysteria and sensationalism in the antivivisectionist cause that were to force the R.S.P.C.A. into an arm's-length relationship and to discredit the movement in the eyes of a significant proportion of the press and the public. The society repeatedly dissociated itself from emotional accusations and called instead for carefully authenticated charges.[49] Presumably its leadership was entirely satisfied to see separate antivivisectionist organization developing, for extremist pressures might well provide leverage for concrete reforms through the good offices of the moderate majority of the animal protection movement.

The unique features of antivivisection as an issue also made an aggressive policy on the part of the R.S.P.C.A. unlikely. There can be no doubt that there was a wide spectrum of opinion on experimental medicine within the society, from blanket condemnation through enthusiastic support. As *Animal World* noted in 1875, "The R.S.P.C.A. is not so entirely unanimous as to desire the passing of any special legislative enactment on the subject [of vivisection]."[50] Doctors and scientists were active in the society.

[48] Brian Harrison, "Animals and the State in Nineteenth-Century England."

[49] See, e.g., *Animal World* vi (1875), 84. Cf. xvii (1886), 82.

[50] Ibid. vi (1875), 180.

Indeed the R.S.P.C.A. looked confidently toward science to supply information upon which to base animal protection legislation and to develop technology to ease the burdens of animals. Thus the animal protection movement scarcely partook of that hostility toward science which was, as we shall see, so important a part of antivivisectionism. Furthermore, experiments on living animals were carried out exclusively by highly educated members of a prestigious profession and not by the working-class offenders who made up the bulk of the society's concern. Finally, women played a very active and decisive role in the antivivisection movement, whereas the later Victorian R.S.P.C.A., despite its very substantial female support, provided much less opportunity for women like Cobbe to take on a public leadership position within its ranks.

There were, of course, many enthusiastic antivivisectionists within the R.S.P.C.A. and more than once over the last decades of the century ginger groups within the organization were to prod the society toward more aggressive action or to attempt to purge allegedly provivisection officers.[51] The difficulties faced by the leaders of the R.S.P.C.A. in maintaining cohesion in the society were exemplified at its annual meeting in 1885. From the same platform upon which the president of the society had just rebuffed antivivisectionist charges of foot-dragging, the Bishop of Carlisle made what was taken to be a plea for even greater moderation by the R.S.P.C.A. toward experimental medicine. The Bishop finished by saying "Do not let us, by any wanton interference, drive our scientific men out of this country. . . ."[52]

From the mid-seventies onward, antivivisectionists were disgusted by the R.S.P.C.A.'s failure to commit its substantial power wholeheartedly to the battle against vivisection. They frequently took public occasion to goad, criticize,

[51] Here and elsewhere in this discussion I am greatly indebted to the advice of James Turner of Harvard University, whose research on the Victorian R.S.P.C.A. may be expected to shed further light on its history.

[52] *Animal World* xvi (1885), 119–120, 130.

or castigate the larger, more powerful, and more prestigious society with regard to its policy on the subject of their special interest.[53] Lord Aberdare, when president of the R.S.P.C.A. in 1879, voted against legislation proposing total abolition, thereby enraging antivivisectionists.[54] One of their periodicals had this to say in 1881:

> To us it seems a dreary farce to blow a trumpet of humanity and collect immense funds for a society with a score of royal and noble names at its head and then do nothing, or worse than nothing, to prevent cruelties so far transcending all others in the degree and duration of the agonies they cause, as to make the brutalities of the street and the shambles seem humane by comparison.[55]

A survey of R.S.P.C.A. branch societies in 1881 found almost none that were actively antivivisectionist.[56] Mutual criticism between the society and antivivisectionists continued in the nineties.[57]

Certain leaders of the antivivisection movement, notably one Charles Adams, urged the development of good relations with the R.S.P.C.A. Adams pointed out that without the society's establishment of the precedent of humane treatment for animals, an antivivisection agitation could scarcely be a practical possibility.[58] He tried to evolve a modus vivendi by portraying the two movements as concerned with the existence of cruelty in different social classes: the R.S.P.C.A. was fighting "cruelty in fustian" (by

[53] *Animal World* vii (1876), 34, and G. R. Jesse, *Man's Injustice to Animals. The Brown Animal Sanatory Institution* (London, 1888), pp. 93–96; *Anti-Vivisectionist* vi (1879), 542–543.

[54] *Spectator* lii (1879), 947–948; *Zoophilist* i (1881), 102–103.

[55] *Zoophilist* i (1881), 105.

[56] Ibid., 129–132, 169–172.

[57] See, e.g., E. G. Fairholme and W. Pain, *A Century of Work for Animals. The History of the R.S.P.C.A., 1824–1934* (2nd ed., London, 1934), pp. 197–198.

[58] *Verulam Review* iii (1892–1893), 139–145. Cf. *Champion* (1885), 228–231.

working people) while antivivisectionism was fighting "cruelty in broadcloth" (by educated members of the middle or upper classes).[59]

Despite such efforts, the antivivisectionists' envy and frustration at what they conceived to be the misuse of magnificent resources by the R.S.P.C.A. resulted in the kind of carping dissension that reduced cooperation to a minimum. The failure of the R.S.P.C.A. to support antivivisectionists in either their legislative attempts or their appeals to the public was a very serious blow to their prospects of success. To the general public, animal protection meant the R.S.P.C.A., and no other society. By withholding its endorsement from total abolition, the R.S.P.C.A. played a sort of buffer role, with the general public on one side, and the antivivisectionists on the other.

In the spring of 1875, then, the exigencies of the practical politics of antivivisection were beginning to drive in the wedge that was to separate the society from organized antivivisection, nascent in the activities of Frances Power Cobbe and George Hoggan, among others. When Cobbe's prestigiously supported memorial to the R.S.P.C.A. failed to move that organization to her satisfaction, she was forced to consider a separate society, formed specifically to combat vivisection. She realized that such a body would encumber her:

> I abhorred societies, and knew only too well the huge additional labour of working the machinery of one, over and above any direct help to the object in view. I had hitherto worked independently and freely, taking always the advice of the eminent men who were so good as to counsel me at every step. But I felt that this plan could not suffice much longer, and that the authority of a formally constituted society was needed to make

[59] *Report of the Annual Meeting of the Victoria Street Society for Protection of Animals from Vivisection* (London, 1882), p. 5. Cf. *Verulam Review* viii (1899–1900), 217–230, esp. 224. Harrison gives a similar analysis of the class orientation of the R.S.P.C.A. B. Harrison, "Religion and Recreation in Nineteenth Century England," *Past and Present* xxxviii (1967), 116–119.

> headway against an evil which daily revealed itself as more formidable.[60]

In addition to the management and knowledge that she and Hoggan could provide, Cobbe knew her society would need the countenance of eminent men in order to attract membership and subscriptions. Without such support, there would be no point in trying to form a society. She succeeded in attracting the Archbishop of York and the eminent reformer Lord Shaftesbury (Figure 4). The latter was to become the first president of the society, its parliamentary spokesman, and, with Cobbe, a major force in its near-hegemony over organized antivivisection. As Cobbe later put it,

> Lord Shaftesbury never joined the Victoria Street Society; it was the Society which joined Lord Shaftesbury. There was a day in November, 1875, when, having telegraphed his readiness to support the project of Dr. Hoggan and Miss Cobbe, he, in fact, founded the Society. It was around him and attracted in great part by his name, that the whole body eventually gathered.[61]

Throughout her involvement with the society, Cobbe maintained for it a prestigious group of vice presidents, who might or might not actively participate. Among them at various times were certain members of the aristocracy, poets Browning and Tennyson, members of parliament James Stansfeld and A. J. Mundella, Lord Chief Justice Coleridge, and religious leaders such as Cardinal Manning.[62] Their role was primarily to attract public interest and confidence. Although Shaftesbury was closely involved in the week-to-week operation of the society, it was Cobbe's opinion that the "actual work done by Lord Shaftesbury

[60] F. P. Cobbe, *Life of Frances Power Cobbe* (Boston and New York, 1894), ii. 586.

[61] F. P. C(obbe), "Lord Shaftesbury and our Society," *Zoophilist* v (1885), 114. Cf. ibid., 587.

[62] *Home Chronicler* (1876), 64; *Zoophilist* i (1881), No. 7. See also *Life of Frances Power Cobbe*, ii. 590.

Figure 4. The Earl of Shaftesbury (1801–1885), parliamentary spokesman for antivivisection, whose prestige did much to legitimize the movement in the eyes of the public. *Courtesy of the Wellcome Trustees.*

for the Anti-Vivisection Cause, great as it was, constituted only a small part of the support he gave it by lending it the authority of his great character, and infusing into it his lofty and religious spirit."[63]

Cobbe's Society for the Protection of Animals Liable to Vivisection became the most important and politically influential of the antivivisection societies, largely on the strength of Cobbe's success in enlisting famous names. It became generally, and finally officially, known as the Victoria Street Society. Its political activity, carried on by Shaftesbury, Coleridge, and Mundella, was only one part of its overall program, which program will be considered in detail later. For the moment, the point is that the bulk of the work of the society—and of hundreds of other societies of similar structure though widely varying aims—was performed by a group of people generally quite different from those well known personalities who chaired its public meetings and were listed in advertisements and flysheets. The day-to-day and week-to-week operation of the society fell to a few paid employees, notably the secretary, to the honorary secretary or secretaries, and to the executive committee, a small corps of enthusiasts whose gratuitous labors were the heart and soul of the movement.

For the Victoria Street Society, Cobbe was honorary secretary from 1876 to 1884. She later became president and was a dominant figure on the executive committee until at least the mid-nineties. George Hoggan was an honorary secretary from the inception to 1878, when he left the society amicably following its adoption of a total abolition policy. As for the executive committee, Cobbe's account of its original composition in her autobiography provides an excellent sense of the eminently respectable but relatively obscure figures who served in that capacity.[64] By March of 1876, the society had taken offices in Victoria

[63] F. P. C(obbe), "Lord Shaftesbury and our Society," *Zoophilist* v (1885), 115.

[64] F. P. Cobbe, *Life of Frances Power Cobbe* (Boston and New York, 1894), ii. 589. Leslie Stephen was a member of the committee for a very short time.

Street, an event symbolizing the existence and operation of an influentially supported organization pledged to fight vivisection more intensely and directly than the R.S.P.C.A.

Despite its imposing backing and preeminent political influence, the Victoria Street Society found itself by no means alone in the field of organized antivivisection. In February 1875, George Jesse (1820–1898) had founded the first specifically antivivisectionist society. Jesse, of Macclesfield, Cheshire, was the author of a *History of the British Dog*, and a retired civil engineer of prodigious idiosyncracy. His Society for the Abolition of Vivisection never became more than a vehicle for one man's quixotic tilting at various vivisectional windmills, but its early appearance attracted support that subsequently migrated to larger, more democratic, and more effective metropolitan organizations. Public sentiment on the subject had only to be tapped.[65]

The founding of the Victoria Street Society in late 1875 was followed by the emergence of new societies in Ireland and Scotland, and two new metropolitan societies: the London Anti-Vivisection Society and the International Association for the Total Suppression of Vivisection. These societies were organized between March and June 1876, in the

[65] For Jesse's early activities, see *Times* and *Morning Post* for 1875, passim; "The Abolition of Vivisection," *B.M.J.* i (1875), 316; "The Royal Commission on Vivisection," *B.M.J.* ii (1875), 25–26; *Spectator* xlviii (1875), 436; "The Abolition of Vivisection Society," *Medical Press and Circular* ii (1875), 35. Jesse made three abortive appearances before the Royal Commission: "Report of the Royal Commission," Q. 4435–4455, 5551–5577, 6418–6551. He apparently weathered court actions to make him account for the subscriptions his society had received (approximately £2,000 in the first year), brought by former supporter and antivivisectionist M. P., J. M. Holt. See "The Anti-Vivisection Society," *Medical Press and Circular* i (1876), 228, 534. Jesse published any number of newspaper advertisements, letters, and pamphlets between 1875 and 1897. Among his favorite targets were the Brown Institution, the R.S.P.C.A. (see *Animal World* vii (1876), 34), and the Royal Commission (see *Extracts From and Notes Upon the Report of the Royal Commission on Vivisection, Refuting its Conclusions* (London, 1876), 2 parts). Other antivivisection societies made things as difficult for Jesse as they could. See *Anti-Vivisectionist* vi (1879), 354.

midst of political controversy over the report of the Royal Commission and government bill.[66]

The Irish society was never terribly powerful and at times became moribund[67]—there was relatively little experimental activity in Ireland in any case. The Scottish society, located in Edinburgh, an important center of experimental medicine, ultimately became enthusiastic and vociferous.[68] Owing to its relative isolation, however, it had little parliamentary influence.

Both the London and International societies were well financed and active from the beginning, and both stood uncompromisingly for the total abolition of vivisection. The thinking of these organizations is illustrated well in a speech by one of the founders of the International Association, made in ignorance of the nearly simultaneous foundation of the London society. The R.S.P.C.A. had failed to stand for total abolition, said the Rev. George Collins,

> And then there is the society in Victoria Street. Well, in regard to that I will only say they go in for a compromise, [i.e., restriction rather than total abolition] and I hate compromises of any kind . . . Lastly, there is the society of which Mr. Jesse is the head and body, and everything in one, and moreover he is found to be so impracticable that it is quite impossible to work with him.

According to Collins, therefore, there was clear need for an independent total abolition society.[69] Another speaker insisted that the society not so much as affiliate with any other not committed to total abolition.[70]

Thus, the report of the Royal Commission on Vivisection,

[66] *Dates of the Principal Events connected with the Anti-Vivisection Movement* (London, c. 1897), pp. 2–3.

[67] *Zoophilist* iii (1884), 246.

[68] See, e.g., *Fourth Annual Conference of the Scottish Society for the Total Suppression of Vivisection* (Edinburgh, 1883).

[69] *The International Association for the Total Suppression of Vivisection. Minutes of Proceedings at a Meeting Convened for the Purpose of Inaugurating the Above Association* . . . (London, 1876), p. 12.

[70] Ibid., 26.

(as it became known), emerged in early 1876 into an atmosphere of indignant letters to the editor, feverish committee room debate, and increasing political organization on behalf of the antivivisection movement. The R.S.P.C.A.'s moderation had spawned specialist societies that were not to be so hesitant to press the case against experiments on living animals.

1875–1876: THE ROYAL COMMISSION

The Royal Commission faced the challenge of finding a way to regulate vivisection that would prevent abuse and calm public disquiet while permitting maximum progress in scientific research. The importance of its deliberations to experimental medicine was clear to the *B.M.J.*: "What further services [physiology] may yet render it is impossible to say; but whether its future development will go on in England or will be confined to continental countries where laboratories are built, professorships endowed, and the expenses of experiments defrayed by the governments, will in no small degree be determined by the Royal Commission. . . ."[71] British physiology, slowly developing after a generation of neglect, scraping up support from begrudging institutions, was now struggling for its life in an increasingly polarized debate over the morality of its means. The situation was clearly outlined in a letter written by Sir Edward Frankland (1825–1899) to his fellow chemist Lyon Playfair, on the presentation of the latter's bill:

> Physiological and biological research of all kinds is at a scandalously low ebb in this country. It is no exaggeration to say that more of such researches are made in the laboratories of Germany in one academical session than in those of the United Kingdom in several years. Experimental research in some branches of science, chemistry for instance, may possibly remunerate the discoverer, but in physiology this can never be the case. The physiological investigator is the most disinterested worker and he requires and deserves all the encouragement which

[71] *B.M.J.* ii (1875), 26. I have slightly altered the punctuation in this passage.

> the state can afford. At no previous period has physiological research promised so much toward the relief of human suffering, as it does at the present moment. . . . The incentives to physiological research in this country are at present so feeble as to cause this most important branch of experimental science to be almost entirely neglected, but if the enactments in your Bill become law, how many young or even old physiologists will be found brave enough to face a criminal prosecution the result of which would depend upon the opinion of a magistrate or jury as to whether a certain experiment upon an animal was or was not likely to lead to an important discovery . . .[72]

Given its mission of mediation between medical and humanitarian opinion on such a complex problem, the composition of the Commission was crucial. Home Secretary Richard Cross attempted a balance of scientific expertise, practical medical experience, judicial wisdom, and humanitarian zeal. He was advised in the task of choosing the commissioners by the fourth Earl of Carnarvon (1831–1890), son of a previous president of the R.S.P.C.A. whose antivivisection speeches had drawn medical fire thirty years before, who was himself a vice president of the society (Figure 5). Carnarvon advised "It is, from my point of view, essential that the partisans of Vivisection sd. be well represented on the Commission: but it wd. be the greatest mistake to allow them to have a numerical preponderance."[73]

Cross chose T. H. Huxley as the representative of science, after Burdon Sanderson had refused to serve.[74] Huxley had served on several Royal Commissions, had been an original member of the London School Board, and was widely recognized as the leading public spokesman for British science. Articulate and forceful, Huxley could be relied upon to make a strong case for research (Figure 6).

[72] Frankland to Playfair, 20 May 1875, Playfair Papers #280.

[73] Carnarvon to Cross, 27 May 1875, Brit. Mus., Addit. MSS. 51268 (Cross Papers) ff.97–99.

[74] L. Huxley, *Life and Letters of Thomas Henry Huxley*, i. 439, and Cross to Burdon Sanderson, 10 June 1875, Burdon Sanderson Papers, MS. Add. 179/2 f.23.

For chairman of the Commission, Cross chose Lord Cardwell (1813–1886), a vice president of the R.S.P.C.A., and a highly experienced statesman who had held a number of cabinet positions, including Secretary for War under Gladstone from 1868 to 1874. During that period, military commissions by purchase were abolished. Cardwell had a reputation as a dispassionate and practical man.

To fill out the Commission with men uncommitted to either side, Cross chose Sir John Karslake (1821–1881) and Lord Winmarleigh (1802–1892), both veteran politicians with no record, as far as I am aware, of activity in animal protection. Some members of the press took them to be representative of a common parliamentary type whose position with regard to vivisection was thought to be of considerable interest: the fox-hunting, grouse-shooting country gentleman.[75]

Another member of the commission supposed to represent a middle view on the issue was a liberal politican and educational reformer, W. E. Forster. Forster was, however, a vice president of the R.S.P.C.A.,[76] whose Quaker background and basic predispositions made him very strongly opposed to painful experiments in all but the rarest cases.[77] While a cabinet member under Gladstone, he had taken steps to reduce painful experiments on animals performed under his authority and one such (misinformed) step drew him into a clash with Huxley over a course of lectures by the latter.[78]

The commissioner most clearly responsible for putting

[75] As Carnarvon wrote to Cross: "A master of the hounds would be a good element added if you can find one on your back benches." See note 73, and L. Huxley, *Life and Letters of Thomas Henry Huxley*, i. 437.

[76] Besides Carnarvon, Cardwell, and Forster, other vice presidents of the R.S.P.C.A. included Shaftesbury and W. H. Smith, leading Tories. *Animal World* vii (1876), 111.

[77] T. W. Reid, *Life of the Right Honourable William Edward Forster* (4th ed., London, 1888), i. 84, ii. 7–8. Cf. *B.M.J.* ii. (1876), 227–228.

[78] Ibid. and L. Huxley, *Life and Letters of Thomas Henry Huxley*, i. 430–433.

Figure 5. The Earl of Carnarvon (1831–1890), the only antivivisectionist in Disraeli's cabinet. *Courtesy of the Wellcome Trustees.*

Figure 6. Thomas Henry Huxley (1825–1895), the dominant public figure in later Victorian science. *Courtesy of the Wellcome Trustees.*

forward the claims of antivivisection was R. H. Hutton. As Carnarvon put it in his letter to Cross, "Mr. Hutton—who carried on I believe most temperately and ably the controversy at the London University and who probably knows more *against* vivisection than anyone else."[79]

The final member of the Commission was John Eric Erichsen (1818–1896), Professor of Surgery at University College, London. Erichsen had once performed experiments upon living animals in the course of his research and was unquestionably a proponent of the practice, but Carnarvon and Cross do not seem to have been aware of this when Erichsen was appointed. At least in his letters to Cross, Carnarvon clearly described Erichsen as an opponent of vivisection[80] and it would seen that Erichsen was intended as a practical medical antidote to the virus of the Huxleian passion for research. He turned out to be no antidote, but he did provide a certain balance, for, according to Hutton in 1889, "we do not regard Mr. Erichsen as having simply represented the interests of the scientific vivisectors on the Commission, where his attitude was more moderate and far less pronounced in that sense than Professor Huxley's."[81]

The care taken in the selection of commissioners was vindicated in the smooth operation of the Commission and the lack of any internal conflict important enough to be reflected in the minutes of evidence in the resulting blue book. The Commission met between 5 July and 15 December 1875, examining fifty-three witnesses and some written submissions, and asking over 6,000 questions. Although, before he was appointed a commissioner, Hutton had forecast that the Commission would be a "total failure" without the power to obtain compulsory evidence under oath,[82] no difficulties of this nature seem to have arisen. The wit-

[79] See note 73.

[80] See Carnarvon to Cross, 27 May 1875 and 9 June 1875, Brit. Mus., Add. MSS. 51268 (Cross Papers), ff.97–99, 100–102.

[81] *Spectator* lxii (1889), 699.

[82] Ibid., 679. Hutton was probably thinking of his difficulties as a member of the Senate of the University of London in obtaining information as to activities within the Brown Institution.

nesses who appeared included a number of prominent physicians and surgeons, two veterinarians, a few antivivisectionist medical men, about a dozen more or less practicing physiologists, and a few private citizens interested in the question, including J. M. Holt, M. P., and G. R. Jesse.

The task of the Commission in its cross-examination of witnesses was to establish the extent of experimentation on living animals taking place in Great Britain, the amount of cruelty that might be taking place in connection with this experimentation, and the best means of preventing such cruelty. As indicated previously, British physiology, under Burdon Sanderson at University College London, Foster at Cambridge, and at certain other medical schools in London and Edinburgh, had begun a resurgence. Virtually all the witnesses agreed that vivisection was on the increase in Britain, and some of the testimony by those in the best positions to know is of interest in this regard. John Burdon Sanderson provided for the Commission an excellent description of organized research and teaching in physiology. He enunciated an ideal for every medical school, in which the professor would be, like himself and Foster, a fulltime physiologist and not a medical practitioner. Opportunities for laboratory research would be open to qualified men:

> I mean by a school of physiology, not the teaching of students by lectures, but the association of workers under the direction of a head, such persons not being students in the ordinary sense, but men who are themselves devoted to science, either for their whole lives, or for a certain definite period, men who are intending earnestly to engage in research for a certain time. It is in this way, I think, that physiology ought to be studied, and the more it is studied in this way the better. In such schools of physiology it is clear that the work which will be done will be in a great measure experimental, because we cannot make progress without experiments of one sort or another, painful or otherwise.[83]

[83] "Report of the Royal Commission," Q. 2302. See also Q. 2301.

Burdon Sanderson made it clear that his models in such a proposition were the German schools of physiology. This and similar testimony by other physiologists[84] made an immense impression upon the lay commissioners, an impression scarcely modified by testimony as to the small *absolute* amount of vivisection actually taking place in Britain.[85] With visions of a "geometrical rate of increase" of experiments like those in the *Handbook for the Physiological Laboratory* and those condemned in ghastly detail from first hand experience in continental laboratories by Hoggan and De Noë Walker, Hutton pressed Foster:

> 2411. [Hutton] I may take it that it is the intention, is it not, of this book, and of what I should call your school of thought, to introduce the general system of study abroad very much more generally into England than it has hither to been introduced?—[Foster] I think our study is a study entirely of our own; I mean, I think we teach physiology in a way that it is not taught on the continent at all.
> 2412. The experimental method is derived very much from the continent, is it not?—The experimental method is coeval with physiology.
>
> . . .
>
> 2414. Still your object is to extend the experimental method in England, is it not?—My object is undoubtedly to advance physiological science, for which the experimental method is one good means.

Hutton gave the authors of the *Handbook* a thorough cross-examination, and routinely asked witnesses their opinion of it. For example, with witness George Rolleston, Linacre Professor of Anatomy and Physiology at Oxford:

> 1351. [Hutton] Then I understand that your opinion about the handbook is that it is a dangerous book to society, and that it has warranted to some extent the feeling of anxiety in the public which its publication has

[84] See, e.g., Q. 5383, 5420.
[85] See, e.g., Q. 2606–2607.

> created.—[Rolleston] I am sorry to have to say that I do think that is so . . .

The *Handbook* and the evidence as to the expansion of physiology in Britain raised the spectre of incompetent bunglers or unsupervised students enthusiastically resorting to experiments on living animals, and a number of witnesses joined Samuel Haughton in strongly condemning such a possibility. Evidence of clandestine experiments by students or young practitioners not under supervision was given by a few witnesses.[86] This evidence produced no specific cases of serious abuse, but strong assertions of its existence, which further impressed the Commission. Persistent questioning as to the relative care taken over animal pain on the continent and in Britain received wildly varying responses, without any particular correlation to the witness's own views on animal pain. Such questions must be seen against the backgound of antivivisection literature of 1875, in which all the staple examples of experimental cruelty were taken from continental sources. Some British medical men and most physiologists were reluctant to concede that cruelty occurred, even on the continent.

An important question arose in the minds of Hutton and some of his fellow commissioners as to the purpose to which animal experiment was directed. A standard justification of physiological research, reiterated again and again (not least before the Royal Commission) by British defenders of vivisection, was predicated upon the medical utility of the knowledge gained thereby—a defense presumably calculated to appeal to a practical, hard-headed people. Hutton insisted, therefore, on making a distinction between research directed toward medical purposes and the indulgence of mere morbid curiosity in the interests of "pure" scientific research, despite the testimony of certain

[86] "Report of the Royal Commission," Q. 2509–2549, 4916–5183, 5188–5251, 5996–6145. The evidence was given by an elderly practitioner, John Anthony, M. D., a young veterinary surgeon, John B. Mills, and a young physician, W. A. B. Scott, M.D. The last two had Edinburgh backgrounds.

physiologists that one could not tell in advance whether or not a certain line of research might have practical consequences. To Hutton, the problem appeared to be that so much current research seemed not to answer questions but simply to open up new ones. He was convinced that Britain had arrived "at the time when the danger is beginning to occur, when we are actually getting a new school of men who are not employed in diminishing suffering, but who are employed in merely investigating physiology as a pure science."[87] In such a contention, Hutton had the support of some absolutely unimpeachable witnesses. Sir Henry Acland, Regius Professor of Medicine at Oxford and President of the General Medical Council, had this to say on the subject:

> I think that the Commission ought to be aware that the number of persons in this and other countries who are becoming biologists, without being medical men, is very much increasing. Modern civilization seems to be set upon acquiring, almost universally, what is called biological knowledge; and one of the consequences of that is, that whereas medical men are constantly engaged in the study of anatomy and physiology for a humane purpose,—that is, for the purpose of doing immediate good to mankind,—there are a number of persons now who are engaged in the pursuit of these subjects for the purpose of acquiring abstract knowledge. This is quite a different thing. I am not at all sure that the mere acquisition of knowledge is not a thing having some dangerous and mischievous tendencies in it. I know that would be thought by many persons a very *arriérée* and foolish opinion. Nevertheless I am not at all prepared to say that the mere desire to attain so much more knowledge is a good condition of mind for a man . . . now it has become a profession to discover; and I have often met persons who think that a man who is engaged in original research for the sake of adding to knowledge is therefore a far superior being to a practising physician who is simply trying to do good with his knowledge . . .[88]

[87] Ibid., Q. 4403.

[88] "Report of the Royal Commission," Q. 944. See Q. 978–993.

Acland's colleague at Oxford, George Rolleston, also displayed a suspicion of research: ". . . I must just say that with regard to all absorbing studies that [it] is the besetting sin of them, and of original research, that they lift a man so entirely above the ordinary sphere of daily duty that they betray him into selfishness and unscrupulous neglect of duty."[89]

In general, the testimony of antivivisectionist medical men (De Noë Walker, Hoggan, Fergusson, Haughton, and, inevitably, Macilwain) was confined to an attack on claims for the medical utility of experiments on living animals. None of them made a really forceful presentation and Huxley and Erichsen only rarely bothered to undertake careful cross-examination, preventing a real discussion of the antivivisection critiques. Recommendations for measures to be considered to curb abuse of vivisection were either half-hearted and unoriginal or ludicrously impractical, though De Noë Walker did suggest the institution of a board of control, composed of physiologists, pathologists, physicians, and surgeons, to rule on experiments on living animals.[90]

The only lay representatives of antivivisection to testify before the Commission were George Jesse and J. M. Holt, M. P. Jesse's pompous ignorance immediately succeeded in antagonizing the Commission, such that it did not see fit to publish all his testimony *in extenso*. Jesse produced a highly unfavorable climate for the plea of total abolition put forth by Holt. The Commission was more interested in pointing out the irregularities of Jesse's Society for the Abolition of Vivisection to Holt, who was listed as a member of the Society's committee, than it was in listening to his testimony.

The connoisseur of antivivisectionist *élan* must always regret that history was deprived of the spectacle of a rotund, forceful, and articulate Frances Power Cobbe lecturing the eminent commissioners on the evils of scientific experiment on living animals. Cobbe would have acquitted

89 Ibid., Q. 1287.
90 Ibid., Q. 1773–1774.

herself as well as any witness, and better than most, but she did not testify.

The most important testimony by the ostensible spokesmen for the humanitarian party was that given by the secretary of the R.S.P.C.A., John Colam. Capably and systematically he laid his society's case before the Commission. He declared that he did not know a single case of wanton cruelty by British experimenters and felt that they were generally more concerned about animal pain than their continental brethren. He did, however, make a lengthy submission extracting and citing by chapter and verse experiments, which the Society found objectionable, as described in British publications. This submission included a lengthy section from the *Handbook for the Physiological Laboratory*. He described the experiments he had witnessed and suggested that his survey of scientific literature showed vivisection to be a frequent practice in Britain. Furthermore, continental views on vivisection were rapidly making inroads. Arrangements for obtaining animals for laboratories had been investigated by R.S.P.C.A. operatives, who had apparently been unable to determine details as to the source of the animals. Colam submitted a bill for the regulation of experiments on living animals, which the R.S.P.C.A. wished to see made law (although a variety of opinion on the subject existed within its membership). The bill provided for a system of enforcement, borrowed in part from the Henniker and Playfair bills, and involving registration of places, licensure of persons, and inspection in a pattern substantially like that embodied in the subsequent legislation. The principle of the bill was prohibition of all painful experiments on living animals, in addition to prohibition of all vivisection, even under anesthesia, for the purposes of demonstration or of obtaining dexterity. A careful cross-examination of Colam by Huxley, Erichsen, and Cardwell revealed a number of important investigations that the prohibition of all painful experiments would prevent: experiments, like Sir Charles Bell's, of great interest and significance for physiology; experiments in aid of forensic medicine for the conviction of poisoners; and bacteriological and pathological experiments for the purposes of public health. While clearly impressed with the care and comprehensiveness of the society's presentation, and prepared to use parts of it in their report, the commission was ultimately unable to accept the recommendation as to painful experiments and as to experiments under anesthesia for the purpose of demonstration.[91]

Criticism of experiments on living animals by witnesses before the Commission was somewhat more than counterbalanced by the testimony of such physiologists and eminent medical men as Sir George Burrows, Joseph (later, Lord) Lister, Sir Thomas Watson, Sir James Paget, and Sir William Gull as to the morality and medical value of such experiments. Under these circumstances, the differences within the Royal Commission between the Huxley and the Hutton positions might well have been so great as to split completely the Commission's final report. I think it likely that, without the testimony of Emanuel Klein, several members of the Commission would have been entirely unwilling to sign a report recommending legislation of any kind with regard to experiments on living animals.

Klein (1844–1925) was an Austrian trained in Vienna who had come to England in 1871, where he had become, by 1875, Lecturer in Histology at St. Bartholomew's Hospital and Assistant Professor at the Brown Institution.[92] He had been one of the authors of the *Handbook for the Physiological Laboratory*, and had done a good deal of bacteriological research for the Local Government Board under John Simon.

T. H. Huxley had frequently been absent from meetings of the Commission owing to his teaching duties; his absences, according to the *B.M.J.*, were "the subject of dissatisfied comment."[93] It was on such a day that Klein testified

[91] For Colam's testimony and R.S.P.C.A. submissions, see "Report of the Royal Commission," Q. 1515–1702; Appendix iii, Section 5; Appendix iv. See also Q. 4246–4283.

[92] W. Bulloch, "Emanuel Klein 1844–1925," *J. of Pathology and Bacteriology* xxviii (1925), 684–697.

[93] *B.M.J.* ii. (1875), 74.

and by his testimony effectively destroyed any possibility that the Commission would find the scientific and medical interest capable and responsible enough to mind its own affairs with regard to vivisection. He managed to achieve this by the simple but repeated affirmation that as far as experiments on living animals in the course of research went, he bothered with anesthetics only for the sake of convenience—when he wished to avoid the bites and scratches of agonized dogs or cats:

> . . . just as little as a sportsman or a cook goes inquiring into the detail of the whole business while the sportsman is hunting or the cook putting a lobster into boiling water, just as little as one may expect these persons to go inquiring into the detail of the feeling of the animal, just as little can the physiologist or the investigator be expected to devote time and thought to inquiring what this animal will feel while he is doing the experiment. His whole attention is only directed to the making [of] the experiment, how to do it quickly, and to learn the most that he can from it.[94]

Repeated questions by commissioners failed to dredge up any trace of humanitarian consideration in Klein's decisions on the use of anesthetics. Klein's testimony as to the small amount of vivisection he did where anesthetics could be applied (most of his research was histological and bacteriological) only magnified the effects of his cavalier treatment of animal pain.

The impact of Klein's testimony upon opinion within the Commission and upon the public, after the minutes of evidence appeared, cannot be overestimated. He was Frances Power Cobbe's archvivisector incarnate, and experimenting in London under government funding to boot. Huxley's disgust at Klein's evidence was unbounded. He wrote immediately to Darwin and Burdon Sanderson bitter indictments of Klein. Huxley was told by Cardwell and

[94] "Report of the Royal Commission," Q. 3562. See Q. 3538ff.

Forster that Klein's "manner of giving utterance to these views was even more cynical than their substance." "I suspect that I might as well throw up my brief" he wrote, adding that Klein's "practical effect on our deliberation is that the ground is cut away from beneath my feet; the advocates of legislation and restriction are furnished with all the arguments they want. . . ."[95] Seven months later, Huxley wrote to Foster "It is not Hutton who has beaten me, but Klein. He has done more for our enemies than they could have done by their joint efforts, without him, by his wantonly and mischievously brutal talk."[96] Huxley was quite ready to assent to any law that would deal with the likes of Klein.

When his evidence was sent to him in proof, Klein attempted to alter it, "moved" as Hutton later wrote "by the remonstrances of his English colleagues."[97] The Commission refused to accept his alterations, but printed the amended testimony in an appendix.[98] Klein's status as a foreigner was an important factor in the reception of his testimony and spawned a number of xenophobic antivivisectionist pamphlets.[99] The medical and scientific interest excused Klein's testimony as attributable to his unfamiliarity with the English language, but to George Jesse, Klein

[95] Huxley to Burdon Sanderson, 30 October 1875, Burdon Sanderson Papers, MS Add. 179/3 f.30. See also L. Huxley, *Life and Letters of Thomas Henry Huxley*, i. 440, and Darwin to Huxley, 1 November 1875, Huxley Papers 5.322.

[96] Huxley to Foster, 25 May 1876, Huxley Papers, 4.120.

[97] R. H. Hutton, "The Anti-Vivisectionist Agitation," *Contemporary Review* xliii (1883), 514.

[98] "Report of the Royal Commission," p. vii and Appendix ii, section 2.

[99] See, e.g., *Extracts of Evidence Concerning Foreign Physiologists* (London, c. 1876). Klein's evidence was printed "As an illustration of the reasons for alarm at the importation of foreign teachers, and the influence of foreign example in English physiological laboratories. . . ." See also F. P. Cobbe, *Public Money. An Enquiry Concerning an Item of its Expenditure* (London, 1892) and the anonymous *A Foretaste of the Institute of Experimental Medicine* (London, 1898). Klein's evidence was cited and quoted continually in antivivisection literature.

was simply being offered as a scapegoat by his equally culpable fellows.[100]

Home Secretary Richard Cross's memoirs show the importance of Klein's testimony: "I had issued a Royal Commission to inquire into the matter [of vivisection], and their report aroused great indignation in the country, chiefly, as I think, owing to the evidence given by a man of eminence, a German, who said that, in carrying out his experiments, his whole mind was absorbed in the experiment itself, and that he did not regard the pain inflicted on the animal."[101]

The report of the Royal Commission, submitted in January 1876, is a curious and somewhat equivocal document, reflecting differences of opinion between the commissioners. As Hutton wrote in 1885, with regard to the question of the Commission's findings on actual abuse in experiments on living animals in Britain, "The Commission were divided on this point, and they held it to be their duty to pass no opinion on it, but to report the evidence simply, and to point out the great need for restrictions on the practice. We ourselves hold that in the case of one class of experiments at least—Dr. Rutherford's—there was evidence of very great abuse and of serious cruelty."[102] No doubt it was Erichsen and especially Huxley who opposed Hutton on these issues. The report, therefore, after referring to the increase of activity in British physiology, past and anticipated, to the growth of public concern over experiments on living animals, and to the historical development of vivisection as a method, proceeded to a judicious synopsis of the evidence. Colam's failure to find "a single case of wanton cruelty" was mentioned, as were the reassuring

[100] Bulloch (note 91) p. 688 and G. R. Jesse, *Extracts From and Notes Upon the Report of the Royal Commission on Vivisection, Refuting its Conclusions* (London, 1876), i. 32.

[101] R. A. Cross, *A Political History* (Privately printed, Eccle Riggs, 1903), p. 42. Cf. the remarks of the Bishop of Carlisle in *Transactions of the Victoria Street Society for the Protection of Animals from Vivisection* . . . (London, 1880), 13.

[102] *Spectator* lviii (1885), 35. Cf. liv (1881), 527.

statements of the various eminent physicians and surgeons. On the other hand,

> It is manifest that the practice is from its very nature liable to great abuse; and that since it is impossible for society to entertain the idea of putting an end to it, it ought to be subjected to due regulation and control. Those who are least favourable to interference assume, as we have seen, that interference would be directed against the skilful, the humane, and the experienced. But it is not for them that law is made, but for persons of the opposite character. It is not to be doubted that inhumanity may be found in persons of very high position as physiologists. We have seen that it was so in Magendie.[103]

In sum,

> Our conclusion, therefore, is that it is impossible altogether to prevent the practice of making experiments upon living animals for the attainment of knowledge applicable to the mitigation of human suffering or the prolongation of human life:—that the attempt to do so could only be followed by the evasion of the law or the flight of medical and physiological students from the United Kingdom to foreign schools and laboratories, and would, therefore, certainly result in no change favourable to the animals:—that absolute prevention, even if it were possible, would not be reasonable:—that the greatest mitigations of human suffering have been in part derived from such experiments:—that by the use of anaesthetics in humane and skilful hands the pain which would otherwise be inflicted may, in the great majority of cases, be altogether prevented, and in the remaining cases greatly mitigated:—that the infliction of severe and protracted agony is in any case to be avoided:—that the abuse of the practice by inhuman or unskilful persons,—in short the infliction upon animals of unnecessary pain,—is justly abhorrent to the moral sense of Your Majesty's subjects generally, not least so of the most distinguished physiologists and the most eminent surgeons and physi-

[103] "Report of the Royal Commission," p. xvii.

> cians:—and that the support of these eminent persons, as well as of the general public, may be confidently expected for any reasonable measures to prevent abuse.[104]

The report went on to recommend legislation to enforce regulation of experiments on living animals, in line with the above conclusions; it involved administrative machinery like that embodied in the R.S.P.C.A. bill, which had in turn borrowed from both the Playfair and Henniker bills. In doing so it pointed out (i) that the uncertainty of the outcome of a given line of investigation made it extremely undesirable to limit research to "expected discovery of some prophylactic or therapeutic end" and thereby to eliminate "experiment made for the mere advancement of science" (ii) that experiments under anesthesia for purposes of demonstration could be neither cruel nor demoralizing, but highly useful in medical education (iii) that curare should not be considered an anesthetic (iv) that cold-blooded animals such as the (commonly used) frog should be considered in legislation (v) that veterinary vivisection should be considered within the purview of legislation.

This report was simply not strong enough for Hutton. He wrote on 6 January 1876 to the Earl of Carnarvon, Secretary of the Colonies in Disraeli's cabinet, the man who had advised Cross on the composition of the Commission, and the man who was to introduce government legislation on the subject. Hutton pointed out that the report, which was to be issued on 8 January and which was to be unanimously signed, had "the effect of mildly whitewashing the practice." He was anxious to add his own minority report recommending the exclusion of cats and dogs from all experiment whatsoever and was unwilling "to append my name to so very mild and apologetic a production as the general report, without putting in something that represents to my mind more completely the gist of our inquiry." Cardwell, as chairman of the Commission, was anxious that Hutton withdraw his minority report, not, according to Hutton, because he thought it too strong,

[104] Ibid., pp. xvii–xviii.

but because he feared Huxley might introduce a counter report and "because he thinks it may weaken the hand of the government in pressing the bill." Hutton's purpose in contacting Carnarvon was to ask whether Carnarvon felt sufficiently in agreement with Cardwell to recommend withdrawal of the minority report.[105] Apparently this was not the case, for Hutton's brief minority report was appended to the main report, when it was issued on 8 January 1876. In it Hutton recommended the prohibition of all experiments upon cats and dogs (i) because their use motivates an "illicit trade" of pet stealing (ii) because they are probably more sensitive to pain than other animals (iii) because "from the very nature of our relations to these creatures, which we have trained up in habits of obedience to man and of confidence in him, . . . there is something of the nature of treachery as well as of insensibility to their sufferings, in allowing them to be subjected to severe pain even in the interests of science." To Hutton, "where the pursuit of scientific truth and common compassion come into collision, it seems to me that ends of civilization, no less than of morality, require us to be guided by the latter and higher principle."[106] During the sitting of the Royal Commission, Hutton's *Spectator* had remained silent upon the issue of vivisection, but after the report became public, the *Spectator* carried a critique of the report suggesting that it soft-pedaled the case for restriction of vivisection, which the evidence indicated.[107]

Although the *B.M.J.* had predicted that the results of the Royal Commission could not satisfy both parties, and probably would not satisfy either,[108] in fact the cautious wording required in the report to achieve some kind of consensus among the Commissioners meant that both sides

[105] Hutton to Carnarvon, 6 January 1876. P.R.O., Carnarvon Papers, 30/6/19/11. The letter is erroneously dated 6 June 1876, in the P.R.O.

[106] "Report of the Royal Commission," pp. xxii–xxiii.

[107] "The Vivisection Commissioners' Report," *Spectator* xlix (1876), 242–244.

[108] *B.M.J.* ii (1875), 25–26.

could read into the report what they wished. Initial response was generally favorable. For example, most of the medical press and their allies, *Nature*, the *Times*, the *Standard*, and the *Saturday Review*, welcomed the report as exonerating British physiologists from the calumnies circulated about them.[109] The maverick *Doctor* claimed that the Royal Commission had simply failed to uncover the extent of the abuse, but, like most of the humanitarian party at this point, expressed itself as satisfied that the recommendations should become law.[110] One humanitarian reservation concerned the composition of the Commission, Huxley and Erichsen being characterized as "two notorious advocates of vivisection."[111] This complaint was to become a part of the dogma of antivivisection. The R.S.P.C.A. had recommended public hearings, testimony under oath, and employment of counsel, and now complained that the omission of such machinery had left the question of the sources of experimental animals, especially cats and dogs, unclear. According to the society, laboratory attendants should have been called to testify on this and other matters. Furthermore, the R.S.P.C.A. denied the conclusions of the *Times* and the *B.M.J.* that the Commission had failed to uncover any abuse. If this was the case, asked *Animal World*, echoing the animal protectionist viewpoint, "why do the Commissioners ask for legislation?[112] The rationale of the professional press, on the other hand, for the proposal of legislation and its reluctant sanction was the need for concession

[109] *B.M.J.* i (1876), 227–228; *Lancet* i (1876), 285, 317–319; *Medical Times and Gazette* i (1876), 226–227; *Nature* xiii (1876), 321–322; *Times*, 19 February 1876; *Standard*, 19 February 1876; *Saturday Review* xli (1876), 295–296.

[110] *Doctor* vi (1876), 41.

[111] J. B. Firth, *British Friend* (May 1876), quoted in F. P. Cobbe, *Life of Frances Power Cobbe*, ii. 584. Cf. G. Macilwain, *Vivisection. Being Short Comments on Certain Parts of the Evidence Given Before the Royal Commission* (London, 1877), esp. p. 126 ff.; W. Howitt, "Vivisection," *Social Notes* i (1878), 273–276; J. Macaulay, B. Grant, and A. Wall, *Vivisection, Scientifically and Ethically Considered in Prize Essays* (London, 1881), p. 33.

[112] *Animal World* vii (1876), 82–86.

to an uninformed and unreasoning public opinion. A particularly extreme version of this attitude was expressed by Robert Lowe, no longer in sympathy with F. P. Cobbe: The Royal Commission, said Lowe, "acquitted the accused and sentenced them to be under the surveillance of the police for life. We presume that if there had been any justification for this extraordinary proceeding the Commissioners would have given it; but they have offered none, and we therefore can only suppose that it was done to appease the popular clamour."[113]

The R.S.P.C.A. said that it would refuse its support to any measure that permitted painful experiments on living animals.[114] Later, when legislation appeared likely, the R.S.P.C.A. did involve itself to a limited extent, but as an organization it was not to be an important factor over and above its contribution to the recommendations of the Royal Commission. The Society's unwillingness to compromise on legislative measures did not reduce pressure for legislation, however. In 1876, as in 1875, groups of representatives from both sides of the question were ready to take a great deal of lobbying initiative in pursuit of legislation. The difference was that by 1876 the two sides were to be better organized, the issues in contention between them were to be greater, and the conflict was to be more intense.

[113] R. Lowe, "The Vivisection Act," *Contemporary Review* xxviii (1876), 716.

[114] *Animal World* vii (1876), 85.

5. An Act "To Reconcile the Claims of Science and Humanity"

The first public meetings protesting vivisection[1] and the first petitioning against it[2] occurred while the Royal Commission was sitting. For the first time, the topic was treated in the pulpit as well as on the editorial page, and antivivisectionist organizing activity continued apace. In early March 1876, Frances Power Cobbe's new Society for the Protection of Animals Liable to Vivisection (to be known as the Victoria Street Society) issued a statement of its objectives, endorsed the report of the Royal Commission, and called for the government to legislate in accordance with the report.[3] Events were not, however, to prove as simple as all that.

In early 1876, the Conservative government under Disraeli was preoccupied with some of the important pieces of legislation for which they were responsible in the mid-seventies. The only member of the cabinet who had shown any spontaneous interest in the issue of experiments on living animals was Lord Carnarvon, Colonial Secretary, whose correspondence with Home Secretary Cross on the

[1] *Animal World* vi (1875), 98–99.

[2] *Medical Press and Circular* ii (1875), 410–411.

[3] F. P. Cobbe, *Life of Frances Power Cobbe* (Boston and New York, 1894), ii. 588–592. This is as good a place as any to note two articles that serve as general introductions to this period: A. H. Ryan, "History of the British Act of 1876: An Act to Amend the Law Relating to Cruelty to Animals," *Journal of Medicial Education* xxxviii (1963), 182–194, and M. N. Ozer, "The British Vivisection Controversy," *Bulletin of the History of Medicine* xl (1966), 158–167. The former is inaccurate and misleading in certain respects, in particular in its discussion of the motives and tactics of Frances Power Cobbe.

composition of the Royal Commission and with Hutton on the form of its report has been quoted above. Carnarvon was indeed the man ultimately given responsibility for presenting the government's vivisection legislation, but in February and early March 1876 it was not clear that the government intended to act at all upon the report of the Commission. Smith, in his study of Tory social reform during this period, was probably not thinking of vivisection legislation when he wrote,

> It was empirical, piecemeal reform dealing with problems as and when they were pushed into prominence by their inherent size and urgency; by agitation and the pressure of public opinion, by investigation and discussion, and by the exigencies of party politics. . . . Questions . . . were taken up not because Conservative ministers possessed and were anxious to implement a policy for them, but simply because they could scarcely be avoided.[4]

Nevertheless, this is a very apt description of the way in which legislation on the subject was forced upon the government in 1875–1876.

THE GOVERNMENT BILL IN THE HOUSE OF LORDS

Parliamentary questions in the Commons and the Lords in mid-March 1876 as to whether the government could be expected to bring in legislation following the recommendations of the Royal Commission received strictly noncommittal responses from Conservative spokesmen.[5] On 20 March, however, the Victoria Street Society's deputation to the Home Office to urge the necessity of legislation—a deputation including the Earl of Shaftesbury, Cardinal Manning, historian J. A. Froude, and A. J. Mundella, M. P.—received a highly favorable response from Cross. Cross invited the society to submit suggestions for a bill to the

[4] P. Smith, *Disraelian Conservatism and Social Reform* (London, 1967), pp. 257–258.

[5] *R.S.P.C.A. Records* xvi (1876), 2; *B.M.J.* i (1876), 366, 428; *Hansard,* 3rd series, 1876, ccxxvii, 2008 and ccxxviii, 476.

government, which it readily did.[6] According to Cobbe, the government bill introduced by Carnarvon in the House of Lords in May 1876 followed precisely the provisions specified by the Victoria Street Society to the government after the society's deliberations in late March.[7]

It is clear that Carnarvon personally was very strongly opposed to vivisection, no doubt owing to his family background in the R.S.P.C.A., at whose meetings he had spoken as a small boy. He wrote approvingly to his brother Auberon Herbert on the occasion of the latter's enthusiastic attack on vivisection, published in the *Times*.[8] He was also in close contact with R. H. Hutton, to whom he showed a draft of the government bill for Hutton's opinion before its presentation.[9] Hutton sent Carnarvon references on which to base his speech to the Lords on the bill.[10]

The government bill on vivisection was introduced in the House of Lords on 15 May; it was read a second time and debated on 22 May. In moving the second reading, Carnarvon made a lengthy and eloquent speech, which serves as an excellent summary of the antivivisectionist case as it was based on the experience of the preceding two years. He voiced prevailing perceptions of the issue when he saw "on the one side there is a strong sentiment of humanity, on the other are the claims of modern science." His legislation attempted "to reconcile the high laws of modern science with the still higher laws of morality and religion." His particular rationale for legislation was typical: the inadequacy of Martin's Act to deal with an expand-

[6] *Life of Frances Power Cobbe,* ii. 592, 594. Public reaction to the deputation was mixed. See *Spectator* xlix (1876), 295, 359, 391, and *Public Opinion* xxix (1876), 386.

[7] See *Life of Frances Power Cobbe,* ii. 594; F. P. Cobbe, *The Fallacy of Restriction Applied to Vivisection* (London, 1886), p. 3; and *Zoophilist* i (1881), 33–34.

[8] S. H. Harris, *Auberon Herbert: Crusader for Liberty* (London, 1943), pp. 169, 172.

[9] Hutton was pleased with the bill. See Hutton to Carnarvon, 12 April 1876, P.R.O., Carnarvon Papers, 30/6/19/1.

[10] Hutton to Carnarvon, 17 May 1876, Carnarvon Papers, 30/6/19/2.

ing practice of vivisection in Britain; experimentation in private places by students and young practitioners; the adoption of continental models and attitudes as indicated by the *Handbook* and the callous testimony of Klein; the delusive status of curare as an anesthetic; the impracticality and (grudgingly conceded) undesirability of total abolition.[11]

Carnarvon claimed his bill was "founded mainly on the lines of the Report of the Royal Commission"—a statement vigorously contested by the scientific and medical interest when the substance of the bill became public knowledge. The bill's provisions for the mechanics of regulation and enforcement were based on the specimen bill submitted by the R.S.P.C.A. and endorsed by the Royal Commission. Potentially painful experiments on living animals were to be performed only by individuals licensed by the Home Secretary, in premises registered with him and subject to periodic inspection. Although both sides were to urge changes in the details of this arrangement, the principles restricting vivisection embodied in the bill were far more controversial. Vivisection was allowed "with a view only to the advancement by new discovery of knowledge which will be useful in saving or prolonging human life, or alleviating human suffering" despite the Royal Commission's express disavowal of the viability of a distinction between scientific and medical objectives. Dogs and cats were exempted from vivisection completely, in line with Hutton's minority report. Experiments without anesthesia, or those that allowed animals to recover from the effects of anesthesia, or demonstrations under anesthesia were permitted only when certified as necessary by certain specified individuals, such as the presidents of various scientific and medical bodies. The former cases, where pain might occur, would be very rare according to Carnarvon. Experiments to attain manual dexterity or for exhibition to the public were strictly prohibited. Experiments for judicial inquiry were allowed by order of the court.[12]

[11] *Hansard,* 3rd series, 1876, ccxxix, 1001–1013.

[12] *R.S.P.C.A. Records* xvi (1876), 3–4, 18–20.

The presentation of the government's bill "for extending the law relating to cruelty to animals to the cases of animals which for medical, physiological, or other scientific purpose are subjected when alive to experiments calculated to inflict pain" was the beginning of the most important phase of the legislative life of the vivisection issue. The lobbying, deputations and counterdeputations, and editorial salvos that the bill touched off are interesting not only for their impact on the final form of the legislation, which regulates research in experimental medicine in Britain to this day, but for what they reveal about the interplay of interest groups on an issue of relatively small importance to Disraeli's government. These activities also reveal the strength of the commitment to animal protection and the depth of parliamentary and public uncertainty about science and medicine.

Reaction to the government bill, both in and out of Parliament, was widely varied and illustrative of the polarization that had taken place on the issue over the last year.[13] In the Lords, Shaftesbury (now president of the Victoria Street Society) accepted the bill, but complained that the measure "did not go as far as could be wished" since "it could not be denied that the feeling of the country was in favour of total abolition," as was demonstrated by the petitions that had been addressed to Parliament. Shaftesbury went on to justify his call for total abolition by a lengthy speech quoting copiously from various medical sources, and appealing against the "ardent adoration of science" and "the attempted despotism of science over common sense."[14] On the other hand, the Duke of Somerset, probably coached by a medical man, made a number of acute and important criticisms of points at which the bill

[13] For example, Lord Henniker said, in supporting the government measure, "He believed his (1875) bill was all that public opinion would have sanctioned at the time; but during the last year a great advance [in antivivisection sentiment] had been made." *Hansard,* 3rd series, 1876, ccxxix, 1030–1032.

[14] *Hansard,* ccxxix, 1016–1030. Cf. E. Hodder, *The Life and Work of the Seventh Earl of Shaftesbury, K.G.* (London, 1886), iii. 373.

went beyond the recommendations of the Commission.[15] Somerset's criticisms were later reiterated at length by the medical and scientific lobby, and were augmented by the Earl of Airlie[16] and by Lord Winmarleigh.[17] Despite the incongruence of the bill with his Commission's report, the cautious Cardwell refused to endorse Somerset's criticisms of it. He urged the necessity of legislation in the following terms: "We were passing through a scientific revival, and a great start was being given in this country to the practice of vivisection. Everything depended, therefore, whether we should seize the opportunity for wise legislation. If we controlled the system in its infancy, we should have an advantage which might never occur again."[18]

In the country at large, an even wider spectrum of opinion was expressed. Abolitionists, such as George Jesse, rejected the bill out of hand as partially countenancing the evil—a "vivisectors' bill."[19] Restrictionist antivivisectionists like Cobbe were, at this point, both pleased with the bill and optimistic about its fate.[20] Part of the reason for their optimism was the generally favorable response accorded the bill by the lay press. Not unexpectedly, Hutton's *Spectator* was favorable, but so was the *Saturday Review*.[21] Most newspapers joined the *Times* and the *Daily News* in endorsing the bill initially, though the *Morning Adver-*

[15] *Hansard,* ccxxix, 1013–1015.

[16] *Hansard,* 1032.

[17] *Hansard,* 1033.

[18] *Hansard,* 1033–1034.

[19] G. R. Jesse, *Extracts From and Notes Upon the Report of the Royal Commission on Vivisection, Refuting its Conclusions* (London, 1876), i. 35, ii. 6.

[20] *Life of Frances Power Cobbe,* ii. 594–595. Cobbe says in this account that her Victoria Street Society recommended that vivisection be allowed only under complete anesthesia and that this provision was embodied in Carnarvon's bill, as originally presented. As far as the bill goes, Cobbe's account is clearly in error. On antivivisectionist optimism at this point, the *Zoophilist* later wrote, "the friends of the cause began to complain that no larger demands had been made." *Zoophilist* i (1881), 33.

[21] *Spectator* xlix (1876), 672–674; *Saturday Review* xli (1876), 671–672.

tiser and the *Manchester Examiner* wanted stronger steps taken[22] while the *Standard,* for its part, was strongly critical of the bill, on behalf of the scientific interest.[23]

To the R.S.P.C.A., the bill clearly came as something of a surprise, for as late as the May 1876 issue readers of *Animal World* were being told "It is all but certain that the Government do not intend to introduce during the present session any bill relating to the practice of subjecting live animals to experiments for scientific purposes."[24] Nevertheless, when the bill did appear, the society appraised it rapidly and presented it for consideration to a conference of representatives of various branches of the organization. The ways in which the bill advanced on the report of the Commission were welcomed, but the permission to painful experiment under certain circumstances was "a sad extremity." In addition, the society suggested to the government that special protection for cats and dogs be extended to equine species, that provisions for making known results of experiments be instituted to prevent needless repetition, that certification for the right to perform painful experiments should require a character reference from "a layman of eminence or position" rather than the endorsement of a medical or scientific authority. Lord Harrowby, president of the R.S.P.C.A., contacted Carnarvon directly with his society's recommendations.[25]

The medical and scientific press expressed itself as appalled at the government bill. The *Lancet* and the *B.M.J.* were not at all sure that the measure would not in practice mean the end of all research and were especially unhappy with the provisions in which the bill went beyond the report of the Royal Commission, notably the cat and dog clause.[26] *Nature,* speaking for science, was likewise disgusted:

[22] "The Government Vivisection Bill," *Public Opinion* xxix (1876), 670–671.

[23] See *Spectator,* note 21.

[24] *Animal World* vii (1876), 66.

[25] Ibid., 86 and Harrowby to Carnarvon and Colam to Carnarvon, 23 May 1876, P.R.O., Carnarvon Papers 30/6/19/3,4.

[26] *Lancet* i (1876), 784; *B.M.J.* i (1876), 701–702.

> It is but natural to suppose that concomitantly with the rapid advances which have, within the last century or so, been made in our knowledge of scientific method, similar progress has occurred in the theory of legislation. And yet our leading politicians, in introducing the above quoted Bill, are bold enough to advance, as a motive for the legal machinery they are endeavouring to enforce, the idea that there is any real substantiality in the notion that the lengthening of human life can form any direct stimulation to physiological work. In so doing they show how little they are capable of appreciating the spirit of the higher philosopher, whose thoughts and temptations to investigate, however much they may be disguised by secondary motives, are but the involuntary secretion, as it may be termed, of his individual brain. They do not even seem to know that one of the most fundamental of the data of scientific method precludes the possibility of preconceived ideas of any kind forming part of a correctly stated problem.

Nature was very much against any interference with creative scientific research.[27]

Burdon Sanderson, at the centre of the scientists' lobby, had watched with anxiety the proceedings following the publication of the Commission's report. In the face of the 20 March deputation to the Home Office by Cobbe's Victoria Street Society, the scientists had maintained a judicious silence, fearing that opposition to the antivivisection society's activities would only kindle an unproductive public controversy. As G. M. Humphry, Professor of Anatomy at Cambridge, wrote Burdon Sanderson on 27 March ". . . I should almost question the policy of any active movement on the part of the physiologists and should be disposed to leave the matter to the good sense of the House and the Government acting upon the Commission Report."[28] A few days later, Cardwell, chairman of the Com-

[27] "Lord Carnarvon's Vivisection Bill," *Nature* xiv (1876), 65. *Nature* was later reconciled to some kind of legislation in order to reduce public concern over the issue.

[28] Humphry to Burdon Sanderson, 27 March 1876, Library, University College London, Burdon Sanderson Papers, MS Add. 179/2 f.35.

mission, was assuring Burdon Sanderson that any legislation by the government would surely follow the recommendations of the Commission.[29]

In these circumstances, therefore, the scientists were stunned by the severity of Carnarvon's bill when it was read in the Lords on 15 and 22 May. They moved rapidly to reopen lines of communication with parliamentary allies, placing particular emphasis, at this point, on provisions in the bill that went beyond the report of the Commission. As Huxley counseled his friend Michael Foster of Cambridge, these provisions, especially the cat and dog clause, which had been explicitly rejected by six of the seven commissioners, were the vulnerable points of the government bill.[30] Burdon Sanderson and a fellow physiologist, David Ferrier, immediately communicated to a sympathetic Lyon Playfair the great danger to physiological research represented by the cat and dog clause of the bill.[31] According to Playfair, the scientific cause could hope for more in the way of favorable amendments to the bill from the Lords than from the Commons. He therefore advised immediate contact with Cardwell, whose influence as chairman of the Commission was considerable. Time was of the essence as the bill was to be amended in committee of the upper chamber on 29 May. Playfair himself had mobilized Lord Airlie, Lord Morley, Lord Belper, and Lord Winmarleigh

[29] Cardwell to Burdon Sanderson, Burdon Sanderson Papers, MS Add. 179/2 f.37.

[30] Huxley to Foster, 25 May 1876, Imperial College of Science and Technology, Huxley Papers, 4.120.

[31] Ferrier to Playfair, 16 May 1876 and Burdon Sanderson to Playfair, 22 May 1876, Imperial College of Science and Technology, Playfair Papers, #262, 124; Playfair to Burdon Sanderson, 20 May 1876, Burdon Sanderson Papers MS Add. 179/2 f.39. According to Ferrier, "The differences between man and the herbivora are so great that if we were restricted to rabbits many researches of vital importance to physiology and of the utmost value to man would be rendered impossible. I need only mention the physiology of digestion, assimilation and excretion on which experiments on rabbits and guinea-pigs could throw no light."

against the government bill as it stood.[32] Following Playfair's advice, Burdon Sanderson contacted Cardwell, whose failure to point out the extent to which the bill went beyond the report of the Commission had angered Huxley and the scientists. In a letter of 23 May, Burdon Sanderson pointed out to Cardwell the severity of the bill and indicated two modifications of it that the physiologists thought of great importance. The first amendment suggested was that experimentation upon cats and dogs be permitted under the same kind of certification from scientific or medical bodies as the bill required for experiments without anesthetics. The second amendment suggested that legal proceedings involving alleged violations at registered places should not be allowed except when undertaken by inspectors under the Home Secretary. The second provision was intended to prevent interference with research by continual prosecutions instituted by antivivisectionists.[33]

Although frantic activity by the scientists' lobby had succeeded in developing a certain amount of strength in the House of Lords, as well as in presenting its hastily considered demands before Cardwell and Carnarvon, Burdon Sanderson and his allies were greatly relieved when the death of Carnarvon's mother took the Earl out of London. This caused the committee of the Lords on the government bill to be postponed for nearly a month, from 29 May to 20 June.[34] Though the resulting time for lobbying activity allowed both sides to increase representations to the government, it was more important to the medical and scientific interest. Hitherto, through 1875 and early 1876, a handful of medical and biological scientists had had good success in developing parliamentary strength to protect their interests. The time provided by Carnarvon's absence from Lon-

[32] Playfair to Burdon Sanderson, 23 May 1876, Burdon Sanderson Papers, MS Add. 179/2 f.42; Airlie to Playfair, 29 May 1876, Playfair Papers, #510.

[33] Burdon Sanderson to Cardwell, 23 May 1876, P.R.O., Carnarvon Papers, 30/6/19/7.

[34] *R.S.P.C.A. Records* xvi (1876), 49.

don and the severity of the government bill he had presented, however, permitted the scientists and the organs of medical opinion an opportunity to represent the measure as a matter of flagrant disregard for the integrity of the medical profession as a whole. At the beginning of June, the *Lancet* had complained. "It is remarkable that this measure, fraught with such importance to the medical and scientific profession in this country, has hitherto received such scant attention from those whom it most concerns, and there can be no doubt that in medical circles it is regarded, strangely enough, with something akin to apathy."[35] Within the next two months, as we shall see, the leaders of the profession so mobilized professional organizations that opposition to the bill on the part of most medical men was abundantly clear to the public and the government alike. The scientists' lobby now operated with the support of a powerful professional pressure group that, despite the fact that most of its rank and file had never seen an experiment on a living animal, ultimately became more intransigent in their opposition to legislation than some of the physiologists whose interests were most directly threatened.

Before discussing further the development of this pressure group, it is necessary to consider the position in which the government found itself upon postponement of committee in the Lords at the end of May. We have already noted the R.S.P.C.A.'s pressure to strengthen the bill and the scientists' suggestions to weaken it, which were forwarded by Cardwell to the bill's sponsor, Carnarvon. Expecting a bill in accordance with the report of the Royal Commission and, therefore, minimally controversial, the cabinet and especially the Home Secretary, Richard Cross, found themselves with a complex measure of little partisan political importance that was nevertheless the subject of increasingly intense lobbying as it progressed. The measure always had a questionable political future in a crowded parliamentary agenda, especially after the Bulgarian atroc-

[35] *Lancet* i (1876), 830.

ities became a matter of wide public concern. Although Disraeli believed "that the startling contemporary advances of science were primarily significant as dangerous stimuli to the animal restlessness of the masses,"[36] he seems to have cared little for the precise substance of legislation for experiments on living animals. Queen Victoria, strongly against vivisection, had been urging her prime minister to act against it since March,[37] and although Disraeli had informed her secretary that the arguments against vivisection were not worth listening to,[38] he wrote to Her Majesty "The great thing is to pass some Act, and give evidence of the determination of the legislature to control this horrible practice. . . ."[39] Victoria's earnest entreaties against vivisection were a persistent thorn in the side of ministers responsible for the passage of the bill and the administration of the Act.[40]

Within the cabinet itself, there was a good deal of uncertainty about the bill. When Carnarvon's mother's illness removed him from London, he wrote anxiously to Cross about the possible influences that might be brought to bear on the Duke of Richmond and Gordon, who had been left in charge of the bill in the Lords. Carnarvon especially feared amendments proposed by Lord Rayleigh (1842–

[36] *Hansard,* 3rd series, 1867, clxxv, 219–220, cited in C. J. Lewis, "Theory and Expediency in the Policy of Disraeli," *Victorian Studies* iv (1961), 240.

[37] F. J. Dwyer, "The Rise of Richard Assheton Cross and his Work at the Home Office, 1868–1880," (Oxford University B. Litt. thesis 1954), p. 228 n.3. I am grateful to Mr. Dwyer for permission to consult his thesis.

[38] E. Longford, *Victoria R. I.* (London, 1964), p. 406.

[39] W. F. Monypenny and G. E. Buckle, *The Life of Benjamin Disraeli, Earl of Beaconsfield* (London, 1920), v. 483.

[40] See, e.g., F. J. Dwyer, op. cit., 228–229; E. Longford, op. cit., 406; P. Guedalla, *The Queen and Mr. Gladstone* (London, 1933), ii. 152; R. A. Cross, *A Political History* (Privately printed, Eccle Riggs, 1903), pp. 42, 180–181; A. Ponsonby, *Henry Ponsonby, Queen Victoria's Private Secretary* (London, 1942), pp. 48–49; A. G. Gardiner, *Life of Sir William Harcourt* (London, 1923), i. 402–403; A. Hardinge, *The Life of Henry Howard Molyneux Herbert, Fourth Earl of Carnarvon, 1831–1890* (London, 1925), ii. 103–111.

1919), a physicist, and his ally, the influential Secretary for India, Lord Salisbury (1830–1903).[41] Salisbury, later prime minister, had a deep interest in science and was to be president of the British Association for the Advancement of Science in 1894. He strongly opposed Carnarvon's bill.[42]

With the cabinet divided, the scientists' lobby moved rapidly to induce the medical profession to provide ammunition for their parliamentary allies. Burdon Sanderson forwarded to Henry Acland, president of the General Medical Council (G.M.C.), a copy of the physiologists' objections to the bill and of their protest to Cardwell.[43] At its session of the end of May, the G.M.C., an elite body of medical men responsible by statute for medical education and professional qualification, discussed the government bill at length. Throughout their deliberations, the feeling of most of the members of the Council was one of great irritation at the government for presuming to bring in a bill without consulting the one body with any legal responsibility relating to the subject at issue. Certain members of the Council clearly would have liked to defy the legislation in toto. The body as a whole accepted the necessity for legislation only under the duress of public opinion. The single member of the Council advocating the legislation was George Rolleston of Oxford, whose antipathy to the *Handbook for the Physiological Laboratory* had so impressed the Royal Commission.[44] The Council forwarded a strongly-worded memorial to the government with a large number of objections to the bill, including the suggestions

[41] Carnarvon to Cross, 28 May 1876, Brit. Mus., Addit. MSS. 51268 ff.115–118.

[42] A Hardinge, *The Life of Henry Howard Molyneux Herbert, Fourth Earl of Carnarvon, 1831–1890* (London, 1925), ii. 344.

[43] Acland to Burdon Sanderson, 25 May 1876, Burdon Sanderson Papers, MS Add. 179/2d f.44.

[44] "The General Council of Medical Education and Registration. Session 1876," *Lancet* i (1876), 815–825. See also H. C. Cameron, *Joseph Lister. The Friend of Man* (London, 1948), pp. 105–108, and E. B. Tylor, "Biographical Sketch," in W. Turner (ed.), *Scientific Papers and Addresses* (of George Rolleston) (Oxford, 1884), i. pp. lix–lx.

that cats and dogs be made available for research, that experiments be allowed for purposes of the advancement of science and the alleviation of animal suffering, as well as for the alleviation of human suffering, and that inspectors under the bill be "scientifically competent."[45] The British Medical Association (B.M.A.), a national mass membership organization for which any qualified medical practitioner was eligible, soon joined the G.M.C. in criticizing the bill. Under the leadership of Ernest Hart, editor of its *B.M.J.*, the B.M.A.'s Parliamentary Bills Committee indignantly protested the aspersions that Carnarvon's bill allegedly cast upon the profession, and endorsed the G.M.C.'s recommendations.[46] Under the auspices of the B.M.A., Hart sought a meeting with Carnarvon for British teachers of physiology,[47] who submitted a lengthy memorial suggesting that cats and dogs must be made available for research and that research should not be confined exclusively to registered laboratories. The memorial, probably drawn up by Foster and Burdon Sanderson, was signed by all the teachers of physiology in Britain with the exception of Rolleston.[48] Hart organized a large deputation to present the report of the Parliamentary Bills Committee to Richard Cross, while the Presidents of the Royal College of Surgeons of England, the Royal College of Physicians of London, the Royal Society, and the Linnean Society waited upon the Duke of Richmond to similar effect.[49] The *B.M.J.* gave full publicity to these deputations and printed a peti-

[45] "Proof of Memorial of the General Medical Council of the United Kingdom—Report of Committee on Lord Carnarvon's Bill," P.R.O., Carnarvon Papers, 30/6/19/10.

[46] "Proceedings of the Parliamentary Bills Committee," *B.M.J.* i (1876), 707–708.

[47] Hart to Carnarvon, 7 June 1876, Carnarvon Papers, 30/6/19/12.

[48] "Memorandum of Facts and Considerations Relating to the 'Cruelty to Animals Bill,'" Carnarvon Papers, 30/6/19/439.

[49] *B.M.J.* i (1876), 708, 794–798. See also [J. Paget], *Annual Report on the Affairs of the Royal College of Surgeons of England* (London, 1876), p. 13 and "Extraordinary Meeting of the College. June 16, 1876," Royal College of Physicians of London, Annals. I am most grateful to Jeanne Peterson for directing my attention to these sources.

tion form to be signed by medical men for parliamentary submission against the bill. It also supplied petition forms suitable for nonmedical persons.[50] These activities were the harbingers of continual pressure mounted by a variety of medical organizations, which was to plague Cross and Carnarvon for the remainder of the legislative life of the bill.

Such evidence of the political power of a mobilized medical profession made it quite clear that some concessions would have to be made. Carnarvon and his ally in the upper chamber, Lord Portsmouth, were attempting to organize support for the measure before the committee of the Lords on 20 June,[51] but Cardwell, in contact with Burdon Sanderson, was advising compromise.[52] By 20 June, the strength of Carnarvon's position had deteriorated considerably, as the debate made clear.

Although the Lords were treated to a lengthy and enthusiastic speech by Lord Chief Justice Coleridge (of the Victoria Street Society) pleading that Carnarvon's bill be left unaltered, pressure from the Marquis of Landsdowne, Lord Rayleigh, the Duke of Somerset, and Lord Cardwell succeeded in modifying the title of the bill and in amending it to allow experiments for the advance of physiological knowledge or of knowledge for the alleviation of either animal or human suffering. A number of the speakers referred to the dissatisfaction of the medical profession with the measure. Carnarvon was by now quite prepared to make these concessions, which he rationalized both in Parliament and in a letter to Queen Victoria as necessary for effective legal interpretation of the enactment. After withstanding demands by the spokesmen for the medical and scientific interest that some experiments be allowed outside

[50] *B.M.J.* i (1876) 739.

[51] Portsmouth to Carnarvon, n.d. and 10 June 1876, Carnarvon Papers, 30/6/19/8,13.

[52] "Will you please have the kindness to let me know whether the bill as proposed to be amended will satisfy the suggestions you made to me?" Cardwell to Burdon Sanderson, 17 June 1876, Burdon Sanderson Papers, MS Add. 179/2 f.50.

registered places, and by antivivisectionist Lord Henniker that the use of curare be prohibited under any circumstances,[53] Carnarvon faced discussion of the clause excluding cats and dogs from experiment. Although he clearly was most reluctant, he agreed to allow experiments on cats and dogs, under certification by officers of the specified medical or scientific bodies and by the licensee that the object of the experiment would be frustrated if cats or dogs could not be used. A few other minor changes, such as the inclusion of horses, mules, and asses in the same category as the now vulnerable cats and dogs, concluded the changes introduced in the committee of the Lords.[54]

LOBBYING AND COMPROMISE IN THE HOUSE OF COMMONS

Although Carnarvon represented his concessions as "slight" in his report to an anxious Queen Victoria,[55] there can be no doubt that the medical and scientific interest had gained a considerable victory. Carnarvon believed that his concessions would so mollify the medical profession as to secure a relatively smooth passage for his bill through the Commons. He wrote to Disraeli on the day before the debate ". . . I have succeeded in settling the measure fairly to the satisfaction of the various parties interested. There will of course be some discussion . . . , but the principal causes of contention are either removed or lessened and though the Profession or a great part of it, do not like Legislation, they will now accept without much repugnance, if not with gratitude, the Bill."[56] Cross and Carnar-

[53] Curare, a muscular poison frequently used by physiologists, did not, according to Claude Bernard, affect the animal's sensation. Carnarvon's bill specified that curare could not be considered an anesthetic for purposes of the bill, angering the medical profession, who charged the government with legislating a solution to an open question in science. Henniker wished to forbid the use of curare even when a certificate for dispensing with anesthetics was allowed.

[54] *Hansard,* 3rd series, 1876, ccxxx, 105–127.

[55] Carnarvon to Victoria, 21 June 1876, Carnarvon Papers, 30/6/2 f.156. Cf. A. Hardinge, *The Life of Henry Howard Molyneux Herbert, Fourth Earl of Carnarvon, 1831–1890* (London, 1925), ii. 107–109.

[56] Carnarvon to Disraeli, 19 June 1876, Carnarvon Papers, 30/6/11.

von had been under fire not only from the medical profession but also from the Victoria Street Society, in the persons of James Stansfeld and Shaftesbury. Cross, anxious for legislation of some strength, had written, "If then amendments are to be made in the Lords I only trust that the matter will be so fully discussed as to settle it. Discussion in our own house [of Commons] will be absolutely fatal. And if then amendments be made Lord Shaftesbury must speak strongly so as to satisfy his friends."[57] Cross's fears of 21 June for the fate of the bill in the House of Commons provide an interesting benchmark for the impact of the intervention of the medical profession, as a profession, into the struggle, for they contrast markedly with Lyon Playfair's assessment of 23 May 1876. At that time, with the scientists' lobby operating without mass professional support, and faced (before Carnarvon was forced to leave London) with the imminent passage of the bill through the Lords, Playfair had written to Burdon Sanderson, ". . . no time must be lost, as the Bill will pass through the Committee of the Lords on Monday and we have a better chance there than in the Commons."[58] Before professional pressure came to bear, personal influence could be most effectively deployed, on a measure of this sort, in the Lords. After the G.M.C. and the B.M.A. acted, Cross, one of the most knowledgeable politicians in the House of Commons and a man apparently in favor of a stringent bill, felt that discussion in the lower chamber would be "absolutely fatal" to the government's hopes for legislation.

Cross's and Carnarvon's hopes that the amendments made on the bill in the Lords would ensure a calm passage for the bill in the Commons, though echoed by the press,[59] were in the event completely unrealized. Instead, activity on both sides of the question intensified, and came to de-

[57] Cross to Carnarvon, 16 June 1876, Carnarvon Papers, 30/6/9.

[58] Playfair to Burdon Sanderson, 23 May 1876, Burdon Sanderson Papers, MS Add. 179/2 f.39.

[59] "The Cruelty to Animals Bill," *Saturaday Review* xli (1876), 773, 804–805, and "The Amended Vivisection Bill," *Public Opinion* xxix (1876), 798–799.

mand an ever increasing proportion of Cross's attention and powers of mediation.

Antivivisectionists, not unnaturally, were greatly distressed at the weakening of the cat and dog clause. Hutton, whose minority report of the Royal Commission had inspired the clause, quickly wrote to Carnarvon of his "heavy disappointment" at its amendment and hinted at "something like a breach of contract" on Carnarvon's part.[60] During the month of June 1876 a good deal of antivivisectionist organization was taking place besides Cobbe's Victoria Street Society. On 10 June the London Anti-Vivisection Society was founded, to be followed on 21 June by the International Association for the Total Suppression of Vivisection.[61] Both bodies scorned the caution of the R.S.P.C.A. and the compromise of the Victoria Street Society, taking their stand on nothing less than total abolition. An antivivisectionist journal, the *Home Chronicler,* was founded on 24 June and soon adopted a policy of total abolition. By mid-summer, the R.S.P.C.A. noted ". . . about ten new societies have been established in various parts of the Kingdom, for the purpose of carrying on more extreme measures than ourselves;"[62] it also mentioned the initiation of three antivivisectionist journals.[63] Petitioning activity was considerable, the House of Commons reportedly receiving 805 petitions demanding total abolition, with 146,889 signatures, by early August.[64] The R.S.P.C.A.—involved despite itself in a measure of such notoriety in the sphere of animal protection—and the Victoria Street Society could represent their demands to the government as relatively moderate against the background

[60] Hutton to Carnarvon, 25 June 1876, Carnarvon Papers, 30/6/19/18.

[61] *Dates of the Principal Events Connected with the Anti-Vivisection Movement* (London, c. 1897), p. 3, and *The International Association for the Total Suppression of Vivisection* (London, 1876).

[62] *Animal World* vii (1876), 132.

[63] J. Colam (ed.), *Further Proceedings Against Vivisection* (London, n.d.), p. 4.

[64] *B.M.J.* ii (1876), 249.

of this total abolitionist activity. The President of the Victoria Street Society, Lord Shaftesbury, had in fact agreed to the concessions that Carnarvon had allowed in the committee of the Lords and gave the bill his blessing at its perfunctory third reading in the Lords, before presentation to the Commons.[65] No doubt he, too, hoped to avoid further pressure from the medical profession.

Ernest Hart, described by unsympathetic antivivisectionists as the medical profession's "chief wire-puller," had other ideas. His efforts to mobilize the medical profession had achieved a certain momentum. According to his *B.M.J.*, even after alterations in the Lords "the best thing that could happen to the Bill would be, that it should be thrown out altogether. In its present shape it is far less offensive than it was; but we doubt whether it will satisfy anybody."[66] In this mood of militance and buoyed by any number of protests and petitions from a variety of medical and biological associations,[67] Hart determined to maintain as much pressure on the government as possible.[68] The *Lancet*, delighted that ". . . a spirit of downright indignation had at last been awakened" in the profession,[69] and the remainder of the medical press[70] were in a similar mood, but Hart's role was crucial. Referring to antivivisectionist activity, Lawson wrote "The medical profession at first took little notice of this babbling crowd; and we fear that if it had not been for the decided movement of Mr. Ernest Hart it would have been absolutely silent on the matter, and as a consequence we should have had the Act passed

[65] E. Hodder, *The Life and Work of the Seventh Earl of Shaftesbury, K.G.* (London, 1886), iii. 373, and A. Hardinge, *The Life of Henry Howard Molyneux Herbert, Fourth Earl of Carnarvon, 1831–1890* (London, 1925), ii. 109.

[66] *B.M.J.* i (1876), 786.

[67] *B.M.J.* i (1976), 789, 798, ii. 91–92, 121; *Lancet* ii (1876), 100, 169.

[68] *B.M.J.* ii (1876), 24.

[69] *Lancet* i (1876), 930.

[70] *Medical Press and Circular* i (1876), 530–532, and *Medical Times and Gazette* ii (1876), 60–61. See also "Vivisection," *Journal of Science* n.s. vi (1876), 317–336.

Figure 7. Ernest Hart (1835–1898), editor of the *British Medical Journal,* and spokesman for the medical and scientific interest. *Courtesy of the Wellcome Trustees.*

in its original form. . . ."[71] Hart printed a special supplement to the *B.M.J.*, detailing objections to the bill and repeated "An attempt has been made to spread the belief that such concessions as Lord Carnarvon had made are sufficient to satisfy the medical profession. It is necessary to completely undo the effect of that attempt. It is certain that the profession is profoundly dissatisfied with the Bill as it stands."[72] Appealing to the branches of the B.M.A. to support the actions of the Parliamentary Bills Committee of the parent organization, Hart suggested that practitioners should write to their M. P.s and local newspapers since "every individual member of the profession should treat this bill as one which affects his own personal honour, and bestir himself to oppose it."[73] Hart's greatest accomplishment was the second deputation of the Parliamentary Bills Committee of the B.M.A., around 10 July. He led a spectacular deputation of several hundred medical men to the Whitehall chambers of Richard Cross, who now had charge of the bill. There the group, overflowing through adjacent corridors, heard Hart, John Simon, Sir William Jenner, and Samuel Wilks speak strongly and indignantly against the bill and present detailed suggestions for its amendment.[74] In the aftermath of this display of professional power and unity, several medical bodies, such as branches of the B.M.A., called for complete withdrawal of the bill.[75] Robert Lowe, M. P. for the University of London, publicly declared that he would fight any special legislation on the subject whatever,[76] as did Lowe's fellow M. P., J. A. Roebuck.[77] The medical initiative also had the effect of stimulating some of the most influential period-

[71] H. Lawson, "The Vivisection Clamour," *Popular Science Review* xv (1876), 398.

[72] *B.M.J.* ii (1876), 49.

[73] Ibid., 80.

[74] Ibid., 80, 88–91. See also *Life of Frances Power Cobbe*, ii. 595. These four were, of course, among the leading medical personalities of the day.

[75] See, e.g., *B.M.J.* ii (1876), 153–156.

[76] Ibid., 121–122.

[77] *Lancet* ii (1876), 167.

icals to back the profession and its demands: the *Times,* the *Standard,* the *Pall Mall Gazette,* and *Punch* are examples.[78]

This was quite enough to confirm the fears of the Victoria Street Society and the R.S.P.C.A. as to the fate of the bill. Lords Coleridge and Shaftesbury had already warned that the bill might be converted into a measure to protect physiologists.[79] The Victoria Street Society, whatever the influence of its individual representatives, scarcely counterbalanced the B.M.A. The R.S.P.C.A., with somewhat more of a constituency, urged its supporters to write to their M. P.s opposing alteration of the bill and warned that the suggestions of the medical deputation "if adopted will change the present bill into a legislative measure, cunningly designed to repeal in part the present anti-cruelty statute [Martin's Act]."[80] An R.S.P.C.A. deputation on 28 July, including such luminaries as its President, the Earl of Harrowby, the Bishop of Gloucester and Bristol, and Cardinal Manning, urged Cross to stand firm in the face of the demands of the B.M.A. The deputation emphasized the moderation of its own position, noted the mushrooming antivivisectionist activity around the country, and warned ". . . should the Bill now before Parliament be thrown out by opposition, then inevitably an appeal must be made to the country. It will be easy, on one word being spoken by the Committee, to kindle a flame of indignation against vivisection even in the remotest hamlets of this kingdom."[81]

The question now was indeed whether or not any bill could be passed. The threat of the R.S.P.C.A. notwithstanding, Cross informed the society's delegation, "Of course you must be aware that at this period of the session it

[78] "The Deputation to the Home Secretary on Vivisection," *Public Opinion* xxx (1876), 66; *B.M.J.* ii (1876), 222.

[79] *Hansard,* 3rd series, 1876, ccxxx, 110; E. Hodder, *The Life and Work of the Seventh Earl of Shaftesbury, K.G.* (London, 1886), iii. 373.

[80] *Animal World* vii (1876), 98, 117.

[81] Ibid., 132, or J. Colam (ed.), *Further Proceedings Against Vivisection* (London, n.d.), pp. 4–5.

would be absolutely impossible to carry a bill of this kind in the face of opposition from the medical profession and the scientific world. . . ."[82] In the face of the intransigence of the medical lobby, Cardwell was urging further concession upon Cross, "I strongly advise your making such a compromize . . . It would be a thousand pities that the bill should be lost."[83] The government's readiness to negotiate, the product of the opposition of the profession at large, created an opportunity for the scientists' lobby to effect a settlement on its own terms. Some scientists, especially Burdon Sanderson, were dismayed at the medical profession's demands for withdrawal of the bill, but were quite prepared, as representing the interests of those who would be affected in a practical way by legislation, to take advantage of the pressure for compromise within the government produced by Hart's activities. The difficulty was that even within the scientists' lobby, one group, represented by London medical men William Gull and Richard Quain and Edinburgh's Joseph Lister and Robert Christison, shared with the majority of their profession a distaste for any legislation whatever; a second group, including Joseph Hooker, President of the Royal Society, and Michael Foster of Cambridge, was essentially undecided; while a third and most influential group, including Huxley, Burdon Sanderson, and James Paget, preferred settling the matter while the opportunity of favorable terms presented itself.[84] This disagreement over tactics manifested itself in the different instructions sent to the lobby's parliamentary allies. Robert Lowe seems to have been put up to total opposition to legislation by his University College constituents Gull and

[82] J. Colam, op. cit., p. 3.

[83] Cardwell to Cross, 13 July 1876, Carnarvon Papers, 30/6/19/23. Cf. Cardwell to Cross, 28 June 1876, Brit. Mus., Addit. MSS 51271 ff.138–139.

[84] I use the term "scientists' lobby" in loose distinction from professional medical organizations. Obviously all of the men in the lobby were not "scientists" in the strict sense of the word, but the medical men involved all had strong academic connections or academic posts. See the analysis at the end of this chapter.

Quain.[85] Lyon Playfair, M. P. for the University of Edinburgh, was told by Burdon Sanderson that "those who are actually engaged in scientific work . . . want the bill to pass with . . . slight modifications,"[86] and that "We are all of opinion that it would not be advisable to throw the Bill out even if that were possible or rather, that it is not desirable that the Bill should be deferred."[87] By "we" Burdon Sanderson probably meant most of the members of the embryonic Physiological Society, an association formed to protect the interests of research in March 1876, and consisting of the leading men involved in experimental work in London and Cambridge.[88] From his Edinburgh constituents, Robert Christison and James Lister, however, Playfair was being advised that the unanimous verdict of both the Medical Faculty and the Scottish medical profession was for postponement of the bill, and that he was the man to effect that postponement.[89] Christison assailed the G.M.C. for its "mealy-mouthed objections and vain cobbling" over the bill, instead of straightforward rejection.[90] In a second letter, Christison put the situation quite clearly:

> Again however I am importuned from London, and for a special object, which happens to be the same for which

[85] Cardwell to Cross, 13 July 1876, and Quain to Lowe, 11 August 1876, Playfair Papers #598.

[86] Burdon Sanderson to Playfair, 11 July 1876, Carnarvon Papers, 30/6/19/21.

[87] Burdon Sanderson to Playfair, 7 July 1876, Playfair Papers, #126. Paget and Huxley agreed with Burdon Sanderson. Paget to Burdon Sanderson, 11 July 1876, Burdon Sanderson Papers, MS Add. 179/2 f.66. Huxley to Foster, 29 May, 15 June, 4 July 1876, Huxley Papers, #123, 125, 131.

[88] E. A. Sharpey-Schafer, *History of the Physiological Society during its First Fifty Years 1876–1926* (Supplement to *J. Physiol.*, London, 1927), pp. 1–15. Cf. Marshall Hall's 1831 appeal for a physiological society, also predicated upon pressures arising from antipathy to vivisection.

[89] Lister to Playfair, 30 July 1876, Playfair Papers, #451; Christison to Playfair, 2 July 1876–30 July 1876, Playfair Papers, #178–181.

[90] Christison to Playfair, 2 July 1876, Playfair Papers, #178.

> I importuned the Lord Advocate in the letter of which I sent you a copy. This is postponement for another session of Parliament. It seems there is a division of opinion in the big city on this question. Some opponents of the Bill think it best to push it on, and pass it with emendations this year, lest the recent outcry be magnified, and a worse measure brought forward next year. Others, and the greater number, think that during delay the public will be disabused, and recover from their errors and hallucinations, and that all the practicable or necessary good which reasonable people could desire, has been already attained by the extraordinary stir which the question has occasioned. As such was exactly the line taken in my letter to the Lord Advocate, I of course am of opinion that postponement is the right procedure. I am not afraid of worse because worse could scarcely be.[91]

It is important to note that while Christison was probably correct in ascribing his views to the majority, most of that majority was made up of men like Hart, Gull, and Quain whose practical work would be unaffected by legislation and whose opposition to it was based upon its "insult" to the profession. The opposition of Christison and Lister in particular is probably explicable in terms of provincial mistrust of Whitehall[92] and of the failure of the antivivisection movement to penetrate Scotland as rapidly as it had emerged in London. It was, however, Burdon Sanderson, Paget, Huxley, and the Physiological Society who ultimately held the ears of Playfair and Cross, despite Christison's criticism: "The London Physiologists, through timidity, have given very bad advice. Sanderson, one of the most estimable of men, has been too much of a dreamer to pilot in such a storm."[93]

At about this time, Sanderson was writing to Hooker, "My chief fear now is lest Mr. Cross finding himself [beset] so much by certain irreconcilable members of the medical

[91] Christison to Playfair, 19 July 1976, Playfair Papers, #179.
[92] This mistrust was fully justified in the event. See Chapter 7.
[93] Christison to Playfair, 30 July 1876, Playfair Papers, #181.

profession should think it the easiest way out of his difficulty to withdraw the Bill. I hope this will not be the case, as for the sake of science, it is most desirable that it should, if possible, pass with amendments."[94] The government, dismayed indeed by the extremity of professional opinion, did not miss the opportunity for negotiation with the moderate members of the scientists' lobby. Cross wrote to Carnarvon on 18 July 1876, "I am afraid that these Medical men are determined to stop legislation but I am told privately that if you and I were to see one or two of the best of them *without* E. Hart that something might be done. What say you?"[95] The meeting was held on 22 July, Cross and Carnarvon representing the government and Joseph Hooker, Sir James Paget, Michael Foster, and Burdon Sanderson speaking for the scientists' lobby. To Carnarvon the meeting was "not very satisfactory"[96] but Michael Foster, in writing to a scientific colleague, had a different perspective:

> Hooker spoke up capitally and Paget made most of the rest of the rubbing—so I had little to say—and could devote all my energies to be[ing] amused at the state poor Carnarvon was in. They seemed pretty ready to accept any amendment if they could get a promise that the bill should not be opposed in the House and as Paget though he did all he could couldn't promise that I expect the bill will be withdrawn.
> I shan't care much which way it goes.
> Sanderson seems to have got some new maggots in his head—he is a wonderful creature.
> I quietly, or as you grandly call it, diplomatically, suggested to Hooker as we were going up that after all I had some doubts whether it wouldn't be a good thing that the bill should be withdrawn altogether, and he, to my surprise, thought after all it would be a good thing. Paget and Sanderson were *very* strongly of the opposite

[94] Burdon Sanderson to Hooker, n.d., Royal Botanic Gardens, Letters to Lady and J. D. Hooker, iv. 1095.

[95] Cross to Carnarvon, 18 July, Carnarvon Papers, 30/6/9/24.

[96] A. Hardinge, *The Life of Henry Howard Holyneux Herbert, Fourth Earl of Carnarvon 1831–1890* (London, 1925), p. 110.

> opinion—but I managed to draw Paget into statements about the profession not very reassuring to Cross, and there I left it. If the bill is withdrawn, it is not our doing but Providence. It is most absurd to compare the [illegible] timid manner in which Sanderson and I . . . suggested to Cross some slight alleviation of our trouble a month or so ago and the present great desire of Cross to give us all we want consistent with passing some Bill. Carnarvon's sickly smile when Hooker explained how he wholly objected to any Bill and to the whole business was worth seeing . . .[97]

Foster's letter makes amply clear both the indecision about legislation within the scientists' lobby and the impact of organized medical pressure upon the government's willingness to entertain amendment of the bill. After this conference, the modifications in the bill sought by the scientists' lobby were made. The changes were revealed publicly the first time the bill was debated in the Commons, at its second reading on 9 August. Before that took place, Cross was forced to make another concession to medical sentiment: he, Lord Shaftesbury, and Lord Cardwell persuaded J. M. Holt, once a committee member of Jesse's Society for the Abolition of Vivisection, to forget the bill for the total abolition of vivisection that he and three other members of the Commons had introduced on 24 May.[98]

With Holt's bill out of the way, Cross introduced the government bill to the Commons on 9 August with a speech highly conciliatory to extreme opinion on both sides. He announced that the government was willing to amend the bill so that (i) private places, in addition to public laboratories, could be occasionally made legally available for purposes of research experiment, (ii) special certificates to experiment upon cats, dogs, horses, mules, or asses would be necessary only if such experiments were to be performed without anesthesia, (iii) prosecutions of licensees under

[97] Foster to Thiselton-Dyer, 23 July 1876, Royal Botanic Gardens, Letters to W. T. Thiselton-Dyer, ii. 18. Cf. Hooker to Harcourt, 22 July 1876, Harcourt Papers, Stanton Harcourt, Oxfordshire.

[98] Cross to Carnarvon, 2 August 1876, Carnarvon Papers, 30/6/9/26.

the measure could only be undertaken with the permission of the Home Secretary, and (iv) the measure would be applicable only to experiments on warm-blooded animals. These very considerable concessions succeeded in gaining the somewhat reluctant support of almost all of the parliamentary spokesmen for the scientific and medical interest. Dr. Ward, M. P. for Galway (who had given notice of a motion to reject the bill), Sir John Lubbock, entomologist and M. P. for Maidstone, and Lyon Playfair all give the bill a grudging blessing, but joined Robert Lowe in assailing the inconsistency of legislation, its insulting character to the profession, and the ignorance of antivivisectionists, and in upholding the nobility of medical research. Playfair warned Lowe and those who thought with him, "I think that, after the amendments promised by the Government, if this bill be thrown out, agitation will largely increase, and that a reasonable Bill in future will be attainable with more difficulty. The force of public opinion on the subject is underrated by many of my medical friends," and he portrayed opinion in favor of the bill within the ranks of experimental medicine as much more universal than it actually was. For his part, Holt, his own bill now a lost cause, showed himself clearly unmollified by the attentions of Shaftesbury, Cardwell, and Cross. Labelling the amended government measure a "Physiologists' Protection Bill," he pronounced a lengthy exegesis of the report of the Royal Commission in order to demonstrate its bias in favor of vivisection and the necessity for total abolition. For Holt, the government bill was "not even a temporary settlement of the question" since "No compromise has been offered to the opponents of vivisection; all the concessions made have been made to the doctors, all the alterations made tend to secure advantages to physiologists, and every one of them has, in my opinion, made the Bill worse than it was before." Despite Holt on one side and Lowe on the other, the question was put and the bill read a second time.[99]

[99] *Hansard,* 3rd series, 1876, ccxxxi, 886–931.

Hart pronounced himself satisfied with the amendments promised by Cross, which were to be instituted in the committee of the Commons on 11 August. He appealed to the profession to accept the needless but now harmless bill in order to set an example.[100] Hart later claimed that the B.M.A. lobby had at this point "a working majority of upwards of one hundred members" of the Commons and "could by a word, have thrown out the Bill."[101] It may have required persuasion by the prolegislation faction to prevent Hart from trying to do just that, but at any rate Cross was able to effect almost all of his agreements in the committee of the Commons. Cross was without real opposition, except in the form of a few nuisance amendments by Lowe, who was still dissatisfied with the measure. Cross was probably pleased to be stymied by commissioner W. E. Forster, when the latter succeeded in substituting "invertebrate" for "cold-blooded" in a clause specifying animals to which the bill would not apply.[102] This exception, which had the effect of including the frog within the purview of the bill, was the subject of a sharp protest of "breach of faith" on the part of Cross by Playfair and of a certain amount of grumbling in the medical press.[103] It was the only particular in which the Home Secretary failed to live up to the settlement that he and Carnarvon had negotiated with the scientists' lobby.[104]

The government was now pushing the bill through its legislative stages as quickly and as quietly as possible in the dying hours of a very crowded session. After the agreement with the scientists' lobby, the bill was rushed through its second reading in the Commons on 9 August, with Dis-

[100] *B.M.J.* ii (1876), 221–222.

[101] *B.M.J.* ii (1876), 249.

[102] T. W. Reid, *Life of the Right Honourable William Edward Forster* (4th ed., London, 1888), ii. 107. George Rolleston was one physiologist delighted with Forster's restoration of protection to the frog. W. Turner (ed.), *Scientific Papers and Addresses* (of George Rolleston) (Oxford, 1884), i. p. lx.

[103] *B.M.J.* ii (1876), 259, 284; *Medical Times and Gazette* ii (1876), 203–204.

[104] *Hansard,* 3rd series, 1876, ccxxxi. 1147–1152.

raeli pleading for passage at that point on the grounds that members who wished to speak on the subject could do so in the committee. The committee, however, was held only two days later, and took place after midnight, in a meeting largely unattended and unreported.[105] In these circumstances, the onus of decision fell upon the parliamentary spokesmen of the antivivisection movement, who could have blocked the modified bill at its final reading in the Lords. R. H. Hutton preferred postponement of the bill rather than acceptance of it in a form he conceived as advantageous to physiologists.[106] There is evidence that one Victoria Street spokesman, Lord Chief Justice Coleridge, felt the same way.[107] The R.S.P.C.A., with a wide experience of animal protection legislation, undoubtedly gave similar advice, but the decision not to block the bill seems to have been made by the President of the Victoria Street Society, Lord Shaftesbury. He wrote to an unhappy Frances Power Cobbe after the passage of the Act to explain his rationale for accepting it:

> In the bill as submitted to me, just before the second reading at a final interview with Mr. Cross, Mr. Holt and Lord Cardwell being present, some changes were made which I by no means approved. But the question then, was simply "The bill as propounded, or no bill," for Mr. Cross stoutly maintained that, without the alterations suggested, he had no hope of carrying anything at all. I reverted, therefore, to my first opinion, stated at the very commencement of my co-operation with your committee, that it was of great importance, nay indispensable, to obtain a bill, however imperfect, which should condemn the practice, put a limit on the exercise of it, and give us a foundation on which to build amendments hereafter as evidence and opportunity shall be offered to us.

[105] R. Lowe, "The Vivisection Act," *Contemporary Review* xxviii (1876), 722.

[106] *Spectator* xlix (1876), 1001–1003.

[107] *Hansard,* 3rd series, 1876, ccxxx. 109, and *Verulam Review* i (1888–1889), 415.

> The bill is of that character. I apprehended that if there were no bill then, there would be none at any time. No private member, I believed, and I still believe, could undertake such a measure with even a shadow of hope; and there was more than doubt, whether a Secretary of State would again entangle himself with so bitter and so wearisome a question in the face of all science and the antipathies of most of his colleagues.[108]

There was undoubtedly a segment of antivivisectionist opinion that agreed with Shaftesbury. After the changes introduced in the bill in the committee of the House of Lords, the *Home Chronicler* had written, "Is the bill worth having at all in its present form? We should unhesitatingly answer, No, were it not that previous experience in reference to the Factory Acts, for instance, as well as many others, shows how valuable it is to obtain legislative recognition of important principles, which may subsequently be pressed to the correction of details."[109] In any case, the bill was given a perfunctory reading in the Lords on 12 August; as Carnarvon wrote, "under all the circumstances of the case the less that is said on the changes made perhaps the better."[110] The measure received Royal assent on 15 August, to become the Act 39 and 40 Victoria, c. 77. The final passage of the Act received little publicity outside the medical press, public attention having by this time moved on to other issues, notably the Bulgarian atrocities. The only party to the measure even somewhat satis-

[108] Shaftesbury added, "I feel sure that no Secretary of State in any 'Liberal' administration would listen to the proposal [of legislation on vivisection]." *Life of Frances Power Cobbe,* ii. 597–598. Cf. E. Hodder, *The Life and Work of the Seventh Earl of Shaftesbury, K.G.* (London, 1886), iii. 374; F. P. Cobbe, in *Home Chronicler* i (1876), 200; *Second Report of the Society for the Protection of Animals Liable to Vivisection* (London, 1887), 3.

[109] The choice of example probably shows the extent of Shaftesbury's influence. *Home Chronicler* i (1876), 24. Cf. p. 120, and *The International Association for the Total Suppression of Vivisection* (London, 1876), pp. 16–28.

[110] Carnarvon to Cairns, 11 August 1876, Carnarvon Papers, 30/6/6/32.

fied with itself was the government, which had passed a measure enabling it to wash its hands of further legislation on so tendentious and unprofitable a subject. Upon the antivivisection and scientific sides there was only momentary relief, followed by doubts as to the wisdom of past policy and fears as to the forms that practical implementation of the legislation might take.

The Cruelty to Animals Act of 1876, as modified under medical pressures, provides as follows. Any individual wishing to perform experiments upon living vertebrate animals must submit an application to the Home Secretary, endorsed by both a president of one of the eleven leading scientific or medical bodies in Britain (e.g., Royal Society of London, Royal College of Surgeons of Edinburgh, Royal College of Physicians of Dublin) and a professor of medicine or of one of the medical sciences. The individual must perform his experiments at a place registered with the Home Secretary and subject to inspection at any time. Both licensure and registration are at the discretion of the Home Secretary. Licenses are valid for a period of one year only, after which they must be renewed. Experiments "must be performed with a view to the advancement by new discovery of physiological knowledge or of knowledge which will be useful for saving or prolonging life or alleviating suffering" and cannot be performed before the public or for the purpose of attaining manual dexterity. Experiments under anesthesia to illustrate lectures, or to be performed without anesthesia, or allowing the animal to recover from anesthesia, or for the purpose of testing former discoveries, are permitted only when certified as necessary by one of the aforementioned officials. Experiments on cats, dogs horses, mules, and asses without anesthesia can be carried out only under similar certification specifying that the objectives of the experiment would be frustrated if these animals were not used. Curare is not considered to be an anesthetic for the purposes of the Act. The Home Secretary may require reports of all experiments performed by a licensee, and he appoints and directs inspectors under the Act. Penalties under the Act are £50 or less for a first offense and

£100 or less or three months for second and subsequent offenses. No licensee may be prosecuted under the Act without the written permission of the Home Secretary. The precise form of these provisions in the *Public General Statutes* shows clearly the very stringent measure originally introduced by Carnarvon in May, progressively altered by a series of amendments in concession to the scientific and medical interest, to the measure finally forced through the Commons in mid-August 1876.[111]

CONCLUSION

What is the historian to make of the events leading up to the passage of the Cruelty to Animals Act of 1876? Here, it will be convenient to consider this question from two points of view: first, from the point of view of the adoption by the government of a new administrative responsibility, and second, from the point of view of the response of the medical and scientific communities to political pressure.

It has, I hope, been clear from this discussion that legislation to regulate experimentation on living animals was thrust upon an initially uninterested government by powerful and publicity-conscious pressure groups, first the R.S.P.C.A. and then, most importantly, the Victoria Street Society. The ease with which vigilant pressure groups could exploit a single incident until the issue it symbolized became a national cause celebre is graphically illustrated by humanitarian capitalization on the Norwich prosecution. I shall argue that the attitude of the government as a whole toward the issue is best understood as the simple desire to allay the concern of the general public with whatever legislation proved passable. Faced with competing recommendations for legislation, the government made, here as elsewhere, "an empirical response to a more or less pressing problem, undertaken out of necessity, . . . [and] because

[111] "An Act to Amend the Law Relating to Cruelty to Animals," *The Law Reports. Public General Statutes,* 39 and 40 Victoria c. 77.

investigation and discussion . . . made the time ripe for action."[112] The only direct precedent for legislation was the Anatomy Act passed earlier in the century to regulate the supply of cadavers for anatomical dissection in medical schools. Antivivisectionist and governmental proposals for legislation utilized either the existing Inspectors of Anatomy or Inspectors modeled after them, in a provision for field enforcement, which, by the mid-seventies, had ample precedent in other areas of legislation. Besides this particular mechanism of enforcement—or at least surveillance—however, suggestions for legislation had to be developed de novo and with only the most grudging assistance from the group most capable of providing it. Nevertheless, the government, or at least Carnarvon and Cross, proceeded doggedly once stimulated to do so by their perceptions of public opinion and by pressure from organized animal protection. While Carnarvon's long-standing commitment to animal protection provides a clear enough explanation for his involvement, Cross's motivations are somewhat more obscure. His problems with the bill were described by Shaftesbury:

> Mr. Cross's difficulties were very great at all times; but they increased much as the session was drawing to a close. The want of time, the extreme pressure of business, the active malignity of the scientific men, and the indifference of his colleagues, left the Secretary of State in a very weak and embarassing position.[113]

The lack of enthusiasm for legislation from Salisbury and other members of the cabinet makes it surprising that Cross, who was without Carnarvon's emotional involvement in the issue, did not simply drop the bill in the face of medical wrath. Disraeli and Cross were among the Queen's

[112] P. Smith, *Disraelian Conservatism and Social Reform* (London, 1967), p. 218.

[113] *Life of Frances Power Cobbe*, ii. 597.

favorites, and she was undoubtedly pressing them to deal with vivisection.[114] Cross's readiness to compromise important provisions of the bill can only lead us to the conclusion that he shared with Disraeli a simple desire for *some* legislation, *any* legislation, that would shield the government from continual lobbying by humanitarian interests. Cross and Disraeli probably convinced the cabinet, as certain scientists convinced the medical profession, that legislation was preferable to continual harassment at the hands of antivivisectionist zealots. The parliamentary power per se of animal protection was probably less at the root of Cross's persistence than the very high nuisance value of a public opinion aroused by antivivisectionist pamphlets and petitions.

Motivation for settling the issue was described by Robert Lowe:

> Both parties had it must be admitted very strong grounds for making mutual concessions. It was impossible for the Government to carry their measure during the session then drawing to a close without conciliating the very formidable opposition which had been provoked. On the other hand, persons deeply engaged in physiological experiments, and, indeed, the whole medical profession, were equally interested in putting an end to a state of things particularly harassing to busy men continually brought in contact with the public and subject to much vexation, annoyance and misrepresentation. They also had to dread an organized agitation carried on during the

[114] The sheer nuisance value of continual missives from Queen Victoria to the responsible ministers was considerable. Although Longford is clearly overstating the case when she writes "But for her [Victoria], a Vivisection Bill for the humane control of research on animals would not have seen the light of day," Carnarvon was very grateful for her aid: "If we can come to terms with the doctors, this is clearly—to my mind—due to the Queen's influence. She has pressed it strongly on Disraeli." E. Longford, *Victoria R.I.* (London, 1964), p. 406. A. Hardinge, *The Life of Henry Howard Molyneux Herbert, Fourth Earl of Carnarvon 1831–1890* (London, 1925), ii. 110.

> winter with unlimited supplies of money furnished by honest, if mistaken, enthusiasm.[115]

Pressure for a settlement led, as we have seen, to extra-parliamentary negotiations that circumvented or vitiated parliamentary opposition, a procedure pungently characterized by Lowe:

> A practice has sprung up, especially since the advent of power of the present Government, which cannot be too energetically deprecated. It may be described as the art of turning public bills into private ones. A measure is introduced which has two aspects: it involves problems of the largest and most important kind; it raises questions of the highest moral significance. But, besides, it touches and alarms some powerful and well-organized body. Symptoms of a strong and pertinacious opposition arise, the Press takes the alarm, deputations are organized, meetings are called; the Government find it is necessary, if they mean to carry anything, to come to terms. The interest endangered makes its demands, which, after more or less haggling, are agreed to. From this time the measure is looked upon as passed. It is postponed to unearthly hours, when debate is difficult and reporting impossible. The conciliated interest—fearful lest, if the matter be deferred for another Session, a worse thing may befall them, and anxious to realize the fruits of their work in the way of agitation—lend their full support to the Government. The consequence is that the voice of those who would speak for the public is stifled, and there is nothing left but submission to whatever it may please the two high contracting powers to dictate.[116]

An excellent example of this "method of legislation by compact, with no reference to any consideration except the convenience of a Government and an adverse interest,"

[115] R. Lowe, "The Vivisection Act," *Contemporary Review* xxviii (1876), 720.

[116] Ibid., 713. The manner in which the bill was passed was subsequently strongly criticized by antivivisectionists. See, e.g., "M.A., Cambridge" *Vivisection Viewed Under the Light of Divine Revelation* (London, 1877), p. i, and *Some Interesting Evidence Given by Sir George Kekewich* . . . (London, 1908), pp. 5–6.

Lowe went on to say, was the Cruelty to Animals Act of 1876. Cross's strategy of maximizing his control over amendments to the bill and minimizing the publicity given to them was of course successful.

At this point in the discussion, it is possible to demonstrate how similar certain features of the passage of the Cruelty to Animals Act were to the passage of many of the other measures that increased administrative responsibility and contributed to the growth of bureaucracy in the nineteenth-century revolution in government. MacDonagh's model for the process of administrative expansion in nineteenth-century Britain permits us to see how very typical were many of the events of 1875–1876 leading to legislation upon even so unlikely a matter for regulation as scientific experiments.[117]

According to MacDonagh, the process of administrative reform typically took place in five stages. Only the first three stages are relevant here, for reasons that will become clear. In the first stage, the recognition of a social evil (in this case, alleged cruelty in scientific experiments upon living animals) arose "exogenously" (i.e., outside the government; in this case, through the efforts of the R.S.P.C.A. up to the Norwich trial, December 1874). A public outcry over the "intolerable" evil resulted, and demands "to legislate the evil out of existence" were heard. Then,

> As the threat to legislate took place, the endangered interests, whatever they might be, brought their political influence into action, and the various stages of inertia, material and immaterial, came into play. Almost invariably there was compromise. Both in the course of the drafting of the bill, when trade interests often "made representations" or were consulted, and in the committee stage in Parliament, the restrictive clauses of the proposed legislation were relaxed, the penalties for their defiance whittled down and the machinery for their enforcement weakened.

[117] O. MacDonagh "The Nineteenth Century Revolution in Government: A Reappraisal," *Historical Journal* i (1958), 52–67. See also his *A Pattern of Government Growth 1800–60* (London, 1961), pp. 320–350.

> Nonetheless the measure, however emasculated, became law. A precedent was established, a responsibility assumed, the first stage of the process was complete.[118]

In the second stage, the inadequacy of the law to deal with the evil against which it was directed became evident, probably due to the vigilance of those who had originally demanded legislative intervention. The legislation was usually ineffective because of the lack of practical experience with the problem and to the lobbying of the endangered interests before passage through Parliament. As a result of the revelations of this stage, amending legislation was passed to ensure some kind of enforcement of the regulations and to appoint special officers or inspectors responsible for such enforcement. In the third stage, the intensive experience of these executive officers in dealing with the evil brought further consolidating and centralizing legislation.

The similarity between the first stage of MacDonagh's model and the events culminating in the Act of 1876 is clear enough; it highlights the typicality of those events. But two further points should be made. The first is that outside the scientific and medical interest, there was never much doubt that some kind of inspection would have to be involved in the regulation of vivisection, and, indeed, the first and only Act provided for the appointment of inspectors. According to Parris, "After about 1835 a demonstration effect came into existence between different branches of the central administration. The example of the first enforcement officers had set the pattern, and it became normal to appoint them with the first incursion into a new field."[119] In the case of the regulation of vivisection in the mid-seventies, legislative provision for inspectors seems to have been more or less routine. A second point, which I shall mention here and leave to be detailed in the following chapter, is that the regulation of vivisection never reached MacDonagh's third stage, that of consolidat-

[118] *Historical Journal* i (1958), 58.

[119] H. Parris, "The Nineteenth Century Revolution in Government; A Reappraisal Reappraised," *Historical Journal* iii (1960), 33.

ing legislation resulting from the practical experience of inspection. Given that the primary motive of Disraeli and Cross in passing legislation was simply to reduce the general level of public concern over the issue by using the Act to spike the guns of the antivivisectionists, the subsequent lack of interest in additional legislation is not unexpected. As the following chapters will make clear, successive governments shared this attitude.

The reaction of various members of the scientific and medical community to the political threat from antivivisection is an instructive index to the relative degrees of professionalization enjoyed by different segments of that community. It will be convenient to assume, with Friedson, that the extent of autonomy within a sociopolitical system is one crucial parameter in assessing the degree of professionalization.[120] For purposes of analysis here, we may begin with the British medical profession as a whole and go on to distinguish within it significant parts of a nascent profession, the scientists.

As a result of socioeconomic changes within the society as a whole, of secularization with its increased emphasis on bodily health, of progress in medical practice such as anesthesia and antisepsis, and of state recognition in the form of legislation, British medicine in the seventies was a profession of increasing unity and power. It enjoyed a high degree of autonomy, which was ceded to its leadership by legislation, especially from the fifties onward, and was based upon an impressive array of institutions: hospitals, medical schools, Royal Colleges, the governing General Medical Council, and the mass membership British Medical Association. The profession had achieved a good deal of collective political experience, originating with the issues of public health and professional qualification in the forties and fifties.[121]

[120] For this discussion I am greatly indebted to Eliot Friedson's *Profession of Medicine. A Study of the Sociology of Applied Knowledge* (New York, 1970), pp. 1–84.

[121] See e.g., M. J. Peterson, "Kinship, Status and Social Mobility in the Mid-Victorian Medical Profession," (University of California (Berkeley) Ph.D. thesis, 1972); J. L. Brand, *Doctors and the State*

In contrast to the strength and entrenched position of the medical profession as a whole was the position of those individuals within it who considered themselves medical scientists, who wished to expand opportunity in medical science, and whose ultimate goal was the establishment of an at least semiindependent scientific profession. This group had a low degree of autonomy in the mid-seventies, for the institutionalization of experimental medicine as such had only just begun. Insofar as they considered themselves scientists rather than healers, their political strength as an occupational group or nascent profession was negligible. Physicists, chemists, geologists, and astronomers did not share a meaningful professional identity with biological scientists, since virtually none of them rallied to the side of experimental medicine. In any case, the political experience of various scientists was far less extensive than that of medical men, and MacLeod has shown that it was often less successful: "The necessary combination of sophisticated political judgment with scientific integrity was not generally available or widely appreciated. The academic expert still inclined toward the tragic hero rather than the statesman of science."[122]

(Baltimore and London, 1966); H. H. Eckstein, *Pressure Group Politics: The Case of the British Medical Association* (London, 1960); E. M. Little, *History of the British Medical Association, 1832–1932* (London, 1932); P. Vaughn, *Doctors' Commons. A Short History of the British Medical Association* (London, 1959). Cf. R. M. MacLeod's excellent case studies, "The Edge of Hope: Social Policy and Chronic Alcoholism, 1870–1900," *J. Hist. Med.*, xxii (1967), 215–245; "Law, Medicine, and Public Opinion: The Resistance to Compulsory Health Legislation, 1870–1907," *Public Law* (1967), 107–128, 189–211; "The Frustration of State Medicine, 1880–1899," *Medical History* xi (1967), 15–40; and A. F. Beck, "Issues in the Anti-Vaccination Movement in England, *Medical History* iv (1960), 310–321.

[122] R. M. MacLeod, "Science and Government in Victorian England: Lighthouse Illumination and the Board of Trade, 1866–1886," *Isis* lx (1969), 5–38. The quotation is from p. 38. See also MacLeod's, "Government and Resource Conservation: The Salmon Acts Administration, 1860–1886," *J. Brit. Stud.* vii (1968), 114–150, and "The Alkali Acts Administration, 1863–84: The Emergence of the Civil Scientist," *Victorian Studies* ix (1965), 85–112.

It is in this context that the interaction among the scientists' lobby, the medical profession as a whole, and the government during the antivivisection debates must be considered. The scientists' lobby consisted of three main elements. First, there were the medical scientists who utilized experiments upon living animals in the course of their research. Here the leaders were John Burdon Sanderson of London and Michael Foster of Cambridge, physiologists, toxicologist Robert Christison of Edinburgh, and Joseph Lister, the pioneer of antisepsis, also of Edinburgh. They spoke for a handful of generally younger men who were also committed to experimental medicine, such as Thomas Lauder Brunton, a pharmacologist, Edward Schafer, a physiologist, and Emanuel Klein, a histologist, whose immediate interests were likewise directly threatened. A second group within the scientists' lobby consisted of elite medical men who did not themselves perform vivisection, but who were initially believed sympathetic to experimental medicine: Sir John Simon, eminent sanitarian and administrator; the prominent London practitioners, Sir William Gull and Sir James Paget; Sir Henry Acland, Regius Professor of Medicine at Oxford; George Rolleston, Professor of Anatomy and Physiology at Oxford; and a few others. A third group was made up of nonmedical biologists who did not, in general, perform vivisections, but who were sympathetic to experimental medicine. These included Charles Darwin, William Sharpey (retired Professor of Anatomy and Physiology at University College, London), T. H. Huxley, and botanist Joseph Hooker, who was President of the Royal Society at the time (Figure 8). (To a certain extent these groupings are arbitrary. Rolleston and Sharpey could be placed in either of the last two groups.) The Parliamentary spokesman for the scientists' lobby was Lyon Playfair, a chemist who was at the time Liberal M. P. for the universities of Edinburgh and St. Andrews. Playfair often represented medical and scientific interests in Parliament.

It is important to note that the composition of the scientists' lobby was both diverse and fluid. These characteristics

Figure 8. Sir James Paget (1814–1899), John Burdon Sanderson, and Thomas Henry Huxley—a triptych from the medical and scientific elite. *Courtesy of the Wellcome Trustees.*

are in themselves indicators of the relative weakness of medical science. The variety of institutional, intellectual, geographic, and professional identifications of members of the lobby did not aid unity. At the most basic level, members of the lobby differed in the extent of their commitment to animal experiment. George Rolleston was fairly rapidly extruded from the lobby; he was a classic example of an anatomist having tenure of a chair in anatomy and physiology, who had been left far behind by continental advances in vivisectional physiology. Rolleston's intellectual identification with anatomy left him with a considerable distaste for animal experiment, which he expressed to the Royal Commission and to the G.M.C.[123] Rolleston and his Oxford colleague Henry Acland refused to sign a petition drawn up by Huxley on the grounds that its claims for the medical utility of vivisection were too extravagant. Acland's ambivalence toward the issue of animal experimentation points up another somewhat divisive element in the scientists' lobby: variations in the strength and nature of professional identification. Most of the members of the scientists' lobby had had some medical training, but their degree of professional commitment to medicine varied greatly. While Joseph Hooker, T. H. Huxley, and Charles Darwin saw themselves as biological scientists pure and simple, Sir Henry Acland and George Rolleston were profoundly unwilling to see an independent profession of biological investigators emerge. Acland told the Royal Commission that "I am not at all sure that the mere acquisition of knowledge is not a thing having some dangerous and mischievous tendencies in it."[124] His suspicion of research carried out in isolation from the demands of medical practice was echoed by Rolleston and by Andrew Clark, who deplored any legislation that would "tend to establish a class distinction between professional physiologists and general practi-

[123] E. B. Tylor, "Biographical Sketch," in W. Turner (ed.), *Scientific Papers and Addresses* (of George Rolleston) (Oxford, 1884), i. pp. lix–lx.

[124] "Report of the Royal Commission," Q. 944.

tioners."[125] Sir Henry Acland was president of the G.M.C. and a man with a reputation as a progressive medical educator. Andrew Clark was an eminent physician and a confidant of fashionable London. Although Rolleston, Acland, and Clark did not want to see experiments on living animals abolished, their differences with the core of the scientists' lobby, which centered around Burdon Sanderson, provided a gap in the ranks of the scientists' lobby that could be exploited by antivivisectionists. Implicit in the demands of the scientists' lobby was support for the professionalization and institutionalization of experimental medicine. Insofar as some members of the lobby refused to countenance the possibility of this partial loss of dominance by the practicing members of the profession, they were a liability to the lobby.

The final tension within the scientists' lobby was, as we have seen, over the advisability of accepting any legislation whatsoever. To some extent, this issue arose from geographic and institutional factors. Scottish opposition to legislation may be attributed to the fact that antivivisection did not become really active in Edinburgh until after 1876. Hence the Scots were not initially as aware as the English of the threat represented by the movement. Some of the strongest opposition to legislation arose from medical men who had no academic connections and nothing at stake in terms of their daily work. Since many of these men were not active in the scientists' lobby at all, but were simply members of the larger profession of which the lobby was a miniscule part, it is time to confront the problem of the relations between the lobby and the profession.

Thus far I have been concerned to show that the existence of antivivisectionist sentiment within medical ranks was not the only source of disunity within the profession vis-à-vis the issue of experimental medicine. Considerable variation of opinion on various aspects of the vivisection question existed even within a relatively small group at the top of the medical profession and medical education.

[125] *B.M.J.* ii (1876), 154.

As this variation of opinion manifested itself during initial attempts to organize the scientists' lobby in the spring of 1875, those at the center of the lobby began to perceive clearly the small number of reliable allies upon whom they could count to present a united front on behalf of experimental medicine. Huxley, Burdon Sanderson, Foster, Paget, and Simon succeeded in dominating the lobby because they were on the scene in London and because they had the ear of certain key individuals, notably Lyon Playfair. The lobby found itself, however, in an extremely difficult position when Carnarvon introduced his unexpectedly severe bill in the Lords in May 1876. Time was of the essence, and legislators could be expected to be profoundly ignorant of the issues involved. As the *Journal of Science*'s editorial put it, "In such matters we should dread the interference of the British government more than any other." Should the bill be altered satisfactorily, "it will be perhaps the first time that the British legislature has dealt advantageously with a scientific subject."[126] In these circumstances and on the advice of Playfair, the lobby turned to a relatively conciliatory attempt to gain certain key concessions, an attempt involving the nonpublic deployment of personal influence within the House of Lords and the cabinet on the part of a few individuals. Playfair believed that rapid effort to affect the course of legislation in the Lords was crucial, since the tactics apparently dictated by the comparative weakness of the scientists' lobby could be more effective in the upper chamber than in the Commons. This style of political activity seems not to be simply a last resort for a numerically weak pressure group, however. According to Haberer, such insider politics is typical of the community of science: "When involved in politics, the scientist-leader has usually preferred to exercise personal influence with political leaders, to operate in or near the inner councils of government."[127]

The delay in the progress of the bill consequent upon

[126] *Journal of Science* n.s. vi (1876), 332.

[127] J. Haberer, *Politics and the Community of Science* (New York, 1969), p. 303.

the fatal illness of Carnarvon's mother allowed the activation of political pressures drastically different from those mounted by the scientists' lobby. The backgoround to Ernest Hart's decision to mobilize the membership of the British Medical Association against the bill is unclear, though his campaign in the pages of the *British Medical Journal,* in the Parliamentary Bills Committee of the B.M.A., and in the lobbies and corridors of Westminster was by no means atypical of others he fought in the interests of the profession. Despite the fact that most practitioners had no first-hand experience of experiments on living animals, Hart was able to portray the government bill as casting a slur on the profession. As he later wrote in a self-congratulatory article, "Up to the time the Bill left the House of Lords and was introduced into the House of Commons, the changes made in the Bill in accordance with the representations . . . of other bodies was comparatively slight: and it was only after the voice of the profession had spoken loudly and distinctly in the deputation [organized and led by E. Hart] to Mr. Cross, that the Government consented to make the changes that were called for."[128] The intervention of the medical profession as such, upon the initiative of Hart, introduced a public, agitational style of politics that greatly altered the government's options with regard to the bill. After this intervention, it was Cross, a proponent of the bill, who was pleading that the controversy be laid to rest in the Lords; debate in the Commons—where the strength of the profession could be best exploited—would now be fatal to the bill, according to Cross. Efforts of personal persuasion in Parliamentary lobbies and London club rooms were one thing. Inflammatory leading articles, public meetings, and monster deputations on the part of a prestigious profession were quite another. The aggressive stance of the *B.M.J.* overrode the influence of the handful of medical antivivisectionists and provided the cue for powerful newspapers and periodicals to swing their editorial weight behind the profession.

[128] *B.M.J.* ii (1876), 245.

The general tenor of professional opinion called for rejection of any legislation whatsoever, and as we have seen, some parts of the scientists' lobby agreed. The ultimately dominant group within the scientists' lobby, however, chose to exploit the altered balance of power to achieve legislation that would protect experimental medicine from antivivisectionist harassment. They, therefore, made common cause with moderate antivivisectionists and the government by conducting extraparliamentary negotiations without Hart, which they then, presumably, persuaded him to accept. Thus, the nascent professional group—the scientists—achieved their ends by using a private, concilatory style of politics that enabled them to utilize the power of the established profession without surrendering policy initiative. A very small group of scientists, and the individual medical men who sympathized with them, retained effective control through personal contact with Playfair and members of the government.

Haberer's studies of scientists and politics lead him to suggest that scientists are frequently highly ambivalent toward politics and that they consistently lack the "ultimate commitment" that would preserve the integrity of their community when that integrity is threatened. Thus, scientists are likely to strive to reduce conflict by what Haberer calls "prudential acquiescence" to political pressure.[129] In the case we have been considering, the scientists' lobby practiced the politics of prudential acquiescence in the belief that it had achieved a measure that would protect without restricting experimental medicine. During the next few years, many were to feel that the strategy of acquiescence had failed.

[129] J. Haberer, *Politics and the Community of Science* (New York, 1969), pp. 299–332.

6. The Antivivisection Movement and Political Action after 1876

The historian who sought to characterize the antivivisection movement on the basis of the documentation available for the period 1875–1876 would find little indeed on which to base his analysis. The Victoria Street Society had, of course, played an important role as a pressure group during the lobbying of 1876. Its political representatives, Lord Shaftesbury and Lord Chief Justice Coleridge, were powerful men who capably voiced the society's position. But the intense activity in Parliament in 1876, pushed forward by a variety of interests, dominated the attention and resources of antivivisectionists. Thus antivivisectionism's most enduring landmark, the Act of 1876, came into being before the movement had the time or the opportunity to establish any distinctive public image or mode of operation. It was in the twenty years *after* passage of the Act that the movement's personalities, policies, and ideology emerged to general view.

In this chapter, the evolution of antivivisectionist policy and the program of political action based on that policy will be discussed. Such a discussion will complete the outline of the formal relations between the antivivisection movement and government during the Victorian period. More important, it will provide a backdrop against which to view the administration of the Act by sampling the atmosphere of scrutiny and suspicion within which politicians, bureaucrats, and scientists found themselves interacting. Finally, it will raise in the reader's mind a number of unanswered questions about the administration of the Act. This is intentional, for the reader's temporarily frustrated curiosity is a mere specimen of the movement's sentiments on its continuing failure to gain information and insight into the process of the regulation of vivisection.

POLICY AND TACTICS

The London Anti-Vivisection Society and the International Association for the Total Suppression of Vivisection were founded in June 1876, upon the principle of total abolition of vivisection. Both developed too late to get involved in the lobbying surrounding the Act, a measure that, in any case, they rejected as sanctioning the very evil they were pledged to stamp out entirely. The London and International Societies were, however, indicators of a sentiment that moderate antivivisectionists could use in pressing their case. Unlike George Jesse's one-man show, the Society for the Abolition of Vivisection, based in Cheshire, these two societies were London-oriented, of respectable antecedents, and more or less democratically organized. While by no means matching it in prestige or power as yet, both societies were viable alternatives to Victoria Street.

The equivocal results of its own legislative initiative of 1876, combined with the existence of active total abolition societies in London, put the Victoria Street Society in a difficult situation during the months following the passage of the Act. In a very real sense, it was in competition with the other animal protection societies for public support, in terms of membership and subscriptions. Restrictionist Victoria Street found itself in a policy squeeze: it bore primary responsibility for an Act that was to be subjected to increasingly heavy antivivisectionist criticism, and it faced simultaneous competition from the restrictionist R.S.P.C.A. on the one side and from the total abolition societies on the other.

Frances Power Cobbe, bitterly disappointed at the modifications made in Carnarvon's bill, resulting in the Act of 1876,[1] was in no doubt as to where the future lay for

[1] F. P. Cobbe, *Life of Frances Power Cobbe* (Boston and New York, 1894), ii. 596–599.

Victoria Street's policy. She stayed on as honorary secretary only upon the society's declaration that it would press for total prohibition of painful experiments as its ultimate goal. In 1877, the society gave its support to J. M. Holt's bill for total abolition, a measure developed under the auspices of the International Association.[2]

Victoria Street had not yet, however, pledged itself uncompromisingly to total abolition. An influential segment, including Shaftesbury, was determined to give the Act a fair trial, and tried gamely to appear enthusiastic about the results of the society's lobbying. According to the society's annual report of 1877, the Act "has not failed to act as a drag upon the practice" of vivisection and "has been of great utility."[3] By this time, a few members had quit the society, wanting it to go further in its policy.[4] The former secretary of the society, A. P. Childs, applied additional pressure from his position as editor of the antivivisectionist *Home Chronicler.* From Childs, Victoria Street suffered the likes of a sarcastic article of 1878 entitled "Fashionable Humanity," which twitted an unnamed fashionable and moderate restrictionist society.[5] To Childs and those who thought as he did, acceptance of the Act as a step toward total abolition was a delusory and dangerous strategy. The Act was nothing more than a deliberately contrived screen for the misdoings of vivisectors, and any recognition of its possible worth was merely playing into the scientists' hands.[6]

Cobbe, determined to maintain Victoria Street's predominance in the movement, eventually succeeded in convinc-

[2] Ibid., ii. 599; *Home Chronicler* (1876), 140, 155, 201; *Second Report of the Society for the Protection of Animals Liable to Vivisection* (London, 1887), pp. 4–6.

[3] *Second Report of the Society for the Protection of Animals Liable to Vivisection* (London, 1877), p. 3.

[4] Ibid., p. 6.

[5] *Home Chronicler* (1878), 377–379. Cf. *Home Chronicler* (1876), 232, 268; (1877), 1016.

[6] See, e.g., W. H. Llewelyn, *Vivisection: Shall it be Regulated or Suppressed?* (London, 1880). Llewelyn was secretary of the International Association for the Total Suppression of Vivisection.

ing most of her colleagues of the wisdom of the total abolitionist view. In late 1878, Victoria Street declared unequivocally for total abolition.[7] This decision cost Victoria Street a few members, including Dr. George Hoggan, honorary secretary, and the Bishop of Gloucester and Bristol and the Archbishop of York, both vice presidents.[8] Lord Shaftesbury was, however, convinced by now that total abolition was the only valid policy. Experience under the Act, which will be considered below, had augmented disappointment with its original dilution under political pressure. As Shaftesbury wrote to Cobbe in September, 1878:

> We were right to make the experiment. We were right to test the men and the law, Mr. Cross and his administration of it. Both have failed us and we are bound in duty, I think, to leap over all limitations, and go in for the total abolition of this vile cruel form of idolatry . . .[9]

According to Cobbe, this change brought in "a large batch of fresh recruits of new members who had long resented our previous half-hearted policy—as they considered it to have been."[10] Thus, by 1878, the organized antivivisection movement was uniformly total abolitionist in policy.

Total abolitionist convictions in Victoria Street gained strength through the eighties. Cardinal Manning declared in 1881 that the society had been "entirely hoodwinked" by the Act and its administration.[11] The thinking of all total abolitionists was that the Act of 1876 and, indeed, any restrictionist or regulatory measure must be a delusion and a fraud; inspection of all experiments would be impossible, and in any case the efficacy of anesthetics simply could not be assessed. Hence, the only practical alternative was to stop altogether the use of living animals in scientific

[7] *Home Chronicler and Anti-Vivisectionist* (1878), 73–74; *Life of Frances Power Cobbe*, ii. 604.

[8] Ibid., and *Transactions of the Victoria Street Society for the Protection of Animals from Vivisection during the half-year ending, Dec. 31, 1879* (London, 1880), p. 7.

[9] *Zoophilist* v (1885), 115.

[10] *Life of Frances Power Cobbe* ii. 604–605.

[11] *Report of the Annual Meeting of the Victoria Street Society for Protection of Animals from Vivisection* (London, 1881), p. 6.

experiments.[12] Total abolition had the further advantage of being a simple and clear appeal for the public to support: "Public opinion cannot be elucidated on the subject by men who treat Vivisection as a thing to be sanctioned under restriction."[13]

Advocacy of a restrictionist policy in 1885 by a member of the Victoria Street Society, on the grounds that "repeated disappointments" and "years . . . of almost fruitless labours" had produced "almost total stagnation of the movement" met with a stern rebuff from the leadership.[14]

Total abolition remained the policy of Victoria Street until 1898. At that time, Cobbe's protege, the Hon. Stephen Coleridge, engineered a coup d'etat that deprived his patroness, now living in semiretirement in Wales, of effective control of the executive committee. Under Coleridge's leadership, the National Anti-Vivisection Society, as Victoria Street was renamed, attempted to introduce restrictionist legislative proposals. Coleridge's strategy envisioned the passage of progressively more stringent measures, culminating in total abolition. A furious Frances Power Cobbe founded her own rigidly abolitionist British Union for the Abolition of Vivisection.[15]

[12] See, e.g., *Home Chronicler* (1877), 1016; *The Lord Chief Justice of England on Vivisection* (London, c. 1882), pp. 3–4, 6; W. H. Llewelyn, *Vivisection: Shall it be Regulated or Suppressed?* (London, 1880); F. P. Cobbe, *The Fallacy of Restriction applied to Vivisection* (London, 1886).

[13] *Zoophilist* v (1886), 175.

[14] *Zoophilist* v (1885), 202–204. See also vi (1886), 75–77, 93–94.

[15] *Verulam Review* vii (1897–1898), 217–235, 297–313; *Abolitionist* (Bristol, 1899–19??). Both the National Anti-Vivisection Society and the British Union for the Abolition of Vivisection have offices in London at the present time. I am most grateful to the British Union for permitting me to examine some of its early minute books. Similar permission was denied by the National Anti-Vivisection Society. A severe blow was dealt to the National Anti-Vivisection Society and similar organizations by judgment of the House of Lords in 1947 rendering such societies liable to income tax on the grounds that they are political as well as charitable in purpose. See "National Anti-Vivisection Society v. Inland Revenue Commissioners," *All England Law Reports* ii (1947), 217–241. I am grateful to Mr. R. Murray, solicitor of the R.S.P.C.A, for details in this regard.

Antivivisectionists pursued three avenues of direct political action in the attempt to achieve their goal of an Act banning experiments upon living animals. These were the introduction of appropriate legislation, the attempt to pledge and elect candidates for the House of Commons in support of such legislation, and the circulation of petitions for presentation in Parliament.

Including J. M. Holt's effort of 1876, bills for the total abolition of vivisection were presented in Parliament in every year from 1876 to 1884.[16] Holt introduced bills in 1877, 1878, and 1880 with the backing of the International Association for the Total Suppression of Vivisection. His 1877 bill was resoundingly defeated, 83 to 222,[17] and his bills introduced in 1878 and 1880 were crowded off the legislative agenda. The Victoria Street Society, then still nominally restrictionist in policy, supported Holt's 1877 bill, though not the subsequent one.[18] The society also gave parliamentary aid to the bill that Lord Truro introduced in the House of Lords in 1879. Despite a strong speech by Shaftesbury, now leading an explicitly abolitionist Victoria Street, Truro's bill was defeated in the upper chamber, 16 to 97.[19] In 1881, the Victoria Street Society drafted its own total abolition bill under the guidance of one of its leading figures, Lord Chief Justice Coleridge. This bill was introduced in vain into the House of Commons by Sir Eardley Wilmot in 1881 and by R. T. Reid (later to be a distinguished Liberal politician and Lord Chancellor, at this point he was in his first term as an M. P.) in 1882, 1883, and 1884.[20] Wilmot's efforts were also supported by

[16] *Dates of the Principal Events Connected with the Anti-Vivisection Movement* (London, 1897), pp. 3–6.

[17] *Home Chronicler* (1877), pp. 722–732, 741–743; *Hansard*, 3rd series, 1877, ccxxxiv, 205–248.

[18] *Dates* (note 16), p. 3 and *Home Chronicler* (1878), 72.

[19] *Hansard*, 3rd series, 1879, ccxlviii, 419–436; *Transactions of the Victoria Street Society for the Protection of Animals from Vivisection* (London, 1880), pp. 4–9; *Anti-Vivisectionist* vi (1879), 469; Earl of Shaftesbury, *The Total Prohibition of Vivisection* (London, 1879).

[20] *Zoophilist* i (1881), 34; *Annual Report of the Victoria Street Society for the Protection of Animals from Vivisection* (London,

the International and London Anti-Vivisection Societies.[21] After the 1884 failure, no further serious legislative attempts were made until the turn of the century.

Legislative proposals for total abolition generated little enthusiasm, either in or out of Parliament. The lopsided votes and dropped bills were preceded by strong criticism of the measures from government spokesmen and others. For example, Lord Truro's bill was actively opposed by antivivisectionist Lord Henniker, Lord Carnarvon, Lord Cardwell, Lord Aberdare (president of the R.S.P.C.A. in succession to Lord Harrowby), and the Bishop of Peterborough.[22] Even periodicals sympathetic to antivivisection—the *Morning Advertiser* and the *Spectator*—generally agreed with the *Saturday Review*, the *Times*, the *Standard*, and the medical press that the bills were unnecessary, poorly drawn, and in fact potentially harmful.[23]

Frustrated in their repeated introductions of legislation for total abolition, antivivisectionists made continuing efforts to initiate an effective program of political action by means of electoral pledges. By this tactic, a candidate for Parliament who pledged himself to support total abolition legislation received in turn the voting support of antivivisectionists and perhaps also aid in terms of finances or manpower. Antivivisectionists were never completely sure where they stood on the advisability of the system of electoral pledges. Repeated suggestions of its usefulness, especially in small boroughs where the swing of a few votes could change a result, appeared in the late seventies.[24]

1882), p. 3; *Anti-Vivisectionist* (1882), 66; *The Debate on Mr. Reid's Bill in the House of Commons* (London, 1883).

[21] *Spectator* xliii (1880), 400, 431.

[22] See note 19 and J. C. MacDonnell, *Life and Correspondence of William Connor Magee* (London, 1896), ii. 113–117.

[23] *Public Opinion* xxxi (1877), 545, and xxxvi (1879), 66; *Spectator* l (1877), 555; *Saturday Review* xliii (1877), 540–541 and lv (1883), 434–435; *B.M.J.* ii (1879), 96 and i (1883), 696; *Lancet* i (1881), 469–470. By 1880 Hutton's *Spectator* was prepared to endorse total abolition. *Spectator* liii (1880), 899.

[24] *Home Chronicler* (1877), 474; *Anti-Vivisectionist* vi (1879), 102, 116, 195–196, 262.

The *Home Chronicler* published a list of marginal seats together with an indication of those seats in which the incumbent had voted against the second reading of J. M. Holt's total abolition bill in 1877.[25] The Victoria Street Society instructed its followers:

> If you have a vote . . . give it to a candidate who will heartily oppose Vivisection. You may be tolerably sure that in other questions he will support the cause of the weak against the strong, of religion against materialism, and of right against might.[26]

The *Zoophilist* printed a form letter for constituents to send candidates for office, requesting their position on the total abolition of vivisection.[27]

In the general election of 1880, the International Association was active in certain constituencies. According to the *Anti-Vivisectionist,* it played a role in the outcome of the elections in Cheltenham (always a hotbed of antivivisection and antivaccination sentiment) and in Poole.[28] Antivivisectionists felt that their real electoral triumph came in the general election of 1885. Ernest Hart, so-called "chief wire-puller" of the British Medical Association, editor of the *British Medical Journal,* and the man in large part responsible for orchestrating the profession's resistance to Carnarvon's bill of 1876, ran in 1885 as radical candidate for Parliament in the supposedly radical Mile End division of the Tower Hamlets. Victoria Street and other antivivisectionists concentrated their resources in Mile End, "showing the true character of Mr. Hart as a supporter and defender of Vivisection." Hart was defeated by his Conservative opponent—2,091 votes to 1,442 votes—and Victoria Street made a rather unlikely claim to responsibility.[29] Flushed with this supposed success, the society inaugurated an elec-

[25] *Home Chronicler* (1878), 89.
[26] *Do You Wish to Stop Vivisection?* (London, c. 1880).
[27] *Zoophilist* v (1885), 99.
[28] *Anti-Vivisectionist* vii (1880), 18, 224.
[29] *Zoophilist* v (1885), 61–62, 86, 137, 156.

tions fund intended to provide resources for similar efforts in the future.[30] Political activity of this type, under the aegis of various societies, including the Electoral Anti-Vivisection League, continued into the twentieth century.[31] Electoral pledging never succeeded for the antivivisectionists in the way it did for other movements, such as the agitation against Contagious Diseases legislation.

The basic problem with electoral pledges was that the antivivisection movement had neither the mass support nor the financial resources to make of such a tactic more than a piecemeal tinkering with the electoral events of a very few seats. Furthermore, such activity ran the risk of seeming to identify antivivisection with party politics in a given seat. Most, though not all, of the leadership wanted to follow the lead of the R.S.P.C.A. in this regard, and keep the movement aloof from partisan politics.[32] Finally, some antivivisectionists felt that electoral pledging might distort the political process and perhaps open the way for charlatan candidates to ride a wave of antivivisection fervor.[33] This was somewhat extreme, but it was an issue that, one way or another, all pressure groups had to face. Kitson Clark has written: "To ask a voter to make his vote turn entirely on whether a candidate gives a satisfactory pledge on a particular subject is to ask the voter to throw away his vote except insofar as that subject is concerned, and unless he is fanatically interested in that subject, or has no interest in any other, he will not do this."[34]

[30] Ibid. v (1886), 189.

[31] *Verulam Review* iii (1892–1893), 3, 133; vii (1897–1898), 395–401; and *Third Annual Report of the British Union for the Abolition of Vivisection* (Bristol, 1901), p. 5.

[32] For one exception to this view, see *Animals' Guardian* iii (1893), 171, and cf. *Home Chronicler* (1878), 12; *Zoophilist* ii (1882–1883), 169.

[33] M. A., *Vivisection Viewed Under the Light of Divine Revelation. An Essay* (London, 1877), p. 41; *Anti-Vivisectionist* vii (1880), 113, 177.

[34] G. Kitson Clark, *The Making of Victorian England*, (University Paperback ed., London 1965), p. 199.

In addition to the promotion of total abolition legislation and the attempt to elect candidates pledged to antivivisection, the movement used one further tactic to influence Parliament directly: the petition. Petitions were important for another reason as well. The petition "supplied an occasion, which would otherwise be wanting, of bringing the subject before people, in asking for their signatures, who would otherwise pay no attention to the matter."[35] This aspect of the petition's role was at least as important as its ostensible function as a direct spur to legislation. Antivivisection societies and journals provided forms of petition and precise instructions for their preparation and circulation, in order to move individual enthusiasts to gather signatures in their own neighborhood. Petitioning took place from the mid-seventies onward, varying in intensity depending upon whether or not a bill was before Parliament, and upon the current resources and tactics of the societies. In 1876, with political activity intense but little formal antivivisectionist organization evident, the House of Commons received 805 petitions for total abolition with a total of 146,889 signatures.[36] In 1883, "it would appear that on an average about 6 Anti-Vivisection Petitions were presented to Parliament every day of its sitting this Session, up to the second reading of Mr. Reid's Bill; and that at least 1,000 persons each day attached their signatures to such Petitions."[37] Antivivisectionist fervor sometimes turned the matter of canvassing for petitions into an aggressive, coercive intrusion into households with no interest in the issue: "Persons, children included, who have never given the subject a moment's thought, and who scarcely know what 'vivisection' means, are besieged to 'just sign their names,' and for the sake of quietness they often consent."[38] Such unscrupulous methods were publicized not only by medical journals, but by newspapers and in the R.S.P.C.A.'s

[35] *Anti-Vivisectionist* vi (1879), 37.
[36] *B.M.J.* ii (1876), 249.
[37] *Zoophilist* iii (1883–1884), 81.
[38] *Journal of Science* 3rd series, ii (1880), 502.

Animal World.[39] At least one antivivisectionist petition was rejected by Parliament because its signatures were all forged by the same hand.[40]

In the end, however, a quarter century of antivivisectionist efforts to induce Parliament to take practical steps toward total abolition came to naught. It was not so much that the movement's political tactics were themselves faulty, but simply that, despite indefatigable efforts at public education, which will be discussed in a subsequent chapter, public opinion would not support total abolition of vivisection. A vague satisfaction with the Act of 1876 was expressed by newspapers and lay periodicals. Antivivisectionists were advised to scrutinize the operations of the Act with care, rather than display a simple blundering animosity toward it. Only demonstrative evidence that the Act had failed, most editors seemed to agree, would constitute a warrant for further legislation on the subject.

THE MOVEMENT AND THE ACT

Total abolition bills and observation by animal protection and antivivisection interests must have been on the minds of both the Home Office (H.O.) and the scientists as they participated in the operation of the Act. Victoria Street's last total abolition bill failed in 1884 and their attempt to find a parliamentary sponsor for a similar measue in 1892 was a failure.[41] Lack of success in this area naturally led to a preoccupation with the existing Act. For its part, the R.S.P.C.A. stood aloof from, and occasionally opposed, total abolition legislation. As its president, Lord Aberdare, made clear in his speech in the House of Lords against Lord Truro's abolition bill of 1879, the Society never countenanced a policy of total abolition.[42] It did, however,

[39] *Medical Press and Circular* ii (1875), 410–411, and i (1877), 377; *Animal World* xi (1880), 63; *Public Opinion* xxxi (1877), 545. On the limitations of petitions as indicators of public opinion, see B. Harrison, "The Sunday Trading Riots of 1855," *Historical Journal* viii (1965), 242.

[40] *Zoophilist* vii (1887), 74.

[41] *Verulam Review* iii (1892–1893), 1–5, 129–133.

[42] See note 14, but cf. *Animal World* viii (1877), 67–68.

join the antivivisectionist societies in a continuing critique of the Act from 1876 on,[43] a critique that spawned a huge and repetitive literature appealing to the people of Britain to right the subterfuge constituted by the present administration of the Act.

Of what exactly did the antivivisectionists complain? First, they objected strongly to the Home Office's choice of inspectors: George Busk, 1876–1886, J. E. Erichsen, 1886–1900 (a former Royal Commissioner), and G. V. Poore, who succeeded Erichsen. All three were prominent London medical men from University College, where they numbered among their colleagues several licensees. Bitter and rather personal attacks on them ("Mr. Busk appears to have been remarkable for nothing more than mediocrity . . .")[44] succeeded in drawing from both Busk and Erichsen lively attacks on the antivivisection movement, attacks that the movement could then adduce as further proof of the inspectors' bias.[45]

The R.S.P.C.A. abstained from personalities, but joined in the general criticism of the adequacy of inspection under the Act. The inspectors could witness only a tiny proportion of all the experiments under the Act. They were inspectors of registered places, not of experiments. They had to compile their returns from the reports of individual licensees, the accuracy of which they could have little or no means of assessing. Such inspection would be laughable were it applied to factories, mines, schools, or ships.[46] A major thrust of the society's critique of the Act lay in the extent to which the measure counted upon the good faith of the Home Secretary. Its public pronouncements and memorials to the H.O. demanded that the Home Secretary make full

[43] See, e.g., J. Colam (ed.), *Further Proceedings Against Vivisection* (London, c. 1878); *Animal World* vii (1876), 146–147, 162–163, viii (1877), 18–19, xi (1880), 117, xiii (1882), 116.

[44] *Zoophilist* vi (1886), 4–5.

[45] F. P. Cobbe and B. Bryan, *The Vivisection Returns, 1884. An Inquiry into their Value* (London, 1884), p. 5, and *Lancet* i (1888), 948–949. Cf. H.O. 156/4/252.

[46] These charges were made again and again. See, e.g., *Animal World* xiv (1884), 98.

use of his discretionary powers and that his office publish a report of the administration of the Act, including names and certificates of licensees. The *Animal World* endorsed a call for revelation of the discoveries made by vivisectors during the tenure of their licenses,[47] and expanded that idea after perusing the literature on rabies:

> If all experiments, successful or unsuccessful, could be reported to a general register, to whom a physiologist might apply for information relating to past work, before pursuing an inquiry new to himself, albeit well-known to others whose researches had proven it to be barren, an immensity of suffering would be prevented.
> . . . it is a great pity that some universal senate is not empowered to revise the enactments of active, ambitious commoners, belonging to the parliament of vivisection, every one of whom apparently is permitted to follow his own course, erratic or otherwise, comparing notes with others only after much horrible torture has been committed that might have been avoided under the surveillance suggested. A senate borrowed from the ranks of practising savants might be trusted, because it would be unlikely to steal the brains of a struggling operator, or to glorify itself in borrowed lustre.[48]

What was needed, in any case, was information by which the public could judge how the Act was being administered. The Victoria Street Society also took this tack. An 1877 motion by one of its members, an influential M. P., A. J. Mundella (1825–1897), brought about the Home Office's first report.[49] This report contained only the number of licenses and of certificates issued, and some general comments on the administration of the Act. The number of experiments performed, their type, and the names of

[47] *Animal World* ix (1878), 94.

[48] Ibid. x (1879), 102. One is reminded of the proposals of Marshall Hall and others. Cf. E. Gurney, "A Chapter in the Ethics of Pain," *Fortnightly Review* xxxvi (1881), 796.

[49] *Second Report of the Society for the Protection of Animals Liable to Vivisection* (London, 1877), pp. 4–5; *Parl. Papers 1877,* lxviii. 423–426.

licensees were not included in the report, nor in its successor covering the second year of operation of the Act.[50] Naming the licensees was initially thought likely to provoke potential stigma and antivivisectionist abuse upon experimenters.[51] Criticism of the inadequacy of the reports by the parliamentary spokesmen of animal protection and others[52] was a major factor in inducing the Home Secretary to include fuller information. Inspector Busk was informed in 1878:

> Mr. Cross thinks that it would be extremely desirable in the next return to show the *names* of the licensees. Mr. Cross hopes that you will be able to relieve him of the responsibility of refusing this Return, for the Refusal creates great suspicion and will create greater.[53]

In subsequent annual reports, the number of experiments performed under license and the various certificates, the names of licensees, and the locations of registered places were included. Cross claimed in the House of Commons that the majority of licensees had no objection to their names being published.[54]

These concessions to pressure meant nothing to total abolitionists and pleased others in inverse proportion to the exclusiveness of their concern with vivisection. The reports were couched in a reassuring style, transparently attempting to minimize the reader's concern over the amount of pain involved. This galled Frances Power Cobbe, who was not about to accept the word of a Fellow of the Royal Society, like Busk, on such an issue.[55] Besides

[50] *Parl. Papers 1878*, lxi. 153–160.

[51] See, e.g., *B.M.J.* ii (1878), 67.

[52] See, e.g., *Animal World* viii (1877), 18–19; *B.M.J.* ii (1878), 81; *Anti-Vivisectionist* (1879), 261; J. Colam (ed.), *Further Proceedings against Vivisection* (London, c. 1878), pp. 23–24.

[53] Minute of 19 December 1878. H.O. 156/1/255–256.

[54] *B.M.J.* i (1879), 721. The practice of publishing in the annual report the names of licensees and the locations of registered places was curtailed in the twentieth century in the interests of brevity.

[55] F. P. Cobbe (ed.), *Bernard's Martyrs. A Comment on Claude Bernard's Leçons de Physiologie Operatoire* (London, 1879), pp. iv, vii–viii.

being falsely reassuring, the reports were, according to antivivisectionists, inaccurate because they relied upon the good faith of licensees, intentionally misleading, and frequently so late that they could be of no use for purposes of enforcement. They provided the slenderest possible basis for assessment of the administration of the Act. The reports gave no idea how many applicants for licenses to perform experiments on living animals were rejected, nor did they include information on any licenses revoked for violations.[56] Each report was routinely torn to shreds by the press of the movement, and alternative forms of report were suggested.[57]

None of the commentators on the returns accused them of *over*estimating the amount of vivisection taking place, and the data in them, though allegedly incomplete, nevertheless showed clearly that the growth in size of experimental medicine in Britain during the last two decades of the century was quite phenomenal. The emergence of bacteriology and immunology during this period entailed a great increase in the number of animals used by a single researcher in a typical experimental project. Furthermore, this research involved inoculation and did not require anesthesia; it was therefore carried out under certificate A, which antivivisectionists took to be a sure sign of painful experiment. In a single year, 1896, experiments under the Act increased from 4,679 to 7,500. This kind of growth greatly alarmed antivivisectionists, who appealed in vain for an end to it. In this sense, the Home Office returns were thought to provide their own indictment of the Act.[58]

[56] *Anti-Vivisectionist* (1879), 261.

[57] See, e.g., *Verulam Review* viii (1899), 82–121.

[58] My summary of the antivivisectionist case against the administration of the Act is largely based upon *Animal World* (1883–1900); *Verulam Review* (1888–1900); Brayfytte, "The Inspector's Report," *Animals' Guardian* iii (1893), 171–176; *Mr. Busk's Last Testament* (London, 1886); M. R. C. S., *Twelve Years Trial of the Vivisection Act* (London, 1889); S. E. Clark, *The Delusiveness of the Vivisection Act and of the Inspector's Reports* (London, 1890); H. J. Reid, *The Administration of the Cruelty to Animals Act . . .* (London, 1893); S. G. Trist, *A Bird's-Eye View of a Great Question* (London, 1894);

Besides appeals to the public conscience and the showering of petitions and memorials upon the Home Secretary, critics of the Act had few practical alternatives. Once or twice, they had recourse to a motion to lower the inspector's salary during parliamentary debate over Home Office supply.[59]

Despite great hostility to the administration of the Act, perceived, as it was, as an effort to protect experimental research, only a very small segment of the antivivisection movement was prepared to call for repeal of the measure. The Act was better than nothing—after all, the medical and scientific interest complained about it—and it provided a base upon which one could build. John Colam of the R.S.P.C.A. defended the Act:

> The existence of a statute which compels scientists to hold their hands off animals before going to a Secretary of State to get permission to conduct any kind of an experiment on them, be it ever so trifling, is in itself an admission that vivisectors cannot be trusted and require to be restrained; and such admission, expressed in a solemn law, creates a public opinion on the subject, and establishes a strong deterrent against cruelty, and against the doctrine that a man has a right to do what he pleases against an animal; the Act contrariwise insists that animals have a claim for protection against experimenters, even when the alleged aim be laudable and possibly beneficent.[60]

The Victoria Street Society also rejected repeal of the Act.[61]

The practical tactic of the antivivisectionists that had the greatest impact upon the administration of the Act

---

The Lax Enforcement of the Law on Vivisection (London, c. 1895); B. Coleridge, *Commentary on the Cruelty to Animals Act, 1876* (3rd ed., London, 1896); E. Bell, *Vivisection on the Increase* (London, c. 1896); E. Bell, *Flaws in the Act* (London, n.d.); annual pamphlets by B. Bryan, (1883–1895).

[59] *Animal World* xx (1889), 95; *Verulam Review* ii (1890–1891), 39–44.

[60] *Animal World* xxv (1894), 146–147.

[61] *Zoophilist* v (1885), 64. See also *Home Chronicler* (1876), 220.

was, however, the Victoria Street Society's careful monitoring of the literature of experimental research and routine comparison with Home Office returns. A fairly large number of alleged violations—i.e., researchers apparently performing experiments for which they did not possess appropriate certification—was discovered.[62] In at least two cases, these charges were genuine enough to mobilize the H.O. In the first case, a private medical practitioner, who had knowingly overstepped the limitations of his license, had it revoked.[63] The second case was more complicated, but the evidence of the Home Office's good intentions was to be satisfactory to Victoria Street. An unlicensed veterinary professor inoculated some cattle against certain diseases, and a colleague described the inoculations in part of a published report of "experimental work." At one point the H.O. was preparing to press charges. The matter became somewhat of a cause celebre when the veterinarian defended himself by saying that his inoculations did not come within the purview of the Act, as they were intended for the welfare of the cattle involved. Proposals were made for amendment of the Act to deal specifically with cases of this kind, but eventually the whole affair blew over without practical result. The man at the center of it was licensed by the H.O. shortly thereafter.[64]

In summary, then, the H.O. revealed to the movement little of what were in fact rather complex administrative arrangements under the Act. Antivivisectionist criticism succeeded in obtaining a report that named licensees, an achievement providing some leverage in scrutinizing the operation of the Act, and, as we shall see in a later chapter, some basis for direct pressure upon medical scientists and

[62] See, e.g., *The Veracity of the Parliamentary Returns as Illustrated by the case of Dr. C. S. Roy* (London, c. 1882).

[63] Complaints about this case were received from both the R.S.P.C.A. and Victoria Street. H.O. 156/2/535, 546, 605–607, 637–638. See also B. Bryan, *Breaches of the Vivisection Act* (London, 1888).

[64] *Zoophilist* vii (1887), 124–125; B. Bryan, *Experimenters in Check* (London, 1888); *Lancet* i (1888), 888–889; H.O. 156/3/38, 43–45, 73–74; H.O. 156/4/70–71, 75–78, 88–89, 215, 223–224.

the institutions that supported them. Beyond this, the movement could learn little and do less about the operation of the Act. Struggling with the opacity of the annual reports, antivivisectionists knew only that experiments upon living animals were increasing steadily in number each year. This frustration moved Victoria Street to join its sister societies in what was to become a dogmatic total abolitionist policy. The lesson of antivivisectionist alertness, and of demands that administrative records be open to the public in order to aid the movement's self-appointed role in policing experimental medicine,[65] was not lost upon the H.O. The latter in fact moved decisively against careless or negligent licensees. There were, however, some features of the administration of the Act that, from the bureaucratic point of view, more than justified the confidentiality in which they were held.

[65] H. J. Reid, *The Administration of the Cruelty to Animals Act . . .* (London, 1893), pp. 13–15.

7. The Administration of the Act and the Association for the Advancement of Medicine by Research

> Society is so glad to find a pretext for closing a discussion, so ready to take as granted that a thing is done and that its responsibilities are over, the power of attack and of arresting attention is so much diminished, that it constantly happens that the people against whom the law is directed are the people to profit by it.
>
> —Auberon Herbert, antivivisectionist, in a letter to the *Times*, 17 January 1876

The most important factor in determining the fate of the Cruelty to Animals Act of 1876 was the extraordinary degree of discretionary power that the Act conferred upon the Secretary of State for the Home Office (H.O.). Quite simply, the Act allowed, or rather forced upon, the Home Secretary all of the important decisions of administration: at the most basic level, the Home Secretary had to decide whether or not to license an applicant and possibly whether or not to allow him the requested certificates. He made these decisions notwithstanding the endorsement of the application by the statutorily specified officers of certain scientific and medical bodies.

The difficulties of such a task for a politician had not been overlooked by the parties to the controversy over legislation. As Cross himself put it to a deputation of the British Medical Association in July 1876, before the Act was passed, "As the Bill stands, I think the only person to be pitied is the Secretary of State."[1] Medical witnesses before the Royal Commission had insisted that the Home

[1] *B.M.J.* ii (1876), 91.

Secretary be competently advised, and for such a role they envisioned not just some qualified medical practitioner, but a man of professional eminence with a special knowledge of anatomy and physiology.[2] As the *Times* asked ". . . is it not ridiculous to suppose that the Secretary of State would be a better judge of the necessity of an experiment than a benevolent physician who had bestowed a whole life-time upon the investigation and the cure of disease?"[3] Robert Lowe agreed: "In the Bill a medical degree is treated as of no value at all, and for it is substituted a license from a Secretary of State. Of course his knowledge can only be secondhand, and his responsibility merely nominal. And yet to him is entrusted the power of closing the path of knowledge. . . ."[4] From the other side of the question, antivivisectionist M. P., J. M. Holt, had asked Cross during debate over the bill,

> Is my right hon. Friend prepared to undertake to sit in judgment on the theories and proposals of scientific men? Is he prepared to say that this or that theory shall not be tested by experiment? When I know that he is already over-worked with the business of his Department I am convinced that he cannot undertake this additional labour, and that he must rely upon the opinions of the scientific men whom he may consult. . . . If he question the decision of one of these learned bodies, they will tell him that there are grave scientific reasons which justify the experiment which he questions, that the blame must rest with him if he retard the progress of science, and hinder discoveries calculated to mitigate human suffering. Unless my right hon. Friend is prepared to set at nought the scientific world he cannot help himself.[5]

[2] See, e.g., the testimony of Sir George Burrows, "Report of the Royal Commission on the Practice of Subjecting Live Animals to Experiments for Scientific Purposes," *Parl. Papers 1876,* xli (C.-1397), Q. 165–172.

[3] 12 July 1876. See also the issue of 2 June 1876.

[4] R. Lowe, "The Vivisection Act," *Contemporary Review* xxviii (1876), 718. Lowe had attempted to amend the measure such that it should not regulate experimentation by qualified medical practitioners.

[5] *Hansard,* 3rd series, 1876, ccxxxi, 904–905.

As these predictions make clear, the course taken in the administration of the Act was to depend in large part upon the development of the relationship between the Home Secretary and the expert scientific and medical advisors whom he consulted. This chapter is devoted to an account of that relationship and the public events affecting it; it is undertaken as a case study in the more general issue of the interaction of professional expertise and political power, which is at the center of so much contemporary scholarly interest in science and government.

ADMINISTRATION, 1876–1882

Cross's first task was the choice of an inspector to aid him in implementing the regulatory functions required by the Act. Significantly, the position was not advertised,[6] and Cross's choice was George Busk (1807–1886), F. R. S., Fellow of the Royal College of Surgeons, of the Linnean Society, and of the Zoological Society. Clearly a man calculated to appease the scientific and medical interest, if not appointed as a result of its active intervention, Busk was a well known figure in London scientific circles. He was President of the Anthropological Institute from 1873 to 1874, and a University College colleague of commission member J. E. Erichsen and physiologists J. S. Burdon Sanderson, W. B. Carpenter, and E. A. Schafer. Busk's salary for his part-time duties was £105 in his first year, doubled to £210 per year in 1877 in view of the fact that the work to be done by the inspector had "proved to be very much greater than was anticipated."[7]

Cross, Busk, and Godfrey Lushington, a civil servant, slowly developed procedures for administering the Act, as applications under its provisions began to arrive at the H.O. during the last half of 1876.[8] No application for a

[6] P.R.O., Home Office Papers, 156/1/56.

[7] H.O. 156/1/13–14, 124–125, 147–148.

[8] My account of the administration of the Act is largely based upon Home Office #156 letterbooks. It is a measure of the sensitivity of the vivisection issue that these documents remain under one hundred year restriction and I am most grateful to the H.O. for permitting me to examine the nineteenth-century letterbooks for purposes of this study.

license was entertained unless a prior and formal request by the authorities of the applicant's institution for registration of specified premises had been received. Registration was then granted routinely and the application from the would-be licensee could be passed on to the inspector for "enquiries as to the competence of the applicant and as to the purpose for which the license is required."[9] With Busk's ensuing report in hand, Cross made his decision as to whether or not to accede to the applicant's request for a license. If the license was granted, it was for a duration of one year only. The licensee was required to keep a careful record of all experiments, ready for submission to the H.O. upon request.[10] It was understood that the registered place in which any licensee experimented was subject at any time to the inspection of Busk or an occasionally employed fellow inspector, G. J. Allman.[11]

It is clear that the kind and purpose of the experiments proposed by an applicant were carefully scrutinized, both by Busk and by Cross. The precise nature of Busk's recommendations to Cross on individual applications is unknown, as no incoming H.O. correspondence is preserved in the extant letterbooks. There can be no doubt, however, that in addition to Busk's advice, there was a second important channel of (informal) expert advice being sought by Cross. This channel involved Sir James Paget, and perhaps one or two other eminent metropolitan medical men, who were consulted personally by Cross on occasional cases that arose in the course of the administration of the Act.[12] Such consultations always took place when an applicant requested

[9] H.O. 156/1/40–41. Private premises were occasionally, but very rarely, licensed. Even more rarely, and somewhat later, a very few eminent licensees were permitted to perform experimental inoculations outside of any registered place, when it could be shown that transportation of the experimental subjects (typically farm livestock) to a registered place would be a severe difficulty.

[10] Soon annual reports of all experiments under license came to be required of each licensee. See H.O. 156/1/69–70.

[11] On the employment of Allman, see H.O. 156/1/51. Allman never participated in the advisory function of the inspectorate.

[12] Cross to Paget, 3 and 10 January 1878, H.O. 156/1/170, 178.

a certificate allowing him to dispense altogether with anesthesia or to allow subjects to recover from the effects of anesthesia.[13]

Besides advice, which he requested, Cross was also subject to unsolicited special pleading by political colleagues and representatives of medicine and science. Lord Carnarvon wrote to Cross a vigorous protest against what he took to be the licensure of Emanuel Klein in November 1876.[14] Lyon Playfair, M. P., contacted Cross with reference to an application by a Scotsman.[15] T. H. Huxley sent Cross a lengthy letter on behalf of an experimental pharmacologist, Lauder Brunton, concluding, "If investigators of Dr. Brunton's character and scientific reputation are to be inhibited, the position of those scientific men who, like myself, have defended legislation upon this unhappy subject, will cease to be tenable."[16] J. S. Burdon Sanderson supported the application of a fellow physiologist, Arthur Gamgee, by similar means.[17]

There can be no doubt that Cross attempted to administer the Act as conscientiously as possible. As a result, he refused a significant number of applications for licenses or certificates during the remainder of his tenure in the office of Home Secretary, to April 1880. Some, and perhaps most, of these decisions clearly flew in the face of advice Cross received from Busk and other scientists and medical men through the channels outlined above. The refusals

[13] H.O. 156/1/196–198. Applications for certificates required specific endorsement by officials of statutorily designated medical or scientific bodies before submission to the H.O. For examples of this probably unwanted burden placed upon such officials as Sir Henry Acland, see Bodleian Library, Oxford, Acland MS d.98 ff.1–6, 10–11.

[14] Carnarvon to Cross, 9 November 1876, Brit. Mus., Addit. MSS 51268 ff.119–120; Cross to Carnarvon, 9 November 1876, P.R.O., Carnarvon Papers, 30/6/9 f.28. Klein had not applied; he was licensed somewhat later, upon application.

[15] H.O. 156/1/72.

[16] Huxley to Cross, 11 January 1878, Brit. Mus., Addit. MSS 51271 ff.214–215.

[17] Draft of Burdon Sanderson to Cross, 14 April 1878, The Library, University College, London, Burdon Sanderson Papers, MS Add. 179/3 f.23.

Figure 9. Richard A. Cross (1823–1914), the first Home Secretary to administer the Cruelty to Animals Act of 1876. *Courtesy of the Wellcome Trustees.*

will be considered in some detail later in this section, but for the moment the point is that such flat denial of licenses or certificates was not the only means by which the administration of the Act impeded scientific research. In the first instance, some would-be applicants were faced with institutional governing bodies antipathetic to vivisection and therefore unwilling to request the H.O. to register for vivisection the appropriate premises within the institution. Without such registration, a mere formality once officially requested, the H.O. did not consider any application for licensure of the individual. It is difficult to know in how many instances the intransigence of a governing body prevented a would-be experimenter from being licensed. Occasionally, applicants for licenses would withdraw when informed of the requirement that institutional authorities had first to request registration,[18] and the registration of private premises was extremely rare.[19] For three institutions at least this factor was crucial, and prevented experiments from being carried out: the Gloucester County Asylum,[20] Trinity College, Dublin,[21] and Mason's College, Birmingham.[22] A second difficulty for the applicant involved the reluctance of certain men, who were filling the official positions in the scientific and medical bodies named in the Act for a given year, to sign application forms or certificates for intending researchers.[23] And then, even should the ap-

[18] H.O. 156/1/104–105. See the debate in the Senate of the University of London over registration of the Brown Institution. *Lancet* ii (1876), 873–874.

[19] For refusals to license private premises in two specific cases, see H.O. 156/1/141–142, 375. One case was in London, the second in Aberdeen.

[20] The frustrated researcher was pathologist G. H. Mackenzie. H.O. 156/1/36, 93–94.

[21] See *B.M.J.* i (1878), 453, 495–496, 499, 510, 551, 667; ii (1880), 755; *Medical Times and Gazette* i (1878), 340, 370; *Medical Press and Circular* i (1882), 288–289.

[22] The Professor of Physiology was J. B. Haycraft. *Medical Press and Circular* i (1882), 35.

[23] In requesting his signature, as President of the G.M.C., on an application form, C. A. Ballance informed Sir Henry Acland, "Mr. Savory has refused to sign these papers but I understand from him

plicant avoid outright refusal by the Home Secretary (as the majority did), his application was subject to administrative delays of at least three weeks and typically six to eight weeks. These delays, the result of both bureaucratic inertia and Cross's caution and care in scrutinizing applications, could rob an investigator of his momentum or ruin his experimental material.

Following the general election of 1880, Cross was succeeded as Home Secretary by William Vernon Harcourt (1827–1904), a Liberal. Harcourt (Figure 10) was Professor of International Law at Cambridge and a distinguished public figure whose intellectual antecedents appeared to augur well for the scientific and medical interest. He had been in sympathetic contact with Huxley and Hooker during the controversies over legislation in 1875–1876.[24] In June 1876 Liberal Lord Cardwell had written to Conservative Richard Cross, "I have been told by Sir W. Harcourt and some others on my side of the contempt in which they hold the bill and I presume all the milk sops who are parties to it. . . ."[25] When Gladstone named Harcourt his Home Secretary, therefore, the physiologists had reason for optimism. As Tables I and II make clear, their hopes at first proved quite unfounded.

In each of the years from 1876 to 1882, somewhere be-

---

that since he has been President of the Royal College of Surgeons he has adopted this course on all occasions. I know that he has refused his signature to my friends Messrs. Horsley and Watson Cheyne." Ballance to Acland, 25 May 1886, Bodleian Library, Oxford, Acland MS d.98 ff.1–2. In certain cases, the responsible officials refused to endorse an application because of legitimate doubts as to the individual applicant or the course of experiments proposed. That is, they occasionally exercised the kind of judicial responsibility that the legislation envisioned. See Jenner to Acland, 13 June 1886, Acland MS d.98 ff. 18–19.

[24] Huxley to Harcourt, n.d., c. 1875–1876, I.C.S.T., London, Huxley Papers, 18.3; Hooker to Harcourt, 22 July 1876, Harcourt Papers, Stanton Harcourt, Oxfordshire. I am most grateful to the current Lord Harcourt for allowing me to examine this last archive.

[25] Cardwell to Cross, 28 June 1876, Brit. Mus., Addit. MSS 51271 ff.138–139.

Figure 10. William Vernon Harcourt (1827–1904), Home Secretary in Gladstone's government of 1880, and the man responsible for the arrangements between the Home Office and the Association for the Advancement of Medicine by Research. *Courtesy of the Wellcome Trustees.*

Table I. Applicants Refused Licenses, 1876–1882

Applicant	*Date of Refusal*	*Education/Qualifications*	*Home*	*Remarks*
B. C. Waller	Jan. '77	M. B. Edin. and C. M., 1876	Edinburgh	Subsequently published a number of articles on morbid and microscopic anatomy and gross pathology.
J. MacFie	April '77	M. D. Edin., 1876. Berlin, Vienna.	Glasgow	Surgeon; subsequently published clinical cases.
J. J. Putnam	April '77	——	London	——
D. Foulis	Oct. '78	M. D. Glasgow, 1875. Edinburgh, Leipzig, Vienna.	Glasgow	Licensed April 1881. Subsequently published articles on pathology.
E. C. Stirling	Jan. '80	F. R. C. S., 1874. Cambridge	London	F. R. S., 1893. Knight Bachelor, 1917. Professor of Physiology, U. of Adelaide.
C. T. Dent	Jan. '80	F. R. C. S., 1877. Cambridge	London	Distinguished surgeon, sometime Lecturer in Physiology, St. George's Hospital. Published a number of medical articles.
J. Aitchison	Jan. '80	M. B. Edin., 1875	Newcastle	M. D. Edin., 1884. Commended for dissertation.
G. Turner	July '80	L. R. C. P., M. R. C. S., 1872. Guy's Montpellier, Cambridge	Portsmouth	Medical Officer of Health, Portsmouth. Former licensee.
F. W. Mott	Oct. '80	M. R. C. S., 1880. University College, London, and Vienna	London	Licensed March, 1883. M. D. Lond., 1886. F. R. S., 1896. Knighted 1919. Became a leading neuropathologist.
J. W. McFadyean	March '81	Dicks Royal Veterinary College	Edinburgh	Veterinarian
W. W. Cheyne	June '81	F. R. C. S., 1879. Edinburgh, Vienna, Strasburg	Edinburgh	Former licensee, licensed again considerably later. F. R. S., 1894. Knighted 1908. A leading bacteriologist and surgeon.
E. T. Tibbits	June '81	M. D., Lond., 1869. University College, London, and Dublin	Bradford	Published a number of medical articles and two popular books.
T. R. Fraser	July '81	M. D., Edin., 1862. F. R. S., 1877	Edinburgh	Former licensee, licensed again Sept. 1881. Professor of Materia Medica, Edinburgh and leading toxicologist *at the time he was refused.*
W. Jervis	Dec. '81	——	London	——

N. B. Horizontal line signifies advent of Sir William Harcourt as Home Secretary, April 1880. Table based upon information in Home Office letterbook 156/1, and on various biographical sources, especially the *Medical Directory*.

Table II. Applicants Refused Certificates, 1876–1882

Applicant	*Cert.*	*Date of Refusal*	*Education/Qualifications*	*Home*	*Remarks*
G. H. Mackenzie	A	Jan. '77	M. B. and C. M. Edin., 1873	Gloucester	Senior Assistant Medical Officer, County Asylum, Gloucester. License cancelled June 1877, when registration of the Asylum was removed at the request of the Asylum's Committee of Visitors. Published articles on toxicology and on cases of mental disease.
W. Rutherford	A, E	Jan. '77	M. D. Edin., 1863. Studied on continent	Edinburgh	Professor of the Institutes of Medicine, Edinburgh. Engaged in research supported by grants from British Medical Association. See below.
J. W. Weir	A	Feb. '77	M. D. Glasgow, 1871. Edinburgh	Glasgow	
T. L. Brunton	B	April '77	M. D. Edin., 1868. D. Sc. Edin. 1870. Leipzig, Amsterdam. F. R. S. 1874	London	Lecturer in Materia Medica, St. Bartholomew's Hospital. Allowed certificates A and E, Dec. 1877. Leading pharmacologist. Knighted 1900.
A. Gamgee	B	May '77	M. D. Edin., 1862. F. R. S., 1872	Manchester	Professor of Physiology, Owens College, Manchester. Allowed certificates A and E, February 1878.
A. S. Currie	B	June '77	M. D. Edin., 1877. Glasgow	Glasgow	Assistant to the Professor of Physiology, University of Glasgow. Surgeon, subsequently published clinical cases.
J. Coats	A	Oct. '77	M. D. Glasgow, 1870. Leipzig, Wurzburg	Glasgow	Published a number of articles on pathology.
A. James	B	Dec. '79	M. D. Edin., 1876	Edinburgh	Lecturer in Institutes of Medicine, Edinburgh.
W. Rutherford	A, E	Jan. '80	M. D. Edin., 1863. Studied on continent	Edinburgh	See above. Allowed certificates A, B, and E, December 1880
M. Hay	A, E	June, '80	M. B. and C. M. Edin., 1878. Glasgow, Strasburg, Berlin, Munich	Edinburgh	Toxicologist and pharmacologist. Became Professor of Medical Jurisprudence at Aberdeen.
A. James	A	Jan. '81	M. D. Edin., 1876	Edinburgh	See above.
D. Newman	B	May '81	M. B. and C. M. Glasgow, 1878. Leipzig	Glasgow	Allowed certificate B, Jan. '82. Published papers on anatomy and physiology.
G. F. Yeo	B	Nov. '81	F. R. C. S., 1878. Trinity College, Dublin and Paris, Berlin, Vienna	London	Allowed certificate B, June '84. Professor of Physiology, King's College, London. F. R. S., 1889.

N. B. Horizontal line signifies advent of Sir William Harcourt as Home Secretary, April 1880. Table based upon information in Home Office letterbook 156/1. Certificate A permits experimentation without anesthesia, Certificate B permits experimentation allowing an animal to recover from anesthesia, and Certificate E, issued only in conjunction with A or B, permits experimentation upon dogs, cats, horses, mules, or asses. These arrangements have been only slightly altered to the present day.

tween twenty-five and thirty-five applications (including the necessary annual renewals) for licenses and certificates were received by the Home Office. Almost all of these were from practitioners of experimental medicine, the remaining two or three from veterinarians. Of these, as Tables I and II show, fourteen applicants for licenses and thirteen applicants for certificates were refused the permission requisite for their proposed research between 1876 and 1882. In his approximately forty-two months of administering the Act, Richard Cross was responsible for sixteen of the total of twenty-seven refusals. During the first twenty-six months of *his* administration, Harcourt was responsible for eleven refusals. Clearly then, Harcourt's arrival at the Home Office worsened rather than improved the situation from the point of view of experimental medicine. This darkening picture during the first years of Harcourt's term of office had important consequences that will be discussed later in the chapter.

What was the rationale of the two Home Secretaries for their refusal of licenses and certificates, refusal which in many cases probably contradicted advice received from the inspector? While the process seems to have been capricious to a certain degree, documentary evidence and Tables I and II suggest explicit criteria and implicit cause for rejection of some applications. The stated objective of the experiments, their number, and the degree of potential pain involved were undoubtedly important considerations, as any number of Home Office letters indicate. For example, Harcourt used the criterion of possible utility when he refused toxicologist T. R. Fraser a license in July 1881: ". . . the Secretary of State does not wish it to be taken for granted that the discovery of every new poison is to be the reason for instituting a set of vivisection experiments unless there is some particular prospect of the utility of such experiments."[26] In 1878, Cross had refused to allow two independent investigations on the same subject, in this case, on a certain poison.[27] Harcourt affirmed the principle

[26] H.O. 156/1/385–386.

[27] H.O. 156/1/223–230, 241–242, 246.

in 1880 by using a similar rationale.[28] The principle meant of course that once the necessary permission had been granted to one researcher, any request by other workers to do the same or similar experiments had to be denied. Perhaps this was in the back of Harcourt's mind when he refused to allow physiologist G. F. Yeo a certificate in 1881 on the grounds that "you have had three Certificates of the same kind for similar experiments during the last three years."[29] Yeo could not be allowed a monopoly; surely three years was time enough to solve the problems of cerebral localization!

Examination of the backgrounds of rejected applicants in Tables I and II provides further indicators as to what made for success or failure with the H.O. One is immediately struck with the youth of the rejected applicants—few of them had qualified medically before 1870—by the predominance of Edinburgh or Glasgow (combined with continental) educational experience, and by the predominance of Scottish residence. The most cursory comparison of successful with unsuccessful applicants indicates that the age and residence characteristics are very distinctive of the unsuccessful group. The educational characteristics are less significant, because Edinburgh had been providing Britain with the majority of its practitioners of experimental medicine since the early part of the century. Such results are just what we would expect, given that the formal and informal channels of expert advice available to the Home Secretary were overwhelmingly London-oriented. Young men, without reputation and residing in Scotland, would naturally receive the least enthusiastic support from Busk or Paget. Parliamentary debate had mentioned the difficulty that unknown but worthy investigators might have in obtaining a license,[30] and the Scottish predominance in the list of unsuccessful applicants recalls the antilegislation stance of the Edinburgh medical community.

It is important to point out, however, that even the best-

[28] H.O. 156/1/329.

[29] H.O. 156/1/398.

[30] *Hansard,* 3rd series, 1876, ccxxix. 1015.

laid plans of eminent men could go astray for no very obvious reason. Take the case of Arthur Gamgee, Professor of Physiology at Owens College, Manchester, who in early spring of 1877 determined to undertake certain pharmacological research. For his research he required a license and a certificate B (allowing experimental subjects to recover from anesthesia). He arranged that the signatories to his application would be the University College Professor of Physiology, John Burdon Sanderson, and the President of the Royal Society, Joseph Hooker. Furthermore, his application was accompanied by an explanatory covering letter from Burdon Sanderson.[31] Despite these precautions, Cross refused Gamgee his certificate in May 1877. At this, Gamgee made a personal visit to the Home Secretary in London, where he received assurances that his case would be reconsidered. Gamgee was certain that Cross was stubbornly intruding his own judgment into a matter for scientific men to decide. In any case, neither Busk nor Paget would have opposed the licensure and certification of Gamgee. As the latter wrote to Sir Henry Acland:

> Mr. Cross is, I think, placing himself in a very anomalous position, by assuming the personal responsibility of deciding upon the scientific merits of [an] application for a certificate under the Cruelty to Animals Act, and not taking the advice of those who are best qualified to give it. After seeing Mr. Cross I called upon Mr. Busk who answered me that he had done his best to procure me the certificate which I desired to have. In confidence he read to me the very strong report upon my application

[31] Burdon Sanderson to Hooker, 15 April [1877], Royal Botanic Gardens, Letters to J. D. Hooker, xviii. Burdon Sanderson wrote to Hooker with regard to Gamgee's application, "As I am desirous that Mr. Cross shall have no excuse for making objection to it, on the ground of want of information, I have addressed to him a note, in which I state my reasons. I hear that he is making objections in every instance in which it is possible and am anxious that when the time comes for pressing upon him, our position may be as strong as possible. Dr. Gamgee's research appears to me to be one against which no reasonable Home Secretary could find objection." Burdon Sanderson's letter to Cross is cited in note 17.

> which he had addressed to Mr. Cross and he said that he had expressed great surprise to Mr. Cross that I had been refused.[32]

After at least ten months of effort, Gamgee was finally granted appropriate certificates in February 1878.

In conclusion then, it is certain that the administration of 39 and 40 Victoria c.77 interfered significantly with research in experimental medicine in Britain between 1876 and 1882. The vivisectional approach had become an integral part of the research method of physiology, pharmacology, pathology, and bacteriology; insofar as experiments upon living animals were prevented, advance in these sciences was frustrated. To an important minority of applicants under the Act, the right to perform such experiments was denied. The Act operated with an intrinsic bias against would-be researchers who lacked a reputation in London, usually owing to a combination of youth and residence in Scotland or the provinces. In addition, the Act as administered by Cross and Harcourt occasionally blocked research by relatively well-known workers—Rutherford, Brunton, Gamgee, Yeo, Fraser—on the basis of largely unstated criteria applied inconsistently by Home Secretaries without any knowledge of experimental medicine.

During its first years of operation, therefore, the Act probably did not completely merit the blanket condemnation it received from antivivisection interests. Largely unknown to the critics, the Act did curb to some degree the extent of experiments on living animals during this period. Furthermore, antivivisectionist hostility forced the Home Office to police its licensees with as much rigor as the status of the latter would allow. While Inspector Busk had little choice but to be extremely diplomatic in his tours of inspection,[33] he moved rapidly to investigate specific charges of

[32] Gamgee to Acland, 21 June [1877], Bodleian Library, Oxford, Acland MS d.98 ff.12–15.

[33] "As I have been instructed by Mr. Cross to make a visitation of all registered places, according to the Act, sometime in the course of the present month I should be glad if you would give me an opportunity of being present at one or more of your demonstra-

abuse raised by antivivisectionists.[34] Two instances of this were outlined in the previous chapter, and in a number of other cases, which did not reach the public eye, the H.O. showed itself eager to pounce upon any infraction of the Act it detected, by revoking licenses or refusing renewals.[35] Although J. N. Langley of Cambridge himself reported ten experiments that he had inadvertently performed after the requisite certificate had expired, he only narrowly avoided losing his license for the infraction.[36] Careless or negligent licensees were a luxury that neither the H.O. nor experimental medicine could afford.

THE TACTICS OF THE SCIENTIFIC AND MEDICAL INTEREST, 1876–1882

The lack of sympathy for and sensitivity to the demands of scientific research with which the H.O. administered the Act of 1876 disturbed and demoralized the scientific and medical community. There were bitter regrets by the moderates that they had ever acquiesced in the passage of any legislation.[37] The arrogance of civil servants or politicians overruling the advice of scientific and medical experts was assailed by the medical press in the strongest terms: "It is, on the face of it, a monstrous absurdity that a Minister appointed for political reasons only should have the power to disregard or nullify the opinions of the highest scientific bodies as to the need for a given set of experiments, or the competence of the performer."[38] "Who . . . reverses the decision of the Presidents of the Colleges of

---

tions . . ." Busk to Schafer, 2 January 1877, Wellcome Institute of the History of Medicine, London, Sharpey-Schafer Papers, British Colleagues, #13.

[34] H.O. 156/1/234–235 (31 October 1878) agrees to Busk's suggestion that he be allowed to investigate charges referring to vivisection at Cambridge.

[35] H.O. 156/2/172–173, 180–181; H.O. 156/4/502–504; H.O. 156/5/64–65, 74–78, 86–87, 122, 124–127, 211–212, 240, 263–266.

[36] H.O. 156/1/251–254 (14 December 1878).

[37] See, e.g., *Lancet* ii (1881), 877–878.

[38] Ibid. i (1882), 360.

Physicians and Surgeons upon matters which are purely medical and surgical?" the *Lancet* asked.[39] James Paget recommended that those who administer the Act make a thorough study of experimental medicine.[40] Michael Foster felt betrayed by both Cross and his advisors, the inspectors:

> Allman and Busk seem to me to have no back-bone and be mere tools of Cross, who having got us in his power, now disregards his unofficial promises. I shouldn't wonder if the upshot is that we shall throw up our licenses. Research will soon become a mockery if the present state of things goes on, and we are to be sacrificed in every way to [antivivisectionist] Holt's and Co. ravings.[41]

Of what precisely did the scientific and medical community complain? First and foremost of course were the point-blank refusals of licenses or certificates.[42] It was recognized that the Act weighed heavily against young men, who, though without reputation, were at the most productive and creative points in their careers.[43] Lister wrote in 1898 that this difficulty, plus the reluctance of the H.O. to register private premises, virtually prohibited research by active medical practitioners and would have prevented his own early research on inflammation:

> I was then a young unknown practitioner, and if the present law had been in existence, it might have been difficult for me to obtain the requisite licenses; and even if I had got them, it would have been impossible for me to have gone to a public laboratory to work.[44]

[39] See note 37.

[40] J. Paget, "Vivisection: its Pains and its Uses-I," *Nineteenth Century* x (1881), 929.

[41] Foster to Thiselton-Dyer, 8 March [?1877], Royal Botanic Gardens, Letters to W. T. Thiselton-Dyer, ii f.19. Cf. *B.M.J.* i (1877), 495.

[42] See Tables I and II.

[43] *Journal of Science* n.s. vi (1876), 332, and 3rd ser., iv (1882), 14.

[44] Lister to W. W. Keen, 4 April 1898, New York Academy of Medicine, MS d.90.

It was, however, the rejection of applications from leading research workers that had the strongest impact upon the community as a whole. Among those refused were Edinburgh's Professor of Physiology, William Rutherford, whose research on the secretion of bile, supported by the B.M.A., was transferred to France;[45] T. R. Fraser of Edinburgh and Lauder Brunton of St. Bartholomew's, London, whose toxicological researches were frustrated;[46] and Professor of Physiology, at King's College, G. F. Yeo, who was unable to continue his research in cerebral localization, carried out in conjunction with David Ferrier. Rutherford's trip to the continent was copied by his Edinburgh associate Alexander James, who went to the Low Countries to perform some experiments for which the H.O. had refused permission.[47] Joseph Lister did the same thing and was widely thought to have been refused a license, although I have found no evidence in H.O. records that he bothered to apply for one.[48] According to the *Lancet*, "Many are deterred from even applying for a licence by the dread of being refused."[49]

As noted above, institutional authorities could refuse to have premises registered, thus preventing research, as at Mason's College, Birmingham; Trinity College, Dublin; and the Gloucester County Asylum. In addition, some hospital authorities, fearing that antivivisectionist publicity would reduce charitable donations, prevented staff physicians and surgeons from experimenting upon living animals.[50] Some would-be investigators were deterred from doing research or from publishing their results by the potential damage

[45] *B.M.J.* i (1877), 79.

[46] S. Gamgee, *The Influence of Vivisection on Human Surgery* (London, 1882), p. 29.

[47] *Zoophilist* i (1881), 157.

[48] *B.M.J.* ii (1881), 365–366; J. Simon, *Experiments on Life as Fundamental to the Science of Preventive Medicine* (London, 1881), p. 12.

[49] *Lancet* ii (1881), 877–878.

[50] J. W. Ogle, *The Harveian Oration, 1880* (London, 1881), p. 54; J. Bland-Sutton, *The Story of a Surgeon* (London, 1930), pp. 35–36. I am grateful to Jeanne Peterson for the Sutton reference.

that antivivisectionist opprobrium could do to their medical practices.[51] These last two hindrances to research were the direct result of antivivisectionist tactics for which the annual returns from the H.O. provided vital intelligence.[52]

Besides outright refusal by the Home Secretary or direct or indirect interference with work as a result of the administration of the Act, research was regularly frustrated by the considerable delays that plagued even ultimately successful applicants for licenses and especially for certificates. The delays involved, three to eight weeks, could easily destroy a program of investigation. John Simon said ". . . the limitations under which these licenses are granted, and the trouble, delay, and friction which necessarily to some extent, and in fact often to an intolerable extent, attend the obtaining of any one of them, are apparently little better than prohibition" and he was echoed in almost identical terms by G. F. Yeo: "the delays in granting licenses and the general official procrastination, often amount to practical refusal and prohibition by loss of opportunities."[53]

Under such circumstances, the adherents of experimental medicine were understandably demoralized. As the Professor-Superintendent of the Brown Institution put it in 1881:

> It is my deliberate conviction, as a result of my experience, that these hindrances and obstacles are so numerous and

[51] G. Makins, "Autobiography" (MS Typescript, Royal College of Surgeons of England, n.d.), pp. 44–45. I owe this reference to Jeanne Peterson.

[52] By including names of licensees, the returns permitted antivivisectionists to organize charitable boycotts and other sanctions against institutions associated with licensees. See the following chapter.

[53] J. Simon, *Experiments on Life as Fundamental to the Science of Preventive Medicine* (London, 1881), p. 11; G. F. Yeo, "The Practice of Vivisection in England," *Fortnightly Review* xxxvii (1882), pp. 367–368. According to Yeo, the Act actually intended that certificates should be granted automatically by the H.O. once they had been endorsed by the appropriate scientific or medical authorities. Only the latter had the ability and, therefore, the power to decide whether or not a certain experiment required, say, that the animal be allowed to recover from anesthesia.

> so great as to constitute a most serious bar to the investigation of disease, and even of such remedial measures as would by common consent be for the direct benefit of the animals experimented upon. When to this is added all the annoyance and opprobrium which are the lot of investigators, it is to be wondered at that anyone should submit to be licensed. I have not been engaged in other investigations for the simple reason that, with the present restrictions, and the difficulty of obtaining a license, I regard it as almost hopeless to attempt any useful work of the kind in this country.[54]

To the *B.M.J.* the effect of the law was "to practically arrest the scientific movement in this country, and very nearly to annihilate the reasonable prospect that England can, or will, under such a *regime,* maintain her place among the living forces of scientific progress in Europe."[55]

The first practical steps to rectify the bleak outlook for research were taken as early as June 1877, less than a year after the Act was passed, by the Physiological Society. This society, composed of about thirty of the leading practitioners and supporters of experimental medicine in London and Cambridge, had been founded during the hectic days of lobbying in the spring and summer of 1876.[56] Although it included a number of individuals prominent in the scientists' lobby—for example, Burdon Sanderson, Foster, and Huxley—the society qua society played little role in the activity. It determined to watch closely the operation of the Act and by the following year was prepared to submit to the General Medical Council a memorandum complaining of the refusals and delays involved. Despite strong support for the Physiological Society from John Simon, a member of the G.M.C., the Council declined to approach

[54] Quoted in J. Simon, op. cit., p. 12.

[55] *B.M.J.* ii (1877), 79.

[56] E. A. Sharpey-Schafer, *History of the Physiological Society during its First Fifty Years, 1876–1926* (Supplement to *J. Physiol.*, London, 1927).

the Home Secretary for the moment but decided to keep the matter under observation.[57]

For the next few years, the Physiological Society seems to have held aloof from practical action, perhaps because in 1878 and 1879 there were only two refusals, both to relative unknowns. As a small, elite group of scientists the society had little influence and no power without the backing of the medical profession as a whole.[58] Furthermore, certain influential members seem to have favored a quietist policy of avoiding public clashes with antivivisection at any cost. With the return of Gladstone and the Liberals in 1880, and the appointment of Sir William Harcourt as Home Secretary, the scientists had reason for optimism. As Tables I and II show, however, the rate of refusals increased rather than decreased under Harcourt, and this setback under the new Home Secretary may have been the last straw. In any case, in 1881 the Physiological Society took some important steps.

First, the society produced a memorandum detailing the hindrances upon research for which the administration of the Act was responsible. In addition to the refusals of licenses and certificates, the society knew of six cases in which delays had amounted to a practical refusal, and of five cases in which applicants had been deterred from making formal requests when preliminary overtures had been rejected by the H.O. This information was made public somewhat later in a book entitled *Physiological Cruelty: or, Fact v. Fancy,* a strong and comprehensive presentation of the case for experimental medicine. *Physiological Cruelty* appeared under the pseudonym *Philanthropos,* and was rumored to be the work of the first secre-

[57] *B.M.J.* i (1877), 681, or *Medical Times and Gazette* i (1877), 594–595.

[58] Although some medical spokesmen occasionally mooted legislative action (e.g., *B.M.J.* ii (1881), 948), an attempt to modify the Act would undoubtedly have failed in the face of equally unsuccessful but extreme opinion represented by the total abolition bills. Cf. *Medical Times and Gazette* i (1882), 306.

tary of the Physiological Society, C. F. Yeo.[59] Secondly, the society decisively rejected the policy of sub rosa pleas to the H.O. and avoidance of confrontation with antivivisection by appealing to the public through the columns of the *Nineteenth Century*. Yeo's colleague as secretary, George J. Romanes, wrote to Charles Darwin in 1881:

> The Physiological Society was formed, as you may remember, for the purpose of obtaining combined action among physiologists on the subject of Vivisection. The result in the first instance was to resolve on a tentative policy of silence, with the view of seeing whether the agitation would not burn itself out. It is now thought that this policy has been tried sufficiently long, and that we are losing ground by continuing it. After much deliberation, therefore, the society has resolved to speak out upon the subject, and the "Nineteenth Century" has been involved as a medium of publication.[60]

Romanes was hoping to induce Darwin to write a short piece on behalf of experimental medicine and biology. Although Darwin did not contribute, articles by James Paget, a surgeon, Samuel Wilks, a physician, and Richard Owen, a paleontologist, on "Vivisection: its Pains and its Uses" appeared late in 1881.[61]

By far the most effective presentation of the grievances of British physiologists and pharmacologists came during the heavily attended and widely publicized International

[59] *"Philanthropos," Physiological Cruelty: or Fact v. Fancy* (London, 1883). The Physiological Society memorandum is dealt with on pp. 97, 154. It is not clear how a potential applicant could be deterred without formal application, except possibly through conversation with the inspector. On the memorandum, see also E. A. Sharpey-Schafer, op. cit., pp. 64–65. On Yeo's authorship, cf. *B.M.J.* i (1891), 737. According to the subject catalogue of the New York Academy of Medicine (xxxiv, 276), *"Philanthropos"* is the pseudonym of one Francis Heatherley. I have never run across Heatherley elsewhere in my research.

[60] E. Romanes, *The Life and Letters of George John Romanes* (2nd ed., London, 1896), pp. 121–122.

[61] *Nineteenth Century* x (1881), 920–948. For Darwin's support of experimental medicine, see *Nature* xxiii (1881), 583.

Medical Congress held in London in August 1881. British scientists, and perhaps the Physiological Society as an organization,[62] took advantage of this ready-made platform, avidly covered by the national press, to link their claims with the undoubted prestige of the Medical Congress and the profession. The congress was notable for some classic scientific debates, such as those between Koch and Pasteur in bacteriology and between Goltz and Ferrier over cerebral localization.

Keynote speakers again and again ignored these issues to pronounce panegyrics on the experimental method and denunciations of restrictions upon research. Excellent examples were speeches by the famous German pathologist and politician Rudolf Virchow, by the doyen of public health, John Simon, and by Edinburgh Professor of Materia Medica, Thomas R. Fraser, a rejected applicant.[63] Fraser told his colleagues from other countries that "original investigation is now almost impossible" showing "how hazardous it is to place the progress of science entirely at the mercy of a State official, utterly ignorant of its aims and triumphs," adding "only the other day I experienced the mortification of being refused a license."[64] Michael Foster urged his fellow researchers to resist legislation in their own countries at all costs and described the difficulties under the British Act: "We are liable at any moment in our inquiries to be arrested by legal prohibitions; we are hampered by

[62] Romanes wrote, "I have been very busy this past week with the affairs of the Congress in relation to Vivisection. It has been resolved by the Physiological Section to get a vote of the whole Congress upon the subject, and I had to prepare the resolution and get the signatures of all the vice-presidents of the Congress, presidents and vice-presidents of the sections, and to arrange for its being put to the vote of the whole Congress at its last general meeting tomorrow. The only refusal to sign came appropriately enough from the president of the section 'Mental Diseases.'" E. Romanes, op. cit., p. 121.

[63] *B.M.J.* ii (1881), 198–203, 219–223, 227–229, or J. Simon, op. cit. and T. R. Fraser, *The Action of Remedies and the Experimental Method* (London, 1882).

[64] T. R. Fraser, ibid., p. 12. For Simon's remarks, see the quotations in the discussion above.

licenses and certificates. When we enter upon any research, we do not know how far we may have to go before we have to crave permission to proceed, laying bare our immature ideas before those who are, in our humble opinion, unfit to judge them; and we often find our suit refused."[65] A general meeting of the congress "unanimously" endorsed a statement insisting upon the value of, and necessity for, experiments upon living animals.[66]

THE ASSOCIATION FOR THE ADVANCEMENT OF MEDICINE BY RESEARCH

There can be no doubt that the presentation of its case by the International Medical Congress was a substantial propanganda coup for the Physiological Society and its sympathizers. The wide publicity given the proceedings of the congress reached a tremendous audience, far beyond the circulation of the *Nineteenth Century*, for example. In particular, the publicity facilitated perusal by Frances Power Cobbe and her Victoria Street Society, for whom careful scrutiny of the literature of experimental medicine was by now routine. The names of research workers publishing accounts of experiments upon living animals were collated with annual returns of the administration of the Act published by the H.O., in order to insure that all such experiments were carried out legally under the Act. Cobbe and her associates, in examining accounts of the discussions between David Ferrier (1843–1928) and Friedrich Goltz on issues of cerebral localization,[67] came upon what seemed to them a godsend, an opportunity to counterattack experimental medicine. Descriptions taken from the *B.M.J.* and the *Lancet* referred to monkeys with cerebral lesions, operatively induced by Ferrier in the course of his research, being displayed at the Medical Congress. The H.O. return,

[65] *B.M.J.* ii (1881), 587–588.

[66] Ibid., pp. 526–527.

[67] The scientific issues have recently been discussed by R. M. Young. See "The Functions of the Brain: Gall to Ferrier (1808–1886)," *Isis* lix (1968), 251–268, or his *Mind, Brain and Adaptation in the Nineteenth Century* (Oxford, 1970).

however, indicated that Ferrier did not have the appropriate license or certification to perform such experiments. The Victoria Street Society therefore decided to prosecute him under the Act.

There had been one previous prosecution. Shortly after the Act was passed in 1876, a medical lecturer, Dr. Gustav Adolph Arbrath, was given a nominal fine under the proviso that prohibited advertisement of experimentation on living animals. Warned in time, Arbrath's public lecture (on a sensational poisoning case) had not featured the advertised demonstrations but instead a strong denunciation of the Act. Although Arbrath claimed that it was this denunciation, and professional jealousy, that was responsible for his prosecution, he received little sympathy, even from the medical press.[68] It is perhaps worth pointing out that twentieth-century government inquiries are thus quite wrong in their statements that there has never been a prosecution under the Act; including the Arbrath and Ferrier cases, there have been at least three.[69]

The Ferrier case was a vastly different matter from that of Arbrath. It involved actual "brain-mangling" or "cerebral-slicing" experiments by a very eminent physiologist. The resulting trial, though very short, came in the aftermath of the Medical Congress and was widely reported. The prosecution maintained simply that Ferrier had done that for which he was not properly licensed and certificated, pointing to the reports in the medical press as evidence. The defense called one of the medical reporters—a physiologist—who testified that the actual surgical operation upon the monkeys had been carried out by G. F. Yeo, who as it turned out was appropriately licensed and certificated. Faced with this defense—misleading references in the med-

[68] *B.M.J.* ii (1876), 545; "The First Conviction under the Vivisection Act," *Public Opinion* xxx (1876), 442–443.

[69] "Royal Commission on Vivisection. Final Report," *Parl. Papers 1912–1913,* xlviii (C. -6114), 408; *Report of the Departmental Committee on Experiments on Animals* (London, C.-2641, H.M.S.O., 1965), p. 44. The third case was in 1913. See A. W. Moss, *Valiant Crusade* (London, 1961), p. 155.

ical press notwithstanding—the prosecution changed its tactics. Its new plea that Ferrier required licensure and certification even in order to have observed and tested the experimental subjects after the operation was, however, rejected by the court. According to the judgment, licensure and certification were necessary only if the individual actually participated in the surgical operation, and the prosecution had failed to prove that Ferrier had done so.[70]

The prosecution of Ferrier had an important impact upon both sides in the antivivisection struggle. For the Victoria Street Society, the experience caused a final and total loss of faith in the Act, which the society had done so much to promote; it provided a striking example of the "perjury" to which physiologists would descend; and it was a potent publicity tool. As Victoria Street's organ, the *Zoophilist,* put it,

> But though the prosecution itself was a failure its effect upon the movement was in every way most satisfactory. In the first place it demonstrated with remarkable clearness, the entire futility of the present Act and the hopelessness of any proceedings under it. In the second it showed to what previously unimagined lengths the Vivisectionist *clique* would go in support of their favourite practice. In the third it afforded perhaps the most striking evidence ever yet obtained of the moral results of the practice as regards those who follow it. . . . Both the trial itself moreover, and the rage aroused among the vivisectionists had the highly desirable effect of bringing the subject prominently before the public mind.[71]

The effect of the Ferrier prosecution upon the scientific and medical community was profound. The costs of Ferrier's defense were undertaken by the B.M.A., whose solicitor represented him in court,[72] but the fact that an eminent

[70] *B.M.J.* ii (1881), 836–842. Cf. *Journal of Science* 3rd series, iii (1881), 729–730, and F. P. Cobbe, *Life of Frances Power Cobbe* (Boston and New York, 1894), pp. 615–617.

[71] *Zoophilist* i (1881), 218.

[72] See note 70.

research worker could be so harassed and persecuted under the Act was a great shock. The Physiological Society was clearly out of its depth in dealing with a challenge of this type. Although originally founded in response to the antivivisectionist threat, the society was composed of a mere handful of experimental scientists whose aggregate political and social influence, while not negligible, scarcely compared with that of, for example, the Victoria Street Society. Shortly after it was founded, the Physiological Society had in fact taken up the most fundamental function of a specialist research society: it began to provide a forum for presentation and discussion of original investigations by members. The events of 1881 generated a widespread realization that united action on the part of all sympathizers with experimental medicine was necessary. The Physiological Society, highly satisfactory as a locus for scientific debate, would have to link its public claims with the power of the medical profession as a whole. The threat posed by the administration of the Act and possible further prosecutions demanded a practical and activist rather than a quietist policy: "It is clear that the yielding policy—the policy of compromise, has failed signally in this instance, and that it must be changed for bolder and more spirited measures."[73] The medical and scientific press called for an organized effort in the defense of research.[74] The *Lancet* put it:

> What is to be done? Follow the old well-established rules. The public must be educated. An agitation must be carried on. Scientific men must come freely and boldly forth and make the laity as familiar as possible with their claims, and the reasons urged in support of them.[75]

The *Journal of Science* suggested a lengthy program of action for a "Biological Defence League" including the

[73] *Medical Press and Circular* ii (1881), 447–449. Cf. ii (1879), 216.

[74] Ibid., 553–554; *B. M. J.* ii (1881), 365–366; *J. Sci.* 3rd ser., iii (1881), 733; ibid. 3rd ser., iv (1882), 8–9.

[75] *Lancet* ii (1881), 343–344.

provision of legal advice and protection for licensees, public education, petitioning and protest to pressure the H.O., and steps for "counteracting and frustrating the system of espionage carried on by the Anti-Vivisection Societies."[76]

The result of these and other clarion calls for unity and action[77] emerged in March of the following year with the foundation of the Association for the Advancement of Medicine by Research. The A.A.M.R. was born amid enthusiastic huzzahs from the medical press: at last, a counter to the wealth, organization and persistence of antivivisection! The inaugural meeting, featuring an imposing list of English biological scientists and medical men, was held at the Royal College of Physicians of London. Here, the result of several months' thought by the eminent men who were its promoters was unveiled. The object of the association was to be the promotion of research to advance medical theory and practice on a broad range of different specialties. Clearly, as certain of the inaugural speakers noted, such an enterprise would be incomplete without careful scrutiny of the operation of the Act of 1876, with a view not to its repeal but to its just administration. Membership in the A.A.M.R. was to be open to anyone willing to pay a modest subscription, while the constitution provided that the society should be governed by a blue-ribbon council holding their positions ex officio as the elect of various scientific and medical bodies. The presidency of the society would fall upon the president of the Royal College of Surgeons and the president of the Royal College of Physicians in alternate years. The first council of the society was composed of Sir William Jenner (president of the Royal College of Physicians), Sir William Gull, Sir James Paget, Dr. Farquharson, M. P., Samuel Wilks, Joseph Lister, and physiologists Foster, Burdon Sanderson, Yeo, and P. H. Pye-Smith.[78] By the time the first meeting of the executive committee of the nascent society was held, the A.A.M.R.

[76] *J. Sci.* 3rd ser., iii (1881), 733.

[77] *B.M.J.* ii (1881), 834, 878–879.

[78] *B.M.J.* i (1882), 476–478, 635, or *Lancet* i (1882), 542–544. Cf. *Medical Times and Gazette* i (1882), 305–306. Charles Darwin

had received over £1000 in subscriptions, including a few £100 donations from wealthy metropolitan practitioners. At that meeting, the priorities of the A.A.M.R. were demonstrated by the appointment of a subcommittee to report "on the present hindrances to research due to the working of the Vivisection Act."[79]

In its activities, the A.A.M.R. operated on two very different levels. The first level was that upon which most of its expenditure occurred. This level involved a brief though intense effort at public education and more prolonged, but sporadic, financial support of miscellaneous researches in experimental medicine. This public activity was initiated by the paeans in praise of the experimental method for the advance of medicine, which had constituted the inaugural speeches in March, 1882. It was continued over the next three years by publication in pamphlet form of a number of similar speeches.[80] It is important to note that this attempt to speak authoritatively to the public on behalf of experimental medicine essentially ceased in 1885. Among the actual research that the A.A.M.R. backed from time to time was the expedition of C. S. Roy, J.

---

had been sounded out as to whether he would be willing to run for the presidency of the society. See F. Darwin and A. C. Seward (eds.), *More Letters of Charles Darwin* (London, 1903), ii. 437–439.

[79] *B.M.J.* i (1882), 677.

[80] W. Bowman, *Address in Surgery . . . 1866* (London, 1882); J. Cleland, *Experiment on Brute Animals* (London, 1883); T. R. Fraser, *The Action of Remedies and the Experimental Method* (London, 1882); G. Gore, *The Utility and Morality of Vivisection* (London, 1884); G. M. Humphry, *Vivisection: What Good Has it Done?* (London, 1882); R. McDonnell, *What has Experimental Physiology Done for the Advancement of the Practice of Surgery?* (London, 1882); *Official Circular Addressed to the German Universities by the Minister of Public Instruction Enjoining Safeguards Against Abuses of Vivisection* (London, 1885); L. Playfair, *Speech delivered in the House of Commons on the Second Reading of Mr. Reid's Bill for the Total Suppression of Scientific Experiments upon Animals* (London, 1883); J. Simon, *Experiments on Life, as Fundamental to the Science of Preventive Medicine* (London, 1882); S. Wilks, *The Value and Necessity of Experiments for the Acquirement of Knowledge* (London, 1882).

Graham Brown, and C. S. Sherrington to investigate the outbreak of cholera in Spain in 1885.[81]

The second level of activity of the A.A.M.R. was its all-important role in representing experimental medicine on the political front, in particular with the Home Secretary. As previously noted, Harcourt was sympathetic to science and scientists; one of his first actions as Home Secretary had been to appoint T. H. Huxley as inspector of salmon fisheries over various sportsmen who had applied for the job.[82] It is, therefore, difficult to explain why the administration of the Act of 1876 became more stringent under Harcourt, except perhaps under the hypothesis that he actually paid little attention to the Act and allowed the decisions under it to devolve upon civil servants. This hypothesis is lent credence by the inaccuracies in Harcourt's defense of the Act in Parliament during a debate over one of Reid's total abolition bills,[83] though the only evidence I have run across as to partisan feeling on the part of civil servants has indicated their sympathy with experimental medicine.[84] In any case, Harcourt vindicated himself with the scientists by being extremely receptive to overtures from the A.A.M.R. in June 1882. As civil servant Godfrey Lushington wrote to Sir William Jenner at the time:

> I am directed by the Secretary of State to acknowledge receipt of your letter . . . signed also by Sir James Paget, Bart., and Mr. P. H. Pye-Smith, offering on behalf of the Council of a newly formed Association for the Ad-

[81] C. S. Roy, J. Graham Brown, C. S. Sherrington, *Report on the Pathology and Etiology of Asiatic Cholera* (London, 1887); J. P. Swazey, *Reflexes and Motor Integration: Sherrington's Concept of Integrative Action* (Cambridge, Mass., 1969), p. 12.

[82] R. M. MacLeod, "Government and Resource Conservation: The Salmon Acts Administration, 1860–1886," *J. Brit. Stud.* vii (1968), 114–150.

[83] See F. P. Cobbe, *Comments on the Debate in the House of Commons, April 4, 1883, on Mr. Reid's Bill for the Total Prohibition of Vivisection* (London, 1883).

[84] See Lushington to Acland, 9 April 1891, Bodleian Library, Oxford, Acland MS d.98 ff.26–27.

> vancement of Medicine by Research, to place at his disposal such aid and advice as may be within their province to afford in administering the powers conferred upon him in the matter of the performing of experiments on living animals under the Act 39 and 40 Vic. cap. 77. And I am to acquaint you for the information of the Council of the above mentioned Association that the Secretary of State will be extremely glad to avail himself of their advice and assistance and he would be much obliged if they will be so good as to favour him with their views upon the subject of the present working of the Act in question and in what respect they would recommend an amendment of the law.[85]

The A.A.M.R. eventually offered Harcourt the services of its Council in reporting upon applications received under the Act. Harcourt quickly accepted this proposal and inspector George Busk was informed in February 1883:

> I am directed by Secretary of State Sir W. V. Harcourt to acquaint you that, having been in communication with the Association for the Advancement of Medicine by Research he has given instructions that no application under the Statute relating to experiments on living animals 39 and 40 V. c.77 will be entertained by him unless it has been recommended by the Association; and that consequently in future any such application that may come to this Dept. will be first submitted to the Association for their observations and when returned by them will be sent on in due course to the Inspector.[86]

This arrangement between the H.O. and the A.A.M.R.[87] became standard operating procedure under the Act until it was altered in response to the recommendations of the second Royal Commission on Vivisection in 1913. Under the new procedures, experimental medicine in Britain en-

[85] H.O. 156/1/474–475.

[86] H.O. 156/1/540–541. See also 538–540. Cf. *Lancet* ii (1882), 745–746, which suggests precisely this arrangement.

[87] For an example of an A.A.M.R. recommendation, see H.O. 45/10073/B6079/28.

joyed spectacular growth; the number of licenses increased from 12 in 1882 to 638 in 1913.[88] The substantial alteration in the mode of administering the Act, which effectively transferred decision-making on applications from the Home Secretary to the A.A.M.R., was responsible for greatly facilitating the licensure of would-be researchers.

H.O. records show clearly the change that Harcourt's cooperative arrangement with the A.A.M.R. produced during the remainder of his own and subsequent administrations. The Council of the A.A.M.R. seems to have recommended against licensure or certification on only the rarest occasions. There were only three outright and final refusals of licensure or certification between 1882 and 1890, and none of the rejected applicants had an institutional affiliation.[89] Occasionally, the A.A.M.R. would recommend that an individual be licensed but limited to a small number of experiments,[90] or that an applicant be asked to explain his intentions more clearly.[91] Furthermore, A.A.M.R. participation in, or rather monopolization of, the decision-making process bestowed upon the association a quasiofficial status from which it contested decisions by the Home Secretary that disregarded its recommendations. Good examples are Harcourt's refusal to certificate G. F. Yeo in December 1883, and Henry Matthews' (Home Secretary, 1886–1892) similar action with regard to F. W. Mott in September 1886. In both instances, the Home Secretaries were confronted with indignant protests from the A.A.M.R. Council, requests that a deputation be allowed to explain the Council's reasons for its recommendations, and similar pressure. In both instances, the physiologists ultimately got their cer-

[88] See the annual returns published in *Parliamentary Papers* and Chapter XII.

[89] H. O. 156/2/453–454, H.O. 156/5/429, 501, 509.

[90] H.O. 156/2/340–341, 349.

[91] One applicant was told that the Secretary of State "cannot see his way to complying with your request unless you can give more specific details of the nature and extent of the proposed experiments and of the utility of endeavouring to prove experimentally a fact which he is given to understand has long been established clinically." H.O. 156/3/36.

tificates.[92] The highly anomalous nature of the H.O. arrangement with the A.A.M.R. is brought out in clear relief when we remember that G. F. Yeo was himself a member of the Council of the A.A.M.R. Three other licensees, Foster, Burdon Sanderson, and Pye-Smith, were also members. Licensees were deeply involved in the upper echelons of the A.A.M.R. right up to 1913.

The A.A.M.R. eased the burden that the Act placed upon experimental medicine in other ways. Occasionally, its Council seems to have corresponded directly with researchers concerning their applications and sometimes official records indicate that applications were submitted first to the A.A.M.R. and then passed to the H.O. with the Council's endorsement.[93] The A.A.M.R. published an instructional pamphlet intended especially for applicants, which included application forms.[94] In 1889 and again in 1893, a deputation from the A.A.M.R. strongly protested what it viewed as undue delay in granting licenses and certificates. The H.O. felt that most of the delay owed to lack of foresight on the part of licensees, but procedures were streamlined as much as possible.[95]

The A.A.M.R., then, radically transformed the administration of the Act of 1876, but it did so substantially, and I believe designedly, out of the public eye. The A.A.M.R. undertook a program of public invisibility that was virtually complete by the late eighties; after all, by that time its primary objective of removing the Act's restrictions on experimental research had been essentially achieved. The anonymity of the A.A.M.R. was based upon two factors: first, a decision by a majority of its promoters not to engage in a distasteful counteragitation to antivivisection, and second, the A.A.M.R.'s extraordinarily rapid and successful penetration of the inner sanctum of the H.O.

[92] On Yeo, see H.O. 156/1/671–672 and H.O. 156/2/26–28, 31, 32, 55, 57–58. On Mott, see H.O. 156/2/446, 476, 487.

[93] H.O. 156/2/276–277.

[94] S. Paget, *A Short Account of the Act for the Better Prevention of Cruelty to Animals* (London, 1891).

[95] H.O. 156/4/518–519, 622–642; H. O. 45/10062/B2738.

Even before its foundation, certain medical men envisioned the A.A.M.R. as an organization that would fight fire with fire, meeting antivivisection on its home ground of controversy and aiming for mass membership, especially for all medical practitioners.[96] A second and ultimately dominant group held that the organization should be a vehicle by which the leading members of the medical profession might quietly represent their case to influential members of the government. Ernest Hart, predictably enough an advocate of mass agitation rather than diplomacy by the few, gave the matter a thorough airing in the pages of the *B.M.J.* during the weeks following the inaugural meeting of the A.A.M.R. Hart attacked the association's constitution for failing to provide for representation of the average medical practitioner and thereby ensuring that the A.A.M.R. would be governed by a predominantly metropolitan professional elite.[97] In a confrontation between the tactical approaches that had combined effectively in 1876, the advocates of the discreet lobby held the field. As one such advocate put it in answer to a critic of the A.A.M.R. constitution:

> If Dr. Barnes fear[s] that in the Council discretion will be more apparent than zeal, he must remember that the object of the Association is not popular agitation; it is not primarily controversial. The working physiologists of the three kingdoms have expressly stated that they do not desire (at least for the present) to attempt to abolish the Act, of which we are all ashamed, but to secure its being harmlessly administered. To speak with authority to public opinion, and to bring effectual pressure upon officials, needs other means than those which are suited to the arena of controversy. . .[98]

The second major reason for the A.AM.R.'s policy of a low public profile was the astonishingly rapid and com-

[96] See Brunton's letter to Darwin, 14 February 1882, in F. Darwin and A. C. Seward (eds.), *More Letters of Charles Darwin* (London, 1903), ii. 440.

[97] *B.M.J.* i (1882), 516–517.

[98] Ibid., 599. See also 638–639, 679.

plete success of the lobbying tactics adopted. The A.A.M.R.'s quick accession to a crucial position in the administration of the Act, in what was at best a somewhat questionable arrangement, made further publicity positively undesirable. A brief note in the *Medical Times and Gazette*[99] and an announcement by Harcourt in Parliament of "Additional security" in the administration of the Act were the only public acknowledgments of the H.O.-A.A.M.R. arrangement.[100] It is in this context that the failure of the *B.M.J.*, *Lancet*, and *Medical Times and Gazette* to mention the A.A.M.R at all from 1883 to 1891, when a discreet appeal for funds appeared,[101] begins to make sense. It was an attempt to develop a highly desirable public anonymity for the association. Rhetoric and the occasional afterthought for the financial support of research aside, the A.AM.R. was an organization oriented to a single issue: the administration of the Act of 1876.[102] Having achieved a favorable strategic position with regard to the Act, the A.A.M.R. ceased its efforts at public education and silently attempted to drop out of public sight. When in 1908 it became necessary to renew efforts at public education, Stephen Paget, long the secretary of the A.A.M.R., founded a separate organization, the Research Defence Society, to carry out this task.[103]

[99] *Medical Times and Gazette* ii (1882), 108–109. Note that the *Lancet* and *B.M.J.* did not report this meeting.

[100] *The Debate on Mr. Reid's Bill in the House of Commons, April 4, 1883* (London, 1883), pp. 35–37. Harcourt named the A.A.M.R. and gave a certain amount of detail. See also *Nature* xxvii (1883), 549.

[101] *Lancet* ii (1891), 141, 147; ii (1892), 40. Even in 1892, a fresh appeal against the A.A.M.R.'s anticontroversial tactics was made: "the Society should have in readiness one or more trained lecturers to combat the errors of anti-vivisectionists, and . . . the pamphlets of the Victoria-street Society should undergo a critical examination. . . ." W. J. Tyson, *Lancet* ii (1892), 1138.

[102] Though cf. J. S. Burdon Sanderson to R. Burdon Sanderson, 13 March 1882, National Library of Scotland, Haldane Papers, 5902 f.24. I am grateful to Richard Kenin for bringing this reference to my attention.

[103] *Report of the Departmental Committee on Experiments on Animals* (London, C.-2641, H.M.S.O., 1965), p. 64.

The A.A.M.R.'s policy of shunning the spotlight was an overall success until the early twentieth century, when antivivisectionist pressure achieved a second Royal Commission on vivisection. Indeed, at least one antivivisectionist journal in the early nineties was exulting over what it took to be the demise of the A.A.M.R. Harcourt's parliamentary announcement had, however, named the A.A.M.R. and had included a certain amount of detail about the H.O.'s arrangement with the association.[104] A vocal section of antivivisectionist opinion attacked the arrangement at that time, and did not fail to apprise successive Home Secretaries of their distaste for it. In 1893, parliamentary questioning confirmed that the arrangement was still in force.[105]

It is not difficult to understand the fury the antivivisectionists felt toward the H.O.-A.A.M.R. relationship. One flyer of 1883 charged, "The Act's Administrator, the Home Secretary, has lately formally and publicly delivered into the hands of the vivisectors the responsibility of giving licenses and certificates—thus depriving the wretched animals of the last shred of protection, which a lay representative of Government was intended to give."[106] A pamphlet entitled *Facts and Considerations Showing the Raison d'Etre of the Association for the Advancement of Medicine by Research* asked "*Who* now represents the claims of Humanity in the administration of the Vivisection Act of 1876?"[107] The Scottish Society for the Total Suppression of Vivisection criticized the Home Secretary's action at great length and petitioned him against the A.A.M.R.[108] To

[104] See note 100. Harcourt also presented the A.A.M.R. as an "additional safeguard" in communications with an anxious and thoroughly antivivisectionist Queen Victoria. See A. G. Gardiner, *The Life of Sir William Harcourt* (London, 1923), i. 402–403, and correspondence between Victoria and Harcourt, esp. 27 November 1881, Harcourt Papers, Stanton Harcourt, Oxfordshire.

[105] *Animal World* xxiv (1893), 158.

[106] S. S. Monro, *Dr. Burdon Sanderson and Vivisection at Oxford* (Flyer, London, 1883).

[107] (London, c. 1885), p. 7. Quoted passage slightly altered.

[108] *Fourth Annual Conference of the Scottish Society for the Total Suppression of Vivisection* (Edinburgh, 1883), pp. 5–14, 16–17, 24. See also H. J. Reid, *The Administration of the Cruelty to Animals*

antivivisectionists, the Home Secretary's action was one more reason for a complete lack of faith in the Act of 1876.

In the light of our analysis of the A.A.M.R.'s tactics, it is interesting that antivivisectionists were positively livid at their failure to draw the association into open public controversy. The "ultimate bathos of cuttle-fish cowardice" was demonstrated by the association's disregard of repeated challenges by the Victoria Street Society.[109] An antivivisectionist journal in 1894 characterized A.A.M.R. proceedings as follows:

> Their whole policy has been a policy of concealment; of undermining that which they could not pull down, and attacking in the rear those they feared to face. At open argument they have made no attempt. But they have spent money freely in supplying the more effective weapons of insinuation and misrepresentation which the peculiar position of their supporters enables them so efficiently to wield.[110]

Although antivivisectionist criticism had a significant impact upon the administration of the Act, the really important decisions were being made by the A.A.M.R. The H.O.-A.A.M.R. arrangement lasted until 1913. At that time, the second Royal Commission on Vivisection—notable primarily for a general satisfaction with the current administration of the Act on the part of a majority of its members—recommended that the A.A.M.R. be replaced with regard to the Act by an advisory body selected by the

---

Act . . . (London, 1893); S. E. Clark, *The Delusiveness of the Vivisection Act and of the Inspector's Reports* (London, 1890), pp. 4–8; *Zoophilist* iii (1883–1884), 105.

[109] C. Adams, *The Coward Science* (London, 1882), pp. 245–252. Cf. *Zoophilist* ii (1882–1883), 102.

[110] *Verulam Review* iv (1894), 18. Cf. *Zoophilist* ii (1882–1883), 127, where the A. A. M. R. is described as "carying on with much energy an underground propaganda among members of Parliament and other persons of influence."

Home Secretary from a list of names submitted by the Royal Society and the Royal Colleges of Physicians and Surgeons.[111] None of these persons were to be licensees and their names were to be published. While the recommendation itself could be viewed as a tacit admission of a certain impropriety, the commission was at pains to deny the antivivisectionists' charges of collusion and "improper private relations" between the H.O. and the A.A.M.R.[112]

In his testimony before the second Royal Commission, E. H. Starling, president of the A.A.M.R., agreed that "the object of the whole membership is favourable to the promotion of vivisection" and could not recall any occasion when the association had finally refused to endorse an application for licensure.[113] A perusal of the annual returns from the H.O. makes it clear that experimental medicine in Britain enjoyed tremendous growth under the A.A.M.R. arrangement and under the advisory arrangement, that succeeded it, which was adopted in response to the second Royal Commission's report. Although the bureaucratic machinery was cumbersome and occasional annoyance might result from the Home Secretary's limiting the number of experiments allowed,[114] it is hard to imagine such growth occurring under circumstances in which there was any significant interference with research. By 1913, ". . . all applications, coming as they do from and being recommended by competent persons, are granted. An absolute refusal is the very rarest occurrence."[115] From the late eighties on, the medical and scientific interest seems to have been

[111] "Royal Commission on Vivisection. Final Report," *Parl. Papers 1912–13*, xlviii (C.-6114), p. 464.

[112] Ibid., p. 414.

[113] Ibid., p. 407. The functions of the A.A.M.R. were described as "(1) To advise in the granting of licenses. (2) To protect when necessary, the interests of licensees. (3) To watch proceedings in Parliament affecting the interests of research. (4) To publish and distribute to medical men and others who may desire it, literature on the importance of research and the necessity of experiments on the lower animals."

[114] Ibid., p. 409.

[115] Ibid., p. 406.

generally satisfied with the administration of the Act. I believe this holds true to the present day.[116]

CONCLUSION

The evolution of the controversy and the regulatory procedures involving experiments upon living animals shows clearly certain features of the interplay between professional organization and expertise on the one hand, and political power on the other. A tiny and relatively unsophisticated group of experimentalists turned in both 1875–1876 and 1881–1882 to the powerful medical profession in order to counterbalance the threat posed by a popular agitation and its political consequences. The would-be profession of experimental medicine had converted certain elite physicians and surgeons to the point where experimental physiology, pharmacology, and bacteriology were slowly becoming institutionalized in hospitals and medical schools. The experimentalists, however, lacked their own professional associations and the power and experience that goes with them. Important manifestations of their political innocence were the publication of the *Handbook for the Physiological Laboratory* without qualifications as to anesthesia and as to the readership for which the book was intended, the public advertisements for animals to be used in the laboratories of the University of Edinburgh, the unguarded public statements by E. Ray Lankester as to the "geometrical rate of increase" of experiments that was to be expected, the Norwich fiasco, and Emanuel Klein's testimony before the Royal Commission. From these gaffes, and antivivisectionist capitalization upon them, experimental medicine was saved by the (largely metropolitan) medical leadership. James Paget, William Gull, William Jenner, John Simon, and others like them mobilized the power and prestige

[116] Cf. ibid., p. 458; W. Lane-Petter, *The Research Defence Society* (London, 1957), esp. p. 4; and *Report of the Departmental Committee on Experiments on Animals* (London, C.-2641, H.M.S.O., 1965). This last report provides a good, though not always completely accurate, account of the Act and its administration. It is very useful for twentieth-century procedures.

of the profession through bodies such as the British Medical Association and the General Medical Council. As a result, the antivivisection movement and the Home Secretary ultimately found themselves confronting an established profession rather than a handful of scientists.

One of the important things the antivivisection movement did, then, was to force experimental medicine to develop professional associations, utilizing in this regard the strength of the medical elite. The latter were in a position of unparalleled influence over their brethren the practitioners and over the politicians, as a result of the statutory autonomy and regulatory powers ceded to the professional leadership by the government during the fifties and sixties. Both the Physiological Society and the Association for the Advancement of Medicine by Research initially had only a minority membership of active experimentalists. As experimental medicine gained in strength over the last decades of the century, both associations came to be dominated by experimental researchers rather than by the eminent practitioners and nonexperimental biologists who originally provided leadership and support. The Physiological Society became a learned society whose primary purpose was the communication of research results. By 1913, when the A.A.M.R. reemerged into public view, its president was no longer a famous physician or surgeon, but a professional physiologist, E. H. Starling.

Professional associations play a critical role in the development of a profession, by educating its members as to their interests, by exerting discipline within their ranks, and by utilizing the claims of expertise and service to gain resources and autonomy for the profession. One of the important ways in which these functions can be realized is for the association to present the activities of the profession to different audiences within the society in terms congruent with the interests of these particular audiences. The early failures by the proponents of experimental medicine to recognize the variety of audiences for their utterances are classically exemplified by the events mentioned above. The advent of the A.A.M.R., with its leadership of relatively sophisticated medical men, demonstrates the efficacy of differentiating between various audiences for the profession. To the public, the A.A.M.R. developed, in its early years, an approach that confounded the interests of experimental research with the objective of very practical therapeutic advances in clinical medicine. The Association rarely argued before the public two sentiments fairly widespread among experimentalists and their allies: first, that vivisection was justified for scientific advance; and second, that lay people were unqualified to exercise judgment in a matter involving esoteric professional knowledge. Yet when the A.A.M.R. approached the Home Secretary in private, it was precisely this second line of argument—the claims of unique expertise—that was exploited.

The Home Secretary would have found it extraordinarily difficult to withstand this line of argument in the absence of strong public interest in vivisection, which had, by the early eighties, subsided substantially from its peak in 1876. His only access to expert knowledge was through the inspectors, who owed intellectual and institutional allegience to the medical profession whose metropolitan leaders were experimental medicine's most powerful allies. The conciliation of powerful professional interests was thus clearly preferable to the option of reopening to the public a complex issue of little political significance but potentially enormous nuisance value.

In summary, then, the antivivisectionist critique exploited research literature written for a strictly professional audience. By so doing, the movement was able to capitalize upon apparent conflict between scientific practices and public values, and to launch a serious attack upon the autonomy of British medical scientists. Antivivisectionists in general failed in their attempt to draw medical scientists and practitioners who sympathized with them into public controversies that would have conferred additional status upon the movement. However, to the extent the movement was able to arouse public opinion to demand intervention into the affairs of experimental medicine of nonprofessionals such as the Norwich bench of magistrates, most of the

Royal Commissioners, members of the Commons and Lords, and civil servants in the Home Office, it encroached upon the autonomy of the scientists affected.

The Act of 1876 raised the question of lay power within the scientific community in a particularly explicit and acute form. The administration of the Act placed the handful of medical scientists and their sympathizers among the elite of the medical profession in a very difficult position. There was strong sentiment, as we have seen, for a quietist political stance and an avoidance of public confrontation with antivivisection. The influence of individual initiatives on behalf of experimental medicine, occasionally directed privately toward the Home Secretary between 1876 and 1882, was negligible. Once again, however, the medical scientists and their allies were able to utilize the prestige and power of the medical profession as a whole, through its representative institutions, the Royal Colleges, without surrendering control of policy. The Association for the Advancement of Medicine by Research represented a departure from the policy of near-absolute public silence, which had been first broken by the articles in *Nineteenth Century* and the speeches at the International Medical Congress. On the other hand, the A.A.M.R. was by no means a popular counteragitation to antivivisection. Its backers were medical scientists and leading medical practitioners who decisively rejected the kind of distribution of power beyond their own circle that Ernest Hart had enthusiastically advocated. The scientists could not rely on the judgment of rank and file medical practitioners, most of whom had no practical experience with experiments upon living animals. In the event, the self-perpetuating elite on the Council of the A.A.M.R. succeeded spectacularly in convincing the Home Secretary that it undertook to provide expert advice, which it emphasized as necessary for the administration of the Act, with a mandate from the profession as a whole. Harcourt, sympathetic to the claims of expertise, had considerable discretionary power owing both to the provisions of the Act and to lack of interest on the part of the general public. The administrative arrangement that resulted regained an all-important degree of autonomy for the medical scientists. A classic confrontation between expert knowledge and professional autonomy, on the one hand, and the political structure and (sporadic) public opinion on the other, had ended in victory for the scientific and medical community.

This victory was consolidated by a rigid audience differentiation, with the Council of the A.A.M.R. and its clients, the licensees, operating in almost complete insulation from the fading public facade of the Association. The basic purpose of the A.A.M.R. was to allow medical scientists to choose their problems and carry out their research as they wished, without the application by nonprofessionals of criteria of acceptability deemed irrelevant by the investigators themselves. As we have seen, the A.A.M.R. rarely applied any binding criteria whatsoever to applications that came before it, since it virtually never rejected such applications. The important contrast is with the *public* rationale of the A.A.M.R., during the brief period when that rationale was being enunciated to the Home Office and to the interested citizenry. To these audiences, the A.A.M.R. was portrayed not as the guardian of scientific freedom with reference to the Act of 1876, but rather, as its name implies, as an omnibus organization involved in the undoubtedly beneficent and conveniently vague enterprise of advancing medicine. Even as the A.A.M.R. flaunted the utilitarian potential of experimental medicine before the public at large, it worked to remove the application of utilitarian or any other nonscientific criteria to the research actually being carried out.

8. Anatomy of an Agitation

Victorian society was stirred, chastised, annoyed, appalled, and finally enervated by a bewildering variety of reforming agitations. Victorians of the middle and upper classes in particular volunteered on behalf of myriad causes and crusades: political, social, and religious, national, regional, and local. The resulting philanthropic and melioristic organizations present rich territory for historical research, only a small fraction of which has been systematically explored. The study of these movements, which were both symptoms and agents of social change, offers the possibility of illuminating the hopes and fears of their leadership and of their historically anonymous supporters.

The most frequently investigated aspects of such movements have been the political and administrative manifestations of their activity. Especially during the heyday of the private member at mid-century, parliamentarians were bombarded with letters, memorials, petitions, and deputations from a broad spectrum of earnest and indignant reformers. Such pressure-group tactics formed an important part of the political environment for back-benchers, ministers, and peers alike in the days before party discipline and parliamentary staff arose to shield the individual member. Victorian civil servants often found their activities subject to severe scrutiny and criticism from interested philanthropists. The first seven chapters of this study have been devoted to a consideration of the antivivisection movement in precisely this political and administrative vein of analysis, with special reference to the response of the scientific and medical communities to the political threat.

To confine one's analysis of a reforming agitation to the

most visible results of its relations with the state would, however, seriously impoverish historical perspective. The historian must try to explore carefully not only the configuration, policies, and tactics of agitations, but also the conscious and unconscious motives of the participants. He must read between the lines of the monotonous annual reports, the outraged letters to the editor, and the inflammatory propaganda. He must view any particular movement not simply in terms of its success or failure in achieving its ostensible purposes, but also as a particular manifestation of a widespread social phenomenon. This level of analysis is a complex and difficult task. It is dependent not only upon the historical imagination, but also upon the nature and extent of the evidence available. As the concluding chapters of this study will make clear, such an analysis of the antivivisection movement raises rather more questions than the evidence permits answering.

Attempts to make government more responsive to public priorities during the late eighteenth and early nineteenth centuries resulted in the development of new mechanisms for mobilizing opinion and pressing for legislation. The legacy of these endeavors remains with us in the form of innumerable social action organizations, interest groups, and lobbies. It behooves us to attempt to understand their genesis, their variation, and their liability to the role of vehicle for religious, racial, regional, or intellectual prejudices, economic interests, class aspirations, and personal ambition. Furthermore, crusading movements provide fascinating historical cases for investigating the connection between intellectual or religious currents and social action.

STRUCTURE AND MEMBERSHIP

Who were the antivivisectionists? Beyond the easily recognizable two or three dozen leaders in the movement, I have been unable to locate appropriate data to support a convincing response to this question. Collective biography would require membership lists, or something more than sporadic subscription lists dominated by pseudonyms and names of animals. These have not been forthcoming.[1] One has to rely, therefore, upon considerably less satisfactory approaches—approaches that are unsystematic, impressionistic, and dependent upon the opinions of contemporaries whose personal predilections cannot always be accurately assessed. A sense of the social location of antivivisectionists must await an exploration of the structure of the movement, and the inferences and contemporary comments that may be drawn from it.

The basic institutional structure of antivivisection has been sketched in previous chapters: the major metropolitan societies, Victoria Street, the London Anti-Vivisection Society, and the International Association for the Total Suppression of Vivisection, fringed by an assortment of provincial or specialized societies such as George Jesse's Society for the Abolition of Vivisection, the Scottish Society for the Total Suppression of Vivisection, a similar society in Ireland, the Society for United Prayer against Vivisection, the Friends' Anti-Vivisection Association, the Independent Anti-Vivisection League, and the Church Anti-Vivisection League.[2] As there is little or nothing available in the way of documentation pertaining to these latter societies, the following account will focus primarily on the London societies, and especially on Victoria Street. The London societies dominated the movement, and Victoria Street was preeminent among the London societies.

Indeed, Victoria Street's preeminence was perhaps the most important single factor in the internal politics of the movement. In the domineering and dedicated Frances Power Cobbe, Victoria Street was guided by the most influential of antivivisectionist personalities. With the political power and social prestige of Shaftesbury, Lord Chief Justice Coleridge, Cardinal Manning, sundry Bishops of

[1] The National Anti-Vivisection Society, lineal descendant of the Victoria Street Society, generously made available to me its library of published works. I was refused access, however, to whatever unpublished material the Society may possess in its archives.

[2] As early as 1876, the R.S.P.C.A. estimated that there were about ten antivivisection societies in Britain. *Animal World* vii (1876), 132.

the established church, and others, Victoria Street far overshadowed its competitors. This role left Victoria Street as a prominent target for the criticism of other antivivisectionists.

Cobbe's personal ascendancy was one favorite object of negative comment. According to Edward Maitland of the International Association, Cobbe acted viciously against his friend Anna Kingsford (1846–1888) in order to prevent any possible threat to her own leadership. Kingsford was a stunningly beautiful, wilful, avant-garde mystic, who had taken a medical degree in Paris with the express purpose of combating vivisection. Upon her return to Paris in the mid-eighties, Kingsford found her overtures to a previously friendly Cobbe rebuffed. As Maitland and Kingsford saw it, Cobbe took advantage of her connections in London and of Kingsford's unconventional life-style to pass enough gossip to make the younger woman's position in London virtually untenable.[3] As Maitland noted,

> Of [Cobbe's] real motive I was in no manner of doubt, having sufficient insight into the character of the persecutrix to recognize her as capable of indulging any amount of jealousy of one whose endowments bid fair to make her a formidable rival in the cause with which Miss Cobbe had identified herself; and we had, moreover, ample testimony from others to the same effect.[4]

Other critics were two ex-secretaries of the Victoria Street Society, A. P. Childs, editor of the *Home Chronicler*, and Charles Adams, editor of the *Verulam Review*. Childs and Adams saw Cobbe as uncooperatively monopolizing for herself and her society all support and initiative within the movement and all the resulting public attention. Adams devoted an entire article to disproving the dictum that "Miss Cobbe *is* the movement and the movement is Miss Cobbe" and insisted that "any one who is moved to join may do so not only without submitting himself to her com-

[3] E. Maitland, *Anna Kingsford. Her Life Letters Diary and Work* (London, 1896), i. 443–444, ii. 26, 44, 57.

[4] Ibid. i. 444.

mands, but without any connection or contact whatever, either with her society or herself."[5] Childs maintained a continuing attack upon Victoria Street during its early years of restrictionist policy.[6] Adams wrote of Victoria Street's "old unhappy policy, forever dictated by the fear that the action of some other society might 'take the wind out of the sails of its own,' by which it has been guided from the first."[7] In 1889 he pleaded,

> The Victoria Street Society is in possession of the field . . . we would very earnestly press upon any of its members, in whose eyes the work itself is of more importance than their own glorification as workers, to do all in their power to direct its future operations in a sounder and wiser course. Above all things, to insist that that course shall no longer be directed by the petty personal jealousies and self-seeking which have been so disastrous up to now.[8]

From the institutional origins of the movement to the twentieth century, then, the atmosphere within it was soured by disharmony and disunity. Differences between various societies, publications, and individual workers were ostensibly rooted in questions of policy and tactics. But jealousy, personal ambition, and especially the desire for public attention in a leadership role were at least as important as more substantive issues in motivating the disputes that fragmented organized antivivisection—and any number of other reforming agitations. Discussion of ends and means could never be effectively disentangled from the clash of personality and ego. What was at stake was as much career as conviction.

[5] "The Anti-Vivisection Movement and Miss Cobbe," *Verulam Review* iii (1892–1893), 207–208.

[6] See, e.g., *Home Chronicler* (1878), 377–379.

[7] *Verulam Review* vii (1898), 223.

[8] Ibid. i (1888–1889), 415–416. See also iii (1892–1893), 1–5; *Anti-Vivisectionist* vii (1880), 465. Adams' relations with antivivisectionist organizations were never harmonious: the London and Scottish societies joined Victoria Street in persistently boycotting his *Verulam Review*. See *Verulam Review* v (1895), 182.

The inefficiency, dissipation of effort, and dissension resulting from the institutional and personal separation of antivivisectionists one from the other aroused frequent calls for union of all those in the movement.[9] These dated from the foundation of the major societies in 1876,[10] and intensified after the Victoria Street Society joined the other two major societies in calling for total abolition. A. P. Childs wrote in 1879:

> The formation of a settled Central League, round which the main body of Anti-Vivisectionists should rally, and to which all their means should be devoted, is becoming more and more pressing. We have, at present, one body which will not directly co-operate with another; the other having essential differences in feeling with a third. . . . the liberality of Anti-Vivisectionists keeps at least three Societies in existence, whose work not infrequently overlaps each other's.[11]

Charles Adams, while secretary of the Victoria Street Society, called for a similar federation in 1881, using the temperance reformers' United Kingdom Alliance as his model.[12] These proposals were echoed by others,[13] but their realization was only partial. Negotiations between Victoria Street and the International Association began in 1879 and ultimately resulted in amalgamation of the two societies in 1883.[14] While the International placed some of its mem-

[9] Evidence of internecine warfare within the movement was cheering to the scientific and medical interest. See, e.g., *Journal of Science* 3rd series, iv (1882), 12.

[10] See, e.g., *Home Chronicler* (1876), 155.

[11] *Anti-Vivisectionist* vi (1879), 197.

[12] *Report of the Annual Meeting of the Victoria Street Society for the Protection of Animals from Vivisection* (London, 1881), p. 4.

[13] W. Howitt, "Vivisection," *Social Notes* i (1878), 293; *Anti-Vivisectionist* vi (1879), 84–85, 376–377, 498, 531–532.

[14] *Transactions of the Victoria Street Society for the Protection of Animals from Vivisection during the half-year ending, Dec. 31, 1879* (London, 1880), pp. 3–4; *Eighth Annual Report of the Victoria Street Society for the Protection of Animals from Vivisection, united with the International Association for the Total Suppression of Vivisection* (London, 1883) p. 3.

bers in key positions, Cobbe continued to dominate the new organization. The other major antivivisection body could not be tempted into a similar arrangement. In Cobbe's words, "The London Anti-Vivisection Society, though I expended all my blandishments on it, has never consented to amalgamation. . . ."[15] A truly united movement remained an unfulfilled dream. The cause suffered accordingly.

If cooperation among existing societies was elusive, expansion into the provinces to organize the unaffiliated became an attractive alternative.[16] Mobilizing opinion outside the metropolis' intellectual and social circles was regarded as the key to successful political action. The R.S.P.C.A. had had great success with its provincial branch societies (though they were a financial drain[17]), but, while antivivisectionists found it easy to raise signatures for petitions, they learned the difficulty of establishing and maintaining consistent and active organizations outside London. Public cruelty to animals, of the type with which the R.S.P.C.A. was concerned, took place everywhere, while experimental medicine was confined to London, Oxford, Cambridge, and Edinburgh until the nineties. Victoria Street did succeed in establishing a certain number of affiliates, notably at Bristol and in spas within the orbit of the metropolis, such as Bath, Tonbridge Wells, and Brighton. It may be useful to quote Victoria Street's program for its branch societies, since the suggestions provide a good summary of antivivisectionist activity at the local level. Some of the activities of branch societies could be:

1. Asking for the votes of members of Parliament, and interviewing candidates at elections.
2. Obtaining signatures to petitions to Parliament—house to house visitation recommended.
3. Holding drawing-room meetings wherein some well-

[15] *Life of Frances Power Cobbe* (Boston and New York, 1894), ii. 603.

[16] See, e.g., *Anti-Vivisectionist* xiii (1881), 39–41.

[17] *Home Chronicler* vi (1879), 543.

informed speaker can instruct the company on the facts of the case and answer questions. (N.B. Public meetings for discussion are not advised unless the presence of a thoroughly competent medical friend can be secured).

4. Inducing the local clergy to preach against vivisection.
5. Winning the local newspapers to take up the subject; writing letters in them; and inserting, by way of paid advertisements, any information it is desired to make public.
6. Obtaining permission to place the *Zoophilist* and other publications of the Society in the Reading Rooms of Clubs, Mechanics' Institutes, Hotels, and Railway Waiting Rooms.
7. Engaging Stationers and Newsagents to exhibit such publications in their shops, and arranging that their sale be profitable to the agent.
8. Lastly, a most important use of Branch Societies is to furnish information to the Central Society of events in its own locality, and again to transmit from the Central Society the advice, information, and influence which may serve the cause in the locality.[18]

In addition to the formation of branch societies, and the recruitment of as many notables—starting with the Queen—into the movement as possible, the Victoria Street Society made overtures to many bodies whose preexisting organization and resources made them potentially valuable allies. Among such bodies were branch S.P.C.A.s, Young Men's Christian Associations, Mechanics' Institutes, Temperance Halls, antivaccination societies and religious denominations and sects. Cobbe appealed at one time or another, with varying degrees of success, to the Society of Friends,[19] the Pope,[20] "the humane Jews of England,"[21] and, last but by no means least, the established church.

[18] *Zoophilist* ii (1883), 157–158.

[19] T. B. Clark, *A Christian View of Vivisection* (London, 1889).

[20] *Memorial to His Holiness Pope Leo XIII* (London, c. 1885).

[21] M. Rubens, *Anti-Vivisection Exposed, Including a Disclosure of the Recent Attempt to Introduce Anti-Semitism into England* (Bombay, 1894).

The entire movement joined Cobbe in decrying "the inertia of the clergy" and regretting their failure to commit the church in its corporate capacity to the fight against vivisection.[22] Churchmen were taunted with their loss of moral leadership to nonconformists active in the movement.[23] According to Cobbe ". . . High Church clergymen have exhibited little interest in the agitation for the suppression of scientific cruelty, and have left the work to be done by Evangelicals, Roman Catholics, Broad Churchmen, Jews, and Unitarians."[24] It was not until 1890 that Victoria Street succeeded in establishing its affiliate, the Church Anti-Vivisection League.[25]

The links of the antivivisection movement with other English moral crusades is a second problem that is difficult to discuss with precision, short of a comprehensive, cooperative research program by students of the various agitations involved. It is possible, however, to make some relatively low level generalizations. The original prototype for such agitations, the antislavery movement, was regularly invoked inspirationally by antivivisection spokesmen. Its leaders—in both England and America—were heroes for antivivisectionists as for partisans of any number of other causes. Its methods—circulating tracts, itinerant lecturers, parliamentary pledges, and the rest—had become standard resources for its emulators before mid-century[26] and antivivisection was no exception.

What of antivivisection's links with contemporaneous agitations? In a formal sense, antivivisection cooperated with two other movements: agitation against compulsory

[22] See the many letters in the *Home Chronicler* (1876), passim, and M. Thornhill, *The Clergy and Vivisection* (London, c. 1883); F. P. Cobbe, *The Churches and Moral Questions* (London, 1889); J. Stratton, *The Churches and Vivisection* (London, 1895); G. R. Jesse, *Appeal to the Ministers of Religion* (London, 1897).

[23] *Spectator* lii (1879), 1042–1043.

[24] *Zoophilist* i (1881), 103.

[25] *First Report of the Church Anti-Vivisection League* (London, 1890).

[26] See G. Kitson Clark, *The Making of Victorian England* (University paperback, London, 1965), pp. 38, 198.

vaccination[27] and against the Contagious Diseases Acts.[28] Although both of these agitations had distinctive features that would appeal to concerns not so clearly represented by antivivisection—devotion to individual liberty in both cases and overtly feminist motives in the case of Contagious Diseases legislation—I suspect that the constituencies for these three movements were remarkably coextensive. All three appealed to the same kinds of fears of, and hostilities toward, science and medicine. Antivivisectionist spokesmen often invoked compulsory vaccination and the Contagious Diseases Acts in their indictment of the legislation for which the scientific and medical elites were responsible. One finds the same parliamentary spokesmen—Shaftesbury, Harcourt, Playfair, Forster, Mundella, Stansfeld, Duncan M'Laren, P. A. Taylor—ranged predictably on the three issues. The agitators fought or allied with many of the same medical men, they were supported or opposed by many of the same editors, and they adopted many of the same tactics. Victories achieved against the C. D. Acts and compulsory vaccination were used to bolster sagging antivivisectionist morale toward the end of the century. However, beyond mutual admiration in oratory and propaganda and the odd bit of administrative assistance, the three movements seem to have found remarkably little basis for active and continuing cooperation in practical matters. Much more

[27] On antivaccination, see R. M. MacLeod, "Law, Medicine, and Public Opinion: the Resistence to Compulsory Health Legislation 1870–1907," *Public Law* (1967), 107–128, 189–211; A. F. Beck, "Issues in the Anti-Vaccination Movement in England," *Medical History* iv (1960), 310–321; D. L. Ross, "The British Anti-Vaccination Movement in the Nineteenth Century," (unpublished University of Leicester M.Sc. thesis, 1968). On the agitation against the Contagious Diseases Acts, see G. Petrie, *A Singular Iniquity. The Campaigns of Josephine Butler* (New York, 1971); F. B. Smith, "Ethics and Disease in the Later Nineteenth Century: The Contagious Diseases Acts," *Historical Studies* (Melbourne) xv (1971), 118–135; E. M. Bell, *Josephine Butler. Flame of Fire* (London, 1962); J. L. and B. Hammond, *James Stansfeld. A Victorian Champion of Sex Equality* (London, 1932).

[28] G. W. and L. A. Johnson, (eds.), *Josephine E. Butler. An Autobiographical Memoir* (3rd ed. Bristol, 1928), pp. 105–113.

important was the spiritual affinity. The movements struggled against professional solidarity, political expediency, and bourgeois morality to bring issues of health and medical practice out of the hospital, the clinic,—and dark corners of the Victorian psyche—onto public platforms, into editorial columns, where the voice of the community could be heard. All three saw themselves as arousing the basic moral instincts of laymen against an arrogant coalition of scientists, medical men, and legislators, who were blindly following the dictates of technique into an ethical cul de sac, where beneficent ends failed to justify horrid and repugnant means. All three battled a "materialism which sets the body above the soul, profaning the sacred name of science" and pleaded for "a return to God's laws in place of these brutally impure laws invented and imposed by man."[28] The linkage between these movements is a fascinating phenomenon, mirrored in contemporary America, where the same people are found opposing a number of different health programs: water flouridation, mental health, compulsory vaccination, antirabies inoculation for pets, and so forth.[29]

Another movement with which one would expect antvivisection to have a great deal in common would be vegetarianism. While certain leading antivivisectionists—notably Anna Kingsford, Edward Maitland, and F. W. Newman—were also vegetarians, and while antivivisection is implicit in the idea of vegetarianism, the two movements never achieved a close relationship. Vegetarianism was a movement explicitly dedicated to a reconstitution of society and life-styles, and antivivisection was first and foremost an agitation with a practical, limited, legislative objective. Thus, the modes of operation of vegetarianism and antivivisection were different, with the former less absorbed than the latter in such political activity as lobbying and electoral pledging. Some antivivisectionists, such as Frances Power Cobbe,

[29] J. Marmor, V. W. Bernard, and P. Ottenburg, "Psychodynamics of Group Opposition to Health Programs," *American Journal of Orthopsychiatry* xxx (1960), 330–345, esp. 338. I am grateful to Arthur Viseltear for directing me to this study.

were outrightly hostile toward vegetarianism.[30] I suspect this hostility arose from fears that public identification of the two movements would lend general credence to spokesmen for experimental medicine who claimed that one could not consistently oppose the use of living animals for experiment unless one also opposed their slaughter for the purposes of food. In any case, there were tensions in the arm's-length relationship between antivivisection and vegetarianism, as when Maitland, Newman, and Kingsford protested vehemently against any involvement by organized antivivisection in the question of slaughter house reform.[31]

As to other movements, antivivisectionists made passing allusions in published letters and speeches to any number of causes close to their hearts: sabbatarianism, temperance, social purity, protection of children, feminism, ragged schools, magdalen homes, penal reform, pacifism, . . . ; the list is lengthy.[32] Unfortunately, such information is not available with enough consistency to make tabulation worthwhile. Patterns of interest in reform movements are particularly useful in indicating the latent functions served in common by the movements concerned: for example, the function of antivivisection, antivaccination, and agitation against the Contagious Diseases Acts as vehicles for hostility toward medicine. Even where such patterns of interest cannot be discerned, the phenomenon of simultaneous allegiance to a number of different causes in itself indicates the existence of an important and underinvestigated problem, the moral reform mentality in Victorian England. Antivivisection, like other causes, no doubt attracted a significant number of individuals for whom participation in the movement was in itself at least as important as the possibility of achieving ostensible objectives.

Expansion into the provinces and alliance with existing

[30] See, e.g., E. Maitland, *Anna Kingsford. Her Life Letters Diary and Work* (London, 1896), ii. 116.

[31] *Home Chronicler* (1878), 380, 396.

[32] F. W. Newman is an excellent, if extreme, example: I. G. Sieveking, *Memoir and Letters of Francis W. Newman* (London, 1909).

societies and institutions were appropriate tactics for the antivivisection movement within Britain, but outside the country, where the vast majority of experiments upon living animals were performed, its task became that of reproducing itself. Animal protection arose in England well before it took hold on the continent of Europe. The same was true for antivivisection. But while animal protection came in many European countries to exercise some semblance of the influence that the R.S.P.C.A. wielded in Britain, continental antivivisection never rose to such heights. British missions against French veterinary vivisection during the sixties were mentioned earlier, and from this time onward, as Bretschneider's study makes clear, British citizens and the British example were crucial for continental antivivisection.[33]

Personal intervention was one means of fighting continental vivisection. In 1878, a British veterinarian named Cowie toured France, Italy, and Switzerland at his own expense, laying the case for regulation of vivisection before courts and governments, and leaving translations of the Act of 1876 with the officials whom he waited upon.[34] Anna Kingsford fought vivisection in Paris even as she studied medicine there. She and her companion Edward Maitland were active both in France and Switzerland. They published pamphlets and attempted unsuccessfully to mobilize the Catholic Church.[35]

Maitland and Kingsford were affiliated with the International Association; international institutional links were fostered in an attempt to increase the power of the movement. Here Victoria Street was preeminent, founding a branch society in Paris,[36] and entering into correspondence

[33] H. Bretschneider, *Der Streit um die Vivisektion im 19. Jahrhundert* (Stuttgart, 1962).

[34] W. Howitt, "Mission of Mr. Cowie, M.R.C.V.S., on the Continent," *Social Notes* i (1878), 387–389.

[35] E. Maitland, *Anna Kingsford. Her Life Letters Diary and Work* (London, 1896), i. 79–85, 90–91, 302–304, 321–324; ii. 79, 117–120. Cf. H. Kleffler, *Les missions "humanitaires" des dames anglaises sur la continent. La vivisection, son utilité, sa morale* (Geneva, 1883).

[36] *Dates of the Principal Events connected with the Anti-Vivisection Movement* (London, c. 1897), p. 5.

with a large number of antivivisectionist or sympathetic animal protection societies in Western Europe, America, and India. Victoria Street distributed pamphlets in French, German, Italian, Swedish, and Danish, and published a short-lived French language version of its magazine, the *Zoophilist,* called *Le Zoophile.*[37] Victoria Street's connections with Denmark and Sweden were particularly strong; some Scandinavian antivivisectionists gave the society important financial support, while others were among the leaders of the movement in Britain in the early twentieth century.

Probably the most important aid that British antivivisection gave to its continental allies was simply its example. Europeans modeled their movement after British organization and tactics, which were often referred to as the inspiration and the ideal to be striven for. As German physiologist Ludwig Hermann wrote in 1876 to his British colleague, John Burdon Sanderson:

> The anti-vivisection wave, rising from your country, approaches ours with the velocity of a railway train. We also have societies for prevention of cruelty to animals, whom the glory of their english sisters forbids to sleep.[38]

British antivivisectionists also sought to bring their special concern prominently before the animal protection movement as a whole, by enthusiastic lobbying at the international congresses for the prevention of cruelty to animals. Victoria Street sent its then honorary secretary, Dr. George Hoggan, to the Paris congress in 1878. There he found the scientific and medical interest well entrenched in Parisian animal protection circles.[39] Just as in England, antivivi-

[37] Ibid., p. 6, and F. P. Cobbe (ed.), *Bernard's Martyrs. A Comment on Claude Bernard's Leçons de Physiologie Operatoire* (London, 1879), endplate. See also *Transactions of the Victoria Street Society for the Protection of Animals from Vivisection during the half-year ending, Dec. 31, 1879* (London, 1880), pp. 9–12, 27–34, and *Life of Frances Power Cobbe,* ii. 603–604.

[38] L. Hermann to J. S. Burdon Sanderson, 9 July 1876, The Library, University College, London, Burdon Sanderson Papers, MS Add. 179/2 f.64.

[39] *Public Opinion* xxxiv (1878), 207.

sectionists on the continent never succeeded in developing the support they wished for animal protection societies. They believed that medical scientists were deliberately infiltrating the societies with the intention of sabotaging any efforts to move against vivisection.[40] A Victoria Street survey of German animal protection societies in 1881 disclosed that of those that had taken a position on vivisection, a majority were in favor of the practice.[41] Two years later, another survey showed that of the 506 such societies in the world, only 142 were "decided" antivivisectionists.[42]

Antivivisectionist efforts on the continent met with far less success than in Britain, although at present vivisection is illegal in Liechtenstein and is regulated in Austria, Denmark, Norway, Sweden, Poland, and certain parts of Switzerland. Regulation is also in effect in parts of Australia, the Bahamas, Jamaica, and Argentina.[43] It is quite clear that these statutes are in part the legacy of the original British movement, aggressively expanding its geographical purview. Despite acrimonious debates in the Reichstag, however, no regulation was achieved in Germany, nor in France or Italy. It was these three countries that were the locus of most of the activity in experimental medicine, and hence they were primary targets.

During its first twenty years, the British movement paid only sporadic attention to its American counterpart. To the best of my knowledge, American antivivisection has not been studied in any depth. It seems to have been more fragmented than in Britain, focusing upon state legislatures until the nineties. During the last decade of the century, a homeopath-politician, Senator Gallinger, introduced a number of antivivisection bills into the Congress. Some antivivisectionist sentiment existed because of the activities of the societies and the support of *Life* magazine. The

[40] C. Landsteiner, "The Right Position in the Question of Vivisection," *Animal World* xxvii (1896), 94–95.

[41] *Zoophilist* i (1881), 198–200.

[42] Ibid. iv (1884), 6–7.

[43] G. Lapage, *Achievement. Some Contributions of Animal Experiment to the Conquest of Disease* (Cambridge, 1960), p. 224.

threat from Gallinger—while apparently the most serious during the formative period of experimental medicine in the U.S.—was effectively counteracted by the activities of eminent medical leaders such as William Osler and William H. Welch.[44]

Implicit in any analysis of antivivisection must be the question of its unique success in Britain as compared with comparative failure elsewhere, especially on the continent. The reasons are social, intellectual, and religious (of which only the first need detain us here, for the latter two will be explored in a later chapter). Indeed, the most acute contemporary analyses of the social and intellectual background to the movement were framed in precisely this context. One cannot, now or then, point to a single factor to account for the peculiar flowering of antivivisection in Britain. The movement was very much a product of its later Victorian milieu, very much the result of the coincidence of some quite different forces. Baron Ernst von Weber, a leading German antivivisectionist, accounted for the strength of the British movement,

1. By the religious tendency of the English nation . . . ;
2. By the warm sympathy of the aristocracy and the clergy;
3. By the influence of cultivated women, which is much greater than with us;
4. By the great powers of the press and public opinion.

According to von Weber, Britons were decidedly of the opinion that science must be subject to higher considerations of humanity and morality, while in Germany and France the "superstitious reverence for science" prevailed. He went on to give an appreciative summary of British tactics of agitation.[45]

[44] H. Cushing, *The Life of Sir William Osler* (Oxford, 1925) i. 397–398, 434, 449–451, 520–522.

[45] E. von Weber, *The Torture Chamber of Science* (6th ed., London, c. 1880), pp. 3, 11–14, 28, or see *Animal World* ix (1878), 126.

Of the reasons advanced by von Weber, two are most germane for present purposes: the role of the aristocracy, or rather, what can be inferred about the class relations of the movement; and the role of women. In the absence of membership lists or other reliable indicators, any analysis of the class support for antivivisection must be extremely tentative.

Most rank and file antivivisectionists seem to have come from the urban middle class or the clergy; often, they were professional, semiprofessional, or living on an income from a legacy, pension, or sinecure. They certainly identified themselves with the aristocracy, in the tradition of Tory paternalism (and in sharp opposition to utilitarian ideology):

> In England, for instance, the leaders of public opinion have been, and are still, the aristocracy and the quasi-aristocratic classes. The Anti-Vivisectionist movement having originated among these classes, its chances of success are decidedly good,—the more so, since the class of *savants* which forms the resistant medium, is neither very numerous nor alarmingly popular.[46]

In these terms, then, the average antivivisectionist could picture himself as the "quasi-aristocratic" upholder of moral responsibility and noblesse oblige against the vulgar careerism, materialism, utilitarianism, and even "trades-unionism" of the provivisectionist medical profession.

Conditions were otherwise in Germany. There too, the aristocracy was antivivisectionist (went Victoria Street's analysis), but

> The real leaders of the German people are, at present, the devotees of science [not the classically educated] but the working men of the Laboratory, the explorers, tamers and improvers of Nature. . . . They are . . . looked up to and venerated by every class of society, but more particularly by the great mass of half-educated people whose faith in the omnipotence of science seems boundless, and often expresses itself with unbecoming fervour.

[46] *Zoophilist* ii (1883), 23.

Since the German press was said to share uniformly in the public veneration for science, the task of antivivisection in Germany was difficult indeed. According to the *Zoophilist,* the identification of the aristocracy with political conservatism and religious orthodoxy made it easy for provivisectionists to represent the question in terms of partisan politics. Thus discussion of the question in the Reichstag "degenerated . . . into a regular party skirmish." In France, the movement was hampered by the general indifference of the Parisian literati, who were the leaders of public opinion in that country.[47] The French were unsympathetic to *Le Zoophile.* It was too relentlessly antivivisectionist and too blatantly pietistic in its appeal. As its frustrated editor, Frances Power Cobbe, remarked with some asperity, "The idea is, apparently, yet unknown in France of a serious journal serving as an organ for serious workers in a serious moral or religious agitation."[48] So both Germany and France, the two major centers of experimental medicine, lacked elements that were vital to the comparative success of antivivisection in Britain.

The enthusiastically promoted veneer of aristocracy over an essentially middle-class base corresponded fairly closely to the situation of the R.S.P.C.A., as also did the important role of women. The R.S.P.C.A., however, was much more successful in the provinces than antivivisection. I believe that the middle class partisans of antivivisection were generally those who supported antivaccination and the agitation against the Contagious Diseases Acts, but in one important respect antivivisection differed from its two sister movements. While antivivisection was London-dominated and in partnership with portions of the aristocracy, the other two movements gained crucial support from the provinces and were in partnership with the socially-conscious working class, the labor aristocracy. Experimental medicine was confined until fairly late in the century largely to the major centers of London, Edinburgh, Cambridge, and Oxford, whereas animal protection and compulsory vaccina-

[47] Ibid., 24.

[48] Ibid. iii (1884), 285.

tion had implications everywhere, and the seaports and garrison towns regulated by the Contagious Diseases Acts were all in the provinces. Thus, only antivivisection failed to transcend the metropolis. Furthermore, the animal protection stand of antivivisection was undoubtedly linked with the idea of working class offenses prosecuted by the R.S.P.C.A., while the Contagious Diseases Acts and compulsory vaccination were portrayed by their critics as a scandalous interference with the rights of women and children of the working class. Thus the working class could be mobilized most effectively for the latter movements.

Antivivisectionist failure was not for lack of trying. It was in the course of its search for electoral support that the antivivisection movement discovered the working class and began spasmodic attempts to court the support of the newly enfranchised sections of it. Pamphlets and flyers were directed toward working men, mostly in a tone of righteous and fiery oratory or outright condescension.[49] Meetings held in working class institutions, such as the Artisans' Institute, typically featured speeches interspersed with songs, tea, and cakes.[50] In 1878, the Working Men's Association for the Suppression of Vivisection was founded—largely by clergymen.[51] One difficulty in organizing the working class arose over the question of scheduling meetings. Most working men had neither the time nor the inclination to attend unless such meetings were on Sunday evenings. Sunday gatherings were arranged in the face of loud protests by Sabbatarians within the movement and only when additionally justified by the pious hope that they would prove "a turning point for good" in the lives of working men.[52] The ultimate stumbling block emerged in the profound

[49] See, e.g., R. Barlow-Kennett, *Address to the Working Classes* (London, c. 1885); M. P., *Vivisection in the United Kingdom. An Appeal to Working Men* (London, 1893); *A Burning Question* (London, n.d.).

[50] *Home Chronicler* (1877), 489–491; *Zoophilist* v (1885), 180.

[51] *Anti-Vivisectionist* (1878), 300–301; [M. E. Wemyss], *Licensed Cruelty* (Gloucester, c. 1886).

[52] See *Anti-Vivisectionist* vi (1879), 3–5, and the controversy that follows.

indifference of laborers toward the issue. Despite repeated attempts to represent animal experiment as the cause of the experimentation allegedly carried out upon the laboring poor in charity hospitals,[53] working class antivivisection was no more than an abortive dream. One working class convert reported sadly that he could persuade no more than nine of the one hundred men in his factory to sign an antivivisection petition.[54] Those responsible for the supposed "cruelty in fustian" were ill-disposed to aid the crusade against "cruelty in broadcloth." The priorities of the labor aristocracy no doubt lay elsewhere than in antivivisection, notwithstanding Edward Carpenter's portrayal of animal experiment "as the irrelevant and cruel means by which capitalists sought to cure diseases resulting from their own firms' pollution and disruption of the human environment."[55]

WOMEN AND ANTIVIVISECTION

As any number of contemporaries remarked, the antivivisection movement was sustained by the support of women from the middle and upper classes, from Frances Power Cobbe, to Anna Kingsford, to the titled women Cobbe recruited as Victoria Street patronesses, to the female contributors named in subscription lists. An examination of the prospectuses of antivivisection societies shows that their leadership—patrons, vice presidents, and executive committee members—was anywhere from 40 to 60 per cent female. A great deal more research would be required to establish the relative levels of female participation in various other humanitarian movements. If the testimony of contemporaries is to be trusted, however, levels of female participation in the antivivisection movement would seem to have

[53] See, e.g., note 49 above and *Home Chronicler* (1877), 552, 792–793. A. P. Childs placed advertisements publicizing alleged medical mistreatment of working class patients in the *Industrial Review.*

[54] *Home Chronicler* (1877), 542.

[55] B. Harrison, "Religion and Recreation in Nineteenth Century England," *Past and Present* xxxviii (1967), 119, referring to E. Carpenter, "Vivisection and the Labour Movement," *Humanity,* (Nov. 1895), 68.

been among the very highest for movements without overtly feminist objectives.

While 470 of the R.S.P.C.A.'s 739 nineteenth-century legacies were from women,[56] one would have been much less likely to say of the R.S.P.C.A., as was said of antivivisection, that "by far the greater number of workers in the cause were ladies" or that "their cause was a woman's question."[57] According to the *Verulam Review,*

> The enormous majority of the women of England are unquestionably on our side. Were it not for them, in truth, the movement could hardly be carried on at all.[58]

Three-quarters of the audience at an annual meeting of the International Association for the Total Suppression of Vivisection were female.[59] Indeed the movement was frequently stigmatized because so many of its adherents were women.[60]

The presence of women in the movement is also visible in some of the tactics advocated. The correspondence columns of the movement's periodicals rang with appeals to the united strength of women and to the benign influence of their tender sympathies. In particular, women were expected to exercise social sanctions against medical men who refused to commit themselves against vivisection. Some women suggested that their sex could do without doctors entirely. Alternatively, they could enter the profession to strengthen its humane aspects.[61] In any case, women were to stand forth publicly on behalf of the cause. Enough has been said, I think, to show that a very significant number did indeed come forward.

[56] B. Harrison, ibid., 108.

[57] *Home Chronicler* (1878), 412.

[58] *Verulam Review* iv (1894), 113. See also E. von Weber, *The Torture Chamber of Science* (6th ed., London, c. 1880), pp. 3, 28; A. Arnold, *Cruelty is Crime* (London, 1894), p. 3.

[59] *Public Opinion* xxxiii (1878), 814.

[60] M. Thornhill, *The Clergy and Vivisection* (London, c. 1883), p. 6.

[61] See, e.g., E. Blackwell, *Erroneous Method in Medical Education* (London, 1891), pp. 7–8.

The example of women, such as Josephine Butler, speaking at open meetings on subjects of deep contention, such as the plight of "fallen women" and contagious diseases legislation, seemed to many Victorians to constitute a great threat to feminine ideals and a perversion of women's place in society. These ideals could only be further sullied by the movement for the emancipation and enfranchisement of women. By subjecting themselves to the storms and temptations of a public role, women were thought by some to be compromising themselves, to be rendering themselves unfit to play their all-important part as the apotheosis of moral purity. Such critics of the drive for women's rights and of women's participation in public affairs were obviously arguing on the basis both of a stereotype of sexual qualities and a conception of women's social role, which confined her fairly closely to domestic duties. The idealized Victorian household, a haven of peace amid the storm of affairs, reigned over by the angelic maternal spirit, is a familiar enough phenomenon to students of the period. So is the stereotype of sexual qualities that went with it: the intelligent, aggressive, calculating male, conqueror and provider, contrasted with the submissive, pious, and morally superior female, the instinctual, emotional queen of the domestic circle. The antifeminist case was very often rationalized with reference to this popular Victorian mythology.[62]

As with much else characteristic of antivivisection, the important role of women in it may in part be traced to Frances Power Cobbe. Cobbe developed an ideology of female activism that took advantage of contemporary mythology and amounted to a skillfully contrived attempt to draw women into antivivisection (and other philanthropic and moral reform activities) from both the conservative and radical extremes of female activism.

[62] See W. Houghton, *The Victorian Frame of Mind* (New Haven 1957), pp. 451–453; J. A. and O. Banks, *Feminism and Family Planning in Victorian England* (Liverpool, 1964), esp. pp. 25, 46–48, 108–109; and W. Arnstein, "Votes for Women: Myths and Reality," *History Today* xviii (1968), 533–534.

According to two of the movement's leading historians, feminism "may be seen to include literary men and women advocating sex equality, and organized bodies of active workers for the cause."[63] Cobbe qualifies on both counts. During the sixties and seventies, she threw herself wholeheartedly into the movement, writing and lobbying for higher education, enfranchisement, entry into the professions, and legal protection for the rights of married women.[64] Cobbe was not, however, simply a feminist. She was a philanthropist and moral reformer with a pluralistic conception of the potential roles of women.

Cobbe's ideology of female activism began with premises indistinguishable from the sexual stereotypes so cherished by conservatives. She believed that women, although less intelligent and of course physically weaker than men, were more sympathetic, more direct and practical, less hypocritical, more facile in self-expression, and most importantly, more religious and more moral than their masculine counterparts.[65] In consequence, women's mission was the moral and spiritual regeneration of society. So far, so typical. At this point, Cobbe began her indictment of current social mores: the latter did not protect women's pristine virtues but perverted them into "petty social ambition or sordid matrimonial scheming," to which women had been "trained and consigned from girlhood, on the principle of 'keeping women in their proper sphere.'"[66] Indeed, it was precisely because of women's superior moral qualities that they should have wider opportunities outside the home, for in this area of greatest need, such qualities would find an appropriate sphere of usefulness. Women would find their natural place, "in some gratuitous labour of love for the poor, the sick, the ignorant, the blind; for animals; in short, in any cause of humanity; but, above all, in the cause

[63] J. A. and O. Banks, op. cit., note 2, pp. 8–11. Quote from p. 10.

[64] F. P. Cobbe, *Life of Frances Power Cobbe* (Boston and New York, 1894), ii. 525–555; *Duties of Women* (Boston, 1881).

[65] See esp. F. P. Cobbe, "The Fitness of Women for the Ministry of Religion," in her *The Peak in Darien* (Boston, 1882), pp. 199–262.

[66] F. P. Cobbe, ibid., pp. 247–248.

of their own sex, and the relief of the misery of their own sisters." In fact,

> . . . I think that every woman who has any margin of time or money to spare should adopt some one public interest, some philanthropic undertaking, or some special agitation of reform, and give to that cause whatever time and work she may be able to afford; thus completing her life by adding to her private duties the noble effort to advance God's kingdom beyond the bounds of her home.[67]

Cobbe here provided a mandate for everything from discreet parish visiting to impassioned oratory from the dais of Exeter Hall or lobbying in the House of Commons, without explicitly insisting on anything more controversial than the first. Furthermore, if women's unique qualities could be used to justify rather than to deny an expanded role for women in public life, such an argument logically extended to the vote: denial of the franchise to qualified women was the community's moral loss.[68]

What of the corrupting and worldly influences that were abroad, ready to waylay the blameless purity of hitherto sheltered innocence? Cobbe denied outright the antifeminist insistence on the inevitability of such corruption, though she readily admitted the power of the argument if it were valid:

> . . . greatly as I desire to see the enfranchisement and the elevation of women, I consider even that object subordinate to the moral character of each individual woman.

[67] F. P. Cobbe, *The Duties of Women. A Course of Lectures* (Boston, 1881), pp. 186–189. See also Cobbe's "What Shall We Do with Our Old Maids?," *Fraser's Magazine* lxvi (1862), 594–610, and "Female Charity, Lay and Monastic," *Fraser's Magazine* lxvi (1862), 774–788. According to a contemporary assessment by Georgiana Hill, charity and philanthropy were "at bottom women's work" and "women are the mainspring of the humanitarian movement of the nineteenth century." G. Hill, *Women in English Life from Medieval to Modern Times* (London, 1896), ii. 231.

[68] F. P. Cobbe, *Why Women Desire the Franchise* (London, n.d.), p. 9; *The Duties of Women* (Boston, 1881), pp. 5–7.

> If women were to become less *dutiful* by being enfranchised,—less conscientious, less unselfish, less temperate, less chaste,—then I should say: "For Heaven's sake, let us stay where we are! *Nothing* we can ever gain would be worth such a loss". But I have yet to learn that freedom, which is the spring of all the nobler virtues in man, will be less the ground of loftier and purer virtues in women.[69]

The purchase of political rights at the cost of "womanliness" would be far too dear, wrote Cobbe to an American women's periodical; "I think nearly every one of the leaders of our party in America and England agrees with me in this feeling."[70] She warned that confusions of emancipation with looser moral standards or unconventionality in life style were "deadly perils to the whole movement for the advancement of women"[71] whose mission was, after all, "to make society *more* pure, *more* free from vice, either masculine or feminine, than it has ever been before."[72] Thus, Cobbe—and no doubt a number of other feminists—propounded an ideology of female activism designed to woo the moderate and the undecided by emphasizing respectability and a kind of moral hyperconventionality. She based her case on widely held stereotypes of sexual qualities, often used by opponents of female participation in public life, and hitched it to that favorite Victorian shibboleth, moral progress. Cobbe believed that once women made a sincere commitment to the attack on some particular social problem, they would be struck by women's impotence in civil affairs, and be converted to feminism. This was exactly what had happened to her.[73] The intention was

[69] *The Duties of Women*, pp. 11–12.

[70] F. P. Cobbe, *Life of Frances Power Cobbe* (Boston, 1894), ii. 533.

[71] *The Duties of Women*, pp. 8–9. Most of the Victorian feminists agreed. See J. A. and O. Banks, op. cit., esp. pp. 105–106, 133; and C. Rover, *Love, Morals, and the Feminists* (London, 1970) esp. pp. 55–56.

[72] *The Duties of Women*, p. 9, Cf. C. Rover, op. cit., p. 55.

[73] "It was only after I had laboured for some time with my honored friend Mary Carpenter, at Bristol, and learned to feel intense interest

to nudge women into some philanthropic or charitable work, however minor, whence they would be inevitably drawn into moral reform and a newly emerged feminist movement.[74]

Yet Cobbe also developed a rationale to appeal to feminists who were frustrated or disheartened by the relatively slow progress toward the goals of the movement itself. She held out to such women the alternative of other moral reform or philanthropic activity by insisting that "Every woman who works wisely and well for any good public cause (whether the cause directly concern female interests or not) does her share in thus lifting up the womanhood of the nation."[75] Such nonfeminist activism was functional for feminism because it permitted a deemphasis of the noisy demanding of rights, which was a feature of the organized movement, while publicly demonstrating female capabilities, and providing women with unprecedented opportunities for administrative and quasipolitical experience. There is excellent evidence that Cobbe was not alone in her evaluation of public female activity vis-a-vis women's rights.[76] Within the antivivisection movement, Anna Kingsford saw greater opportunity for advancing women's pros-

---

in the legislation which might possibly mitigate the evils of crime and pauperism, that I seriously asked myself . . . *why* I should not seek for political representation as the direct and natural means of aiding every reform I had at heart." *The Duties of Women,* p. 6.

[74] As McGregor has pointed out, "many who contributed to the investigation, definition, and treatment of the social results of industrialism were driven into feminism because they found their usefulness limited and their work frustrated by the socially imposed disabilities of sex." O. R. McGregor, "The Social Position of Women in England, 1850–1914; A Bibliography," *British Journal of Sociology* vi (1955), 56. Cf. C. Rover, op. cit., p. 50.

[75] *The Duties of Women*, p. 11.

[76] According to Klein, "Not a few of these women were aware of the fact that, while they were fighting poverty, slavery, and disease, they were, at the same time, fighting in the cause of women. And many of them had consciously accepted this as a secondary aim . . . they preferred their method of missionary work to the political and journalistic activities of their feminist sisters." V. Klein, *The Feminine Character, History of an Ideology* (New York, 1949), p. 19.

pects than in explicit feminism. According to her companion Edward Maitland:

> Though sympathising to the last in the movement for the enfranchisement of women, she did not long continue to take an active part in it. The reasons for her withdrawal were manifold. One was her conviction that women would more successfully achieve their desired emancipation by demonstrating their capacity for serious work than by merely clamouring for freedom and power. . . . It was by magnifying their womanhood, and not by exchanging it for a factitious masculinity, that she would have her sex obtain its proper recognition.[77]

In one way or another, then, feminism was important for the recruitment of women into philanthropic movements.[78] and these movements provided valuable experience and collective confidence vital to the long term interests of women's rights.[79] In the words of one historian, "the humanitarian interests which formed the starting point of social research, and practical social work itself, actually provided the back-door through which women slipped into public life."[80]

[77] E. Maitland, *Anna Kingsford. Her Life Letters Diary and Work* (London, 1896), i. 19. Note the emphasis on the distinct qualities of either sex. Kingsford disagreed with Cobbe in that she believed overt feminism *did* "harden" or "unsex" women.

[78] See K. Heasman, *Evangelicals in Action. An Appraisal of their Social Work in the Victorian Era* (London, 1962), p. 11 and O. R. McGregor, op. cit., p. 58.

[79] In Georgiana Hill's opinion, "The prominent part taken by women in the modern humanitarian movement is remarkable in two ways. The effect on the women themselves is hardly less striking than the effect on the work in which they are engaged. Women have developed an unexpected capacity for organization, an enterprise in arduous undertakings, and an enthusiasm for difficult, disagreeable, and uncompromising work." G. Hill, op. cit., ii. 227.

[80] V. Klein, op. cit., pp. 16–19. Quotation is from p. 17. See also R. Fulford, *Votes for Women* (London, 1957), p. 19, and cf. O. Anderson, "Women Preachers in Mid-Victorian Britain: Some Reflections on Feminism, Popular Religion, and Social Change," *Historical Journal* xii (1969), 482–483.

Thus social circumstances in general, and Frances Power Cobbe's ideology of female activism in particular, might have been expected to attract women into an antivivisection movement led by Cobbe. There can be no doubt that she herself felt that antivivisection was a cause to which women's unique qualities would make a very important contribution. R. H. Hutton was clearly responding to such a suggestion from Cobbe when he wrote to her:

> I am not a woman's suffrage man and I doubt women being stronger in the matter than men. A least I know one clear [?] and very sensible [?] woman who vehemently defends the practice without being either connected with doctors or herself physiological.[81]

What other factors can be suggested to account for the proportion of women in the antivivisection movement, which so impressed contemporaries? The almost total lack of any scientific education for women certainly contributed.[82] So too did female resentment toward a medical profession that systematically sought to exclude women from its ranks. Cobbe was bitter at what she conceived to be her own mistreatment by doctors,[83] and she saw the profession as "doubly treacherous" to women.[84] The attempt by the profession to maintain masculine exclusivity within it was one of her favorite targets.[85] Many Victorian women were humiliated by the necessity to expose themselves or confide intimate details to male doctors, and they flocked

[81] R. H. Hutton to F. P. Cobbe, 15 March 1875, Henry E. Huntington Library, Frances Power Cobbe Papers, CB.509.

[82] See, e.g., L. E. Becker, "On the Study of Science by Women," *Contemporary Review* x (1869), 386–404; J. Stuart, "The Teaching of Science," in J. E. Butler (ed.), *Woman's Work and Woman's Culture* (London, 1869), pp. 121–151.

[83] *Life of Frances Power Cobbe* (Boston, 1894), i. 310–314.

[84] [F. P. Cobbe], "The Medical Profession and its Morality," *Modern Review* ii (1881), 311, 321–323, 325.

[85] F. P. Cobbe, "What Shall We Do with Our Old Maids?," *Fraser's Magazine* lxvi (1862), 608–609.

to patronize the first female practitioners.[86] These feelings found a natural outlet in antivaccination, antivivisection, and agitation against Contagious Diseases legislation. Finally, women were probably more susceptible than men to the kinds of attitudes toward animals that will be explored in Chapter 11.

That the large and leading role of women in the movement had consequences for feminism was recognized by the feminist *Englishwoman's Review,* which printed a sympathetic article on antivivisection in 1884.[87] In a similar vein, the antivivisectionist *Home Chronicler* ran, under the title "Woman and Woman. A Sketch from Life," a contrived dialogue between a scandalized aunt and her antivivisectionist niece, the former remonstrating with the latter for speaking out in such an "unlady-like and unfeminine" way to men of deep and superior intellect, especially to the Rev. Mr. Plausible, whose views in favour of vivisection are well kown.[88] The *Home Chronicler* was only too willing to foster such a manifestation of the emancipation of women. Frances Power Cobbe perceived that the predominance of women in the movement left it liable to dismissal as so much feminine hysteria. She urged women in the movement not to cry in public nor indulge in mere sentimentality, but rather to denounce sin bravely and forthrightly.[89] Cobbe's strictures and her attempts to fashion a publicly appealing style for antivivisectionist agitation failed; emotionalism, floods of tears, and fainting at antivivisectionist meetings alienated more than it converted. A few proponents of experimental medicine sought to take maximum advantage of popular

[86] B. Harrison, "Underneath the Victorians," *Victorian Studies* x (1966–1967), 254–255; J. Manton, *Elizabeth Garrett Anderson* (London, 1965), pp. 226, 243–244, 264, 286; G. Petrie, *A Singular Iniquity. The Campaigns of Josephine Butler* (New York, 1971), pp. 65–66; E. M. Bell, *Josephine Butler. Flame of Fire* (London, 1962), pp. 45–46.

[87] *Englishwoman's Review* (15 October 1884), extracted in *Zoophilist* iv (1884), 135, under the title "Women's Share in the Anti-Vivisection Movement."

[88] *Home Chronicler* (1878), 150–151.

[89] F. P. Cobbe, *A Charity and a Controversy* (London, 1889), p. 3.

prejudices and aversion to women in public affairs. Elie de Cyon, a physiologist from the continent, crudely impugned the motives of women in the movement in an article commissioned by the *Contemporary Review:*

> Is it necessary to repeat that women—or rather, old maids—form the most numerous contingent of this group? Let my adversaries contradict me, if they can show among the leaders of the agitation one young girl, rich, beautiful, and beloved, or some young wife who has found in her home the full satisfaction of her affections![90]

De Cyon had a similar explanation for the success of antivivisection in Protestant countries and its failure in Catholic ones:

> . . . Catholicism opens to old maids of excited imagination a refuge in its convents. The ecstatic adoration of the Heart of Jesus or the Blessed Virgin offers sufficient food for the mysticism of these disordered minds. An enthusiasm of piety acts as powerful diversion to the explosions of a morbid nervous condition. For want of a similar resource, the Protestant old maids fling themselves into the mysteries of Spiritism, or give themselves up to a fantastic charity rarely directed to any worthy object.[91]

In 1892, repeated attempts by antivivisectionists to have their cause aired before a Church Congress finally bore fruit. Unfortunately, Victor Horsley, a neurosurgeon, turned the occasion into a stunning victory for experimental medicine when he convincingly demonstrated that a major Victoria Street publication, *The Nine Circles of the Hell of the Innocent*, regularly failed to mention the use of anesthetics in the "brutal" experiments it cited chapter and verse from research literature.[92] The book had a preface by Cobbe, whose moral responsibility for the whole produc-

[90] E. de Cyon, "The Anti-Vivisectionist Agitation," *Contemporary Review* xliii (1883), 506.

[91] Ibid., 509–510.

[92] G. M. Rhodes (ed.), *The Nine Circles of the Hell of the Innocent. Described from the Reports of the Presiding Spirits* (London, 1892). A revised edition was later published.

tion rendered her liable to some very rough treatment at the hands of Horsley, no man to mince words. The resulting public controversy very seriously damaged the credibility of both the movement and Cobbe.[93] The spectacle of a man publicly accusing a woman of deliberate falsehood brought home to the public particularly forcefully the new role of women in the nation's affairs. Where now was the superior moral sense of women? How could they retain their fragile and precious femininity in such storms as these? Ernest Hart of the *B.M.J.* took the occasion to lambast Cobbe: was this perversion of truth to be discounted as a "privilege of womanhood"?[94] *Punch* summarized the rather pathetic end to Cobbe's public career, taking off on antivivisectionist distress at Horsley's blunt language:

> *Miss Fanny* (a gentle and most veracious child)—"Yah! you cruel coward! You and your friends skinned a live frog!"
> *Master Victor* (an industrious but very touchy little boy)—"You're a liar! The frog was dead, and *you know it!*"
> *Miss Fanny*—"Boohoo! Whether it was dead or not, you've got no right to call names; 'cos I'm a girl and can't punch your head!"
> *Master Victor*—"Its just because you're a girl that I can't punch yours! You should have thought of that before you called me a coward!"[95]

PUBLIC EDUCATION

The most important prerequisite to any program of antivivisectionist activity was money, and fund-raising was a con-

[93] See *Animals' Guardian* ii (1892–1893), 17–19, 21–34; *Animal World* xxiii (1892), 162; *Spectator* lxix (1892), 514, 583; Cf. *B.M.J.* ii (1892), 817–819, 901–908, 962; *Lancet* ii (1892), 888–889, 908, 1000–1002, 1225–1126; *Nature* xlvi (1892), 557–558.

[94] E. Hart, "Women, Clergymen, and Doctors," *New Review* vii (1892), 708–718.

[95] Quoted in *Public Opinion* lxii (1892), 575. Suggestions that Cobbe was aware of the inaccuracies in the *Nine Circles of the Hell of the Innocent* were about as accurate and as charitable as her sustained assult upon the motives of scientists and doctors.

tinuing concern of the societies. The operation of offices and the employment of a secretary and possibly a clerk or two constituted an overhead on the order of £500 to £700 a year in itself. The second major item of expenditure was for publication of advertisements and movement literature, including, in some cases, a periodical. There was, of course, no limit to the amount of money that could be spent usefully in the area of propaganda. Two major means were used to raise money: direct appeals for subscriptions, donations, and legacies; and fund-raising activities like bazaars. Antivivisection societies circulated precise instructions as to how to provide for the society in one's will, since legal challenges to such provisions—provisions often vaguely worded—were frequently made by heirs. The late seventies were difficult times for the societies. By 1878, Frances Power Cobbe was appealing to Victoria Street's most well-to-do patrons for emergency support, since the society was apparently in imminent danger of collapse.[96] Victoria Street's precarious situation was stabilized by a legacy of £1,350 from a Dutch woman in 1881,[97] and for the remainder of the century the society maintained annual giving at the level of £1,000 to £1,300.[98] In 1882, the London Anti-Vivisection Society had an income of nearly £1,500.[99] Some sense of the amount of literature that could be produced for this kind of money may be achieved by noting that in 1891 the cost of printing, folding, and wire stitching 1,000 sixteen-page, octavo pamphlets was a mere £3-3.[100] Antivivisection societies, in common with most other voluntary societies of the period, published periodic

[96] Marquis of Bute to F. P. Cobbe, 28 January 1878, Henry E. Huntington Library and Art Gallery, Frances Power Cobbe Papers, CB.373.

[97] *Life of Frances Power Cobbe,* ii. 612–613.

[98] These figures are based upon information in the *Zoophilist* and annual reports of the society. It seems likely that Victoria Street's relatively healthy financial position was an important motive for the International Association's amalgamation with it.

[99] *Seventh Report of the London Anti-Vivisection Society* (London, 1883), p. 20.

[100] *Animals' Guardian* i, No. 10 (July, 1891), inside front cover.

lists of subscribers with the amounts given. This practice was generally intended to play upon the desire to establish status on the part of potential donors. A perusal of antivivisectionist lists indicates, however, that the movement gained a great deal of its support in small sums, often less than £5, though it may be that some large donors preferred to remain anonymous, their donations appearing only as indistinguishable parts of the annual balance sheet. Most donors preferred to be anonymous or pseudonymous, or to give in the name of their cat or dog. Women predominate among the relatively few named donors.

The organizational and fund-raising work went to support a variety of activity. The first and primary thrust of antivivisectionist activity was a program of public education combined with political action. The intention was to make vivisection "infamous, then illegal" since hard experience with total abolition bills showed that "to attack a Parliament before converting the Constituencies is to put the bullet into the gun before the Powder."[101] Public education involved a continuing campaign in the press, placards and pamphleteering in the streets and public institutions, holding meetings and providing lecturers, and circulating petitions.[102] Political action, dealt with above, included the introduction of legislation and pressure on the Home Office regarding the administration of the Act, the presentation of petitions in Parliament, and electoral activity, such as the attempt to get antivivisectionist pledges from candidates for office. The second major tactical thrust of antivivisection consisted of an attack upon the practitioners of experimental medicine, their sympathizers, and institutions believed to support them. This approach included the attempt to make things as difficult and unpleasant as possible for proponents of experimental medicine, to cut off funds for them and for their institutions, and to advance antivivisectionist medical alternatives.

[101] *Home Chronicler* (1876), 200–201, 366; *Verulam Review* vii (1897–1898), 228.

[102] See, e.g., *Anti-Vivisectionist* vii (1880), 507; *Do you Wish to Stop Vivisection?* (London, c. 1880).

Public meetings and touring lecturers had been an important fixture in social reform movements since the days of William Clarkson's antislavery rallies. They were adopted enthusiastically by the antivivisection movement. The societies advertised their readiness to supply the services of an expert lecturer free of charge, whenever a hall and a competent chairman could be secured. One of the most active of these speakers was H. Ribton Cooke, who lectured all over England on behalf of either the London or the International society, from 1877 into the early eighties.[103] Cooke had had legal training, a fact that moved the *B.M.J.* to note with satisfaction that he was a "member of the profession whose business it is to advocate a cause without reference to its merits."[104] The societies' failure to induce antivivisectionist medical men to undertake itinerant lecturing left meeting organizers in constant danger of finding their gatherings subverted by the oratorical counterattack of local medical practitioners. This backfiring at antivivisectionist meetings happened from time to time, much to the delight of the medical journals.[105] Cooke described the situation in 1879:

> The present tactics of our opponents may be thus summarized:—Get a popular local man—a fluent speaker and a medical man if possible; send round to all his friends and all the medical men and their assistants, all the chemists, etc., in the town. Let him move an amendment, make bold assertions, praise the profession, abuse all other forms of cruelty, talk as long as ever he can, have a prosy seconder to weary the meeting, interrupt the lecturer at every possible opportunity, carry the amendment, if possible, at a late hour, dispute the authority of the chairman, unless he rules in your favour, and then go noisily home to celebrate the "glorious victory."[106]

[103] Cooke's meetings and speeches are recorded in the successive issues of the *Home Chronicler* and the *Anti-Vivisectionist.*

[104] *B.M.J.* i (1878), 831.

[105] See ibid. and *Lancet* i (1883), 241–242.

[106] *Anti-Vivisectionist* vi (1879), 360.

To combat such opposition, antivivisectionists tried to lay down careful ground rules for such meetings—including the proviso that they should not be held without the presence of an antivivisectionist medical man.[107] Occasionally, meetings were packed by medical students who enthusiastically howled down speakers and prevented any discussion whatsoever. This practice, most common in Edinburgh, was criticized by medical journals as scarcely in the best interests of the profession.[108] Local practitioners were by no means always successful in their exchanges with antivivisectionist speakers. As one frustrated medical man put it, "The speakers chosen are men well-versed in the way of dealing with the gentler sex and with men of illogical minds."[109] Lecturing grew steadily: in 1880, the London Anti-Vivisection Society was proud of holding twelve meetings in three months; by the end of the century, it had an "Anti-Vivisection Van" touring the country, while British Union lectures totaled more than 500 yearly.[110]

The personal contact with potential converts, which public meetings and the circulation of petitions occasioned, was only the smaller part of the antivivisectionists' efforts at public education. Their greatest single medium for information and conversion was the printed word: articles, letters and advertisements in the press, placards, flyers, polemical pamphlets and books, prize essay contests, antivivisection novels and paintings, and the movement's own periodicals. The volume of this heterogenous mass of literature is immense, but an attempt to outline it may be worthwhile.

During the lobbying of 1875 and 1876, little in flyer and pamphlet literature existed. Starting in late 1876 and lasting well into the present century, this kind of material represented the case for abolition of vivisection in its purest and most typical form. The substance of antivivisection

[107] *Home Chronicler* (1877), 776.

[108] *B.M.J.* i (1877), 375, 500.

[109] *Lancet* ii (1892), 963.

[110] *Anti-Vivisectionist* vii (1880), 507; *Against Vivisection . . .* (London, 1899), p. 34; *Third Annual Report of the British Union for the Abolition of Vivisection* (Bristol, 1901), p. 6.

literature will be discussed in the following chapters; here, the point is that publication of pamphlets was relatively inexpensive and, therefore, very widely practiced by both societies and private individuals.[111] For example, Frances Power Cobbe wrote literally dozens of pamphlets—many from articles published originally in the *Zoophilist*—which were circulated by the Victoria Street Society in the tens of thousands every year. In 1885 alone, Victoria Street put out 81,672 books, pamphlets, and leaflets.[112] Such pamphlets and books attempted not only to mobilize readers against vivisection, but also to bring them within the orbit of the particular society responsible for publication. Pamphlets were usually distributed gratuitously though occasionally longer ones (they might remain in soft cover even when over one hundred pages) would cost a few pence. Books and pamphlets varied widely in quality, reflecting the relative ease with which an enthusiast could immortalize his views in print. They ranged from the maudlin to the childish, from vicious personalities to very effective polemic. Certain publishing firms seem to have catered specifically to societies and individuals in certain kinds of movements. E. W. Allen of London published both antivivisection and antivaccination literature.

The effective circulation of pamphlets was difficult to assess and inevitably subject to the attitudes of those into whose hands they fell and to their own physical fragility. Placards on hoardings, passed by hundreds of people every day, had a relatively greater effective circulation.[113] Such placards, frequently including gruesome illustrations, were used from time to time by the movement. Indeed, they are virtually the only public indication of organized antivivisection in Britain today. Placarding was begun in

[111] Cf. B. Harrison, "Drunkards and Reformers: Early Victorian Temperance Tracts," *History Today* xiii (1963), 178–185.

[112] *Eleventh Annual Report of the Victoria Street Society for the Protection of Animals from Vivisection, united with the International Association for the Total Suppression of Vivisection* (London, 1886), p. 16. See also *Life of Frances Power Cobbe* ii. 613–614 and *Animals' Guardian* iii (1893), 49.

[113] *Home Chronicler* (1877), 536, 602.

February 1877, by both Victoria Street and the International Association. Victoria Street's placards featured a life size illustration taken from the instructional textbook of continental physiologist Elie de Cyon. The placards, personally approved by Lord Shaftesbury before their release, caused a sensation among the scientific and medical community.[114] They proved to be the last straw in a style of agitation that alienated many former sympathizers, including such periodicals as the *Morning Advertiser*, the *Globe*, and the *Saturday Review*. As the latter put it:

> It is not perhaps generally known that papers of a most malicious, and in some cases libellous character have been widely circulated, and unwarrantable invasions of domestic privacy have been made. . . Postcards of a bitterly personal and denunciatory tone, inquisitorial visits, and all sorts of aggressive pressure have been used in order to influence weak minds on these questions. In various parts of London blank walls are just now placarded with sensational illustrations of the alleged horrors of vivisection, and appeals to popular passion. There is, for instance, a large picture of a dog and cat as they are supposed to be when arranged for dissection, with the question underneath whether such "Agony"—this word is in large capitals—can be allowed to be inflicted on sensitive and innocent animals for pure selfishness; and there is another bill, in which inquiry is made, "Have you lost your cat?"[115]

Another means of bringing the cause before the public was through fiction and art. Short stories published both as pamphlets or in periodicals were a favorite genre with antivivisectionists, although novelistic treatments of the subject were also available.[116] Short stories typically featured

[114] *Spectator* lvi (1883), 449; *Zoophilist* vii (1887), 138; C. Adams, *The Coward Science* (London, 1882), pp. 34–40.

[115] *Saturday Review* xliii (1877), 540–541; see also *Public Opinion* xxxi (1877), 545.

[116] A few of the latter, none of which I have examined closely, are B. Pain, *The Octave of Claudius* (London and New York, 1897); J. Cassidy [E. M. Story], *The Gift of Life* (London, 1897); L. Graham, *The Professor's Wife* (London, 1881); G. Macdonald, *Paul Faber, Surgeon* (London, 1878).

a cold-hearted German vivisector (Professor Krankherz) whose rampart materialism brought him his just deserts. In Wilkie Collins' "Heart and Science," a novel serialized in *Belgravia,* the wicked vivisector Benjulia committed suicide after being forestalled in the objective of his lifetime's research by two antivivisectionist medical men.[117] Arthur Conan Doyle's "A Physiologist's Wife" excellently caricatured the public stereotype of the morbidly curious physiologist, who makes a mental note to investigate the phenomenon of lachrymal secretion when his wife cries on his shoulder.[118] Such contributions were welcomed by the movement. As the *Animal World* wrote late in the century "Nothing is more common among humanitarians than the expression of a pious hope that English literature may be pressed into our service, and especially that books likely to be read, the most likely being novels, should help to awaken public opinion. . . ."[119] If fiction reached an audience not usually addressed by movement literature, so too did art, whether it was in the form of sensational drawings in the *Illustrated Police News,* cartoons in the likes of *Truth,* or paintings hanging in metropolitan museums and showrooms.[120]

For the historian, perhaps the greatest insight into the policy, tactics, and motivations of Victorian antivivisectionists is provided by an examination of the contents of the periodicals spawned by the movement itself. These were A. P. Childs' *Home Chronicler* (1876–1879) and its successor, the *Anti-Vivisectionist* (1879–1882);[121] Victoria

[117] W. Collins, "Heart and Science," *Belgravia* xlviii (1882), 175–199, 312–333, 438–465; xlix (1882–1883), 54–80, 168–193, 312–341, 443–474; l (1883), 39–69, 160–192, 298–330, 489–508.

[118] A. C. Doyle, "A Physiologist's Wife," *Blackwood's Edinburgh Magazine* cxlviii (1890), 339–351.

[119] *Animal World* xxviii (1897), 190.

[120] For notice of an antivivisectionist painting, see *Zoophilist* v (1885), 102.

[121] The *Home Chronicler* was a weekly. *The Anti-Vivisectionist* became a monthly in late 1880. A. P. Childs died in March, 1881 and the *Anti-Vivisectionist* was briefly edited by H. Ribton Cooke until it ceased publication in the following year.

Street's *Zoophilist* (1881 1915), a monthly edited successively by Charles Adams, F. P. Cobbe, and Benjamin Bryan; Adams' monthly *Champion* (1884–1887)[122] and his *Verulam Review* (1888–?1903); and the London Anti-Vivisection Society's *Animals' Guardian* (1890–19??), a monthly edited by H. J. Reid.[123] The function of these periodicals was twofold: they were intended to convert uncommitted readers and to maintain organization and activity among convinced members of the movement. To obtain the first objective—maximum circulation for the generous doses of polemic and propaganda in each issue—prices were maintained as low as possible, and issues were sent gratuitously to libraries, Temperance Halls, Y.M.C.A.s, Mechanics' Institutes, hotels, police stations, and newspaper offices.[124] Repeated but unsuccessful attempts were made to have the *Anti-Vivisectionist* distributed from W. H. Smith's railway bookstalls.[125] Despite efforts to achieve a wide distribution for the purpose of expanding the movement, a thorough reading of these periodicals leads one to the inescapable conclusion that they were published by and for a small coterie of zealous antivivisectionists. Their substance is too monotonously repetitive, their news items too exclusively selected, their editorial matter too narrowly conceived for them to appeal outside such a well defined group. As the public lifelines of the movement, periodicals such as the *Home Chronicler*, the *Anti-Vivisectionist*, and the *Zoophilist* created an intimacy of concern, a near unanimity of opinion for their readers, giving them a sense of purpose and self-importance. This sense in turn goes a long way toward explaining the extraordinary persis-

[122] I have succeeded in locating only two issues of *Champion*, both in the Bodleian Library, Oxford.

[123] The *Animals' Guardian* carried fairly long articles on the general subject of animal protection, including aspects of vivisection. Since it contained very little in the way of news or editorial matter on the movement, however, it will not be discussed at any length in what follows.

[124] See, e.g., *Anti-Vivisectionist* vii (1880), 194–195; *What Can We Do?* (London, n.d.).

[125] Ibid. viii (1881), 15.

tence not only of such relatively dull literature, but also of the movement itself, and many others like it. The same dedicated group of correspondents wrote again and again to the *Home Chronicler* and the *Anti-Vivisectionist,* to express with the same cliches an indignation that becomes quickly banal with repetition. A movement could provide for them, and a few hundred others like them, an outlet for frustrated emotion, a feeling of moral superiority, and a sense of sharing the concerns of great men like Shaftesbury or Manning. Such a function sustained the periodicals in the face of the general public's indifference. Movement literature never made a profit; on the contrary, it was a severe financial drain.[126]

With the exception of Charles Adams in the *Verulam Review,* the editors of antivivisection periodicals—however substantively narrow their interests—ranged far and wide over many different sources to pluck relevant material.[127] Particularly popular were the correspondence columns of local and national newspapers, the religious press, the medical press, and the field-sports press. From these sources, letters, articles, and editorial matter were freely reprinted. Controversies over vivisection, no matter how obscure their forum, rarely escaped the vigilance of the movement's editors. The *Zoophilist* reprinted literally hundreds of letters that Victoria Street worthies had addressed to an incredible variety of newspapers and journals all over the country. Often included in the wholesale transfers from other periodicals were defenses of vivisection by the scientific and medical interest,[128] as well as a great many news items concerning experimental medicine. Other sources included sermons, accounts of lecture tours and annual meetings, descriptions and illustrations of experi-

[126] See, e.g., *Verulam Review* i (1888–1889), vi.

[127] The *Verulam Review* was composed solely of comparatively long articles on various aspects of the movement, written originally by the editor himself and at most one or two associates.

[128] As Augustine Chudleigh, a regular correspondent to the *Anti-Vivisectionist,* put it: ". . . it is of the greatest importance for us to know our enemy's strongest points, since to underrate them is to court defeat." *Anti-Vivisectionist* vi (1879), 168.

ments in scientific textbooks, scientific or medical disputes in the professional literature, and a certain amount of the literature of ancillary movements. Every issue of the *Home Chronicler* contained a full page illustration of a vivisectional experiment, intended to provide the kind of ghastly shock that would mobilize the uninitiated into a lifetime of dedication to the cause (see Figures 11 and 12).[129] A. P. Childs' use of this tactic was thought in some quarters scarcely likely to increase the popularity of his publication, and the practice was never adopted by other editors. Controversies within science and medicine over either theoretical or practical issues were gleefully publicized by the movement as proof of the inefficacy of the vivisectional method. The *Zoophilist* enlisted antivivisectionist medical men to write original articles in criticism of the alleged triumphs of experimental medicine. Notable targets were research on cerebral localization by David Ferrier and his colleagues and immunological work by Pasteur, Koch, and others. Pasteur's rabies vaccine, involving as it did a disease of great concern to dog lovers, was the subject of innumerable antivivisectionist tirades. The *Zoophilist* went to immense pains to compile a "Pasteur Necrology"—an annotated list of mortalities following treatment at the Pasteur Institute.

Occasionally, material from related movements, notably slaughterhouse reform, humane trapping, and prevention of cruelty to children would appear in antivivisectionist periodicals. The response to such material by the readership of movement periodicals was not sufficiently strong that editors ever attempted to broaden the appeal of their publications by including it on a regular basis. The constituency for antivivisection was a peculiar one. At the very least, that constituency could not be identified, despite a priori similarity, with the constituency for animal protection in

[129] Childs' use of this method compares to the fondness of Temperance periodicals for recounting in detail personal disasters resulting from drink. B. Harrison, "'A World of Which We Had No Conception.' Liberalism and the English Temperance Press: 1830–1872," *Victorian Studies* xiii (1969), 137–138.

1170

THE

HOME CHRONICLER,

With which is incorporated "The Health Chronicle."

No. 56. SATURDAY, JULY 14, 1877. Price 4d.

A VIVISECTOR'S APPARATUS FOR STUDYING THE MECHANISM OF DEATH BY HEAT.

Illustration exactly copied from page 347, of "Leçons sur la Chaleur Animale," by Claude Bernard:—Paris, London, & Madrid, 1876.

LONDON ANTI-VIVISECTION SOCIETY,

MR. HOLT'S BILL.

PUBLIC LECTURES AND MEETINGS, &c.

Bankers, THE IMPERIAL BANK,

LARGE PRINTED HAND-BILL FOR SHOP WINDOWS AND FOR DISTRIBUTION.

CHARLIE AND HIS DOG SHAG.

"A Wonderful and Horrible Thing is committed in the Land . . . Shall I not visit for these things? saith the Lord."—*Jeremiah V.*

SECOND EDITION.

LONDON:
W. KENT & CO., 23, Paternoster Row.
1877.

VERMIN TRAP PRIZES.—First, The Committee of the Royal Society for the Prevention of Cruelty to Animals, have much pleasure in renewing their PREMIUMS for IMPROVED VERMIN TRAPS, designed to supersede the present cruel steel trap used in preserves. The improved traps must be portable, inexpensive, weather proof, easy of adaption to the various requirements of trapping, simple in manipulation, effectual, and humane, killing instantly or capturing without pain. Further particulars may be had of JOHN COLAM, Secretary, R.S.P.C.A., 105, Jermyn Street, London, S.W.

"The opening of living animals has done more to perpetuate error than to confirm the just views taken from the Study of Anatomy."—Sir Charles Bell.

VIVISECTION. BY GEORGE MACILWAIN, F.R.C.S. Price 3s. 6d.

This little book makes clear to the simplest understanding that it was not to Vivisection that we are indebted for any of our great discoveries.

Those who are inclined to buy a copy are requested to MISS WADDELL, Hartfield, Gloucester.

Figure 11. Frontispiece from the *Home Chronicler* of 14 July 1877. Note the advertisements: an appeal for funds for the London Anti-Vivisection Society, an offer of anti-vivisectionist handbills for sale, an animal story complete with biblical allusion, an R.S.P.C.A. prize for improved vermin traps, and George Macilwain's *Vivisection,* which "makes clear to the simplest understanding that it was not to Vivisection that we are indebted for any of our great discoveries." *Courtesy of the British Museum.*

882 THE HOME CHRONICLER. [JULY 14, 1877.

VIVISECTION ILLUSTRATED.

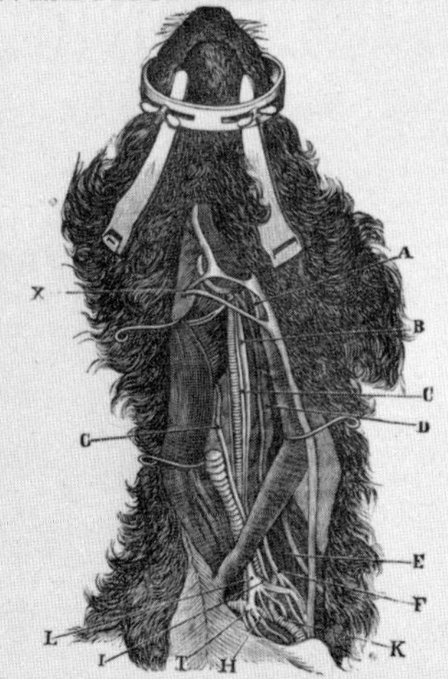

VIVISECTION OF A DOG,

Exactly copied from the Atlas of Professor Cyon's Work,

In reference to which the *Lancet* says,—

"IT IS AN ABSOLUTE NECESSITY FOR ANY WORKING PHYSIOLOGIST."

FROM

THE HOME CHRONICLER,

A JOURNAL ADVOCATING

THE TOTAL ABOLITION OF VIVISECTION.

Published Weekly, Price 4d., 11, Ave Maria Lane.

To ensure THE HOME CHRONICLER being sent direct by Post on the day of publication, send the following to

The Manager of THE HOME CHRONICLER,

12, SYDNEY ROAD, STOCKWELL GREEN, LONDON, S.W.

Enclosed is a Post Office Order, payable to C. ROSS, at Charing Cross, for Five Shillings, for one Quarter's Subscription, in advance, to THE HOME CHRONICLER *(with postage), which I request may be sent weekly to* ____________________

Figure 12. A typical example of antivivisectionist graphics. *Courtesy of the British Museum.*

general, as represented, say, by the membership of the R.S.P.C.A. Hence, any attempt by the antivivisection press or the societies to move into ancillary areas of activity was resisted by the faithful. Protest against a corporate commitment to slaughterhouse reform from vegetarian antivivisectionists Maitland, Kingsford, and F. W. Newman was only the most explicit manifestation of the fact that members of the movement agreed on little but the central tenet: opposition to the use of living animals in experiments.[130] To the rank and file of the movement, such exclusivity of concern was more a gut reaction than a reasoned argument. Nevertheless, the phenomenon had its roots in the variety of motivation and background that individuals brought to the cause.[131] The specialized nature of the potential audience for antivivisection periodicals was further emphasized when the *Anti-Vivisectionist* and later the *Zoophilist* halved their prices (from 4d. and 6d., respectively) in an attempt to increase circulation among medium and lower income groups. Both experiments lasted only a short time, and ended in failure, with prices raised back to their original levels. The only common denominator, besides antivivisection per se, among the rank and file was a sentiment about animals that made letters, articles, and short stories about intelligent dogs, cats, and horses staple fare in the *Home Chronicler* and the *Anti-Vivisectionist.*

While antivivisection periodicals served to unite isolated supporters of the movement at the grass roots level, they had a different function for the leadership. Their existence was partly the result of, and partly the forum for, endemic internecine disputes in which questions of policy and tactics were hammered out in editorials, letters, and articles. A. P. Childs and Charles Adams maintained continual criticism of the conduct of the movement and especially of Frances Power Cobbe and the Victoria Street Society. Although both the *Home Chronicler* and the *Zoophilist* originally portrayed themselves as the organ of the movement

[130] *Home Chronicler* (1878), 380, 396.

[131] For an excellent discussion of many of the issues raised here, see B. Harrison, op. cit., 125–158, esp. 127–132, 152.

as a whole, in practice they became the mouthpieces of the particular society or societies that supported them.[132] Charles Adams frankly described his *Champion* as "the organ of a little band of earnest workers whom personal jealousies had unfortunately excluded from the cooperation of the ordinary Anti-Vivisection Societies."[133] Financial support was crucial because of limited circulation and very sparse advertising revenues. Aside from the announcements of sponsoring societies, movement periodicals contained very little advertising: occasionally, there would be an R.S.P.C.A. announcement or advertisements for books on animals, pet-related products, or humane traps.

Although I have found no precise information on circulation levels, it seems clear that movement periodicals reached only a small proportion of the population. The average citizen constructed his image of antivivisection and its adherents from reports in his newspaper and the journals he read regularly. Antivivisectionists had access to the lay press through letters to the editor, advertisements, and news reporting of their meetings and their political activity. As we have seen, writing letters to the editor was a tactic carried out by the Victoria Street Society as a systematic program, and by hundreds of other antivivisectionists when the mood struck them. The resulting flood of mail upon editorial desks carried its own danger, soon realized, of boring both editors and readers. On antivivisectionist advised fellow workers that "the mere desire to unburthen the mind of its indignation" was an insufficient reason to take pen in hand: "General tirades against the cruelties and abuses of Vivisection now only weary veteran editors, and indispose them for the admission of anything new or really important that may occur in connection with the subject."[134]

If letters to the editor inevitably hastened the rapid onset

[132] The *Home Chronicler* and the *Anti-Vivisectionist* were supported by the London, International, Scottish, and Blackburn societies. *Anti-Vivisectionist* ix (1882), 11.

[133] *Verulam Review* i (1888–1889), 2.

[134] *Anti-Vivisectionist* (1878), 301.

of tedium for all but the converted, advertisement fared little better, and it was prohibitively expensive as well. Although occasionally used by the larger societies to pass information on to members, advertisement was used extensively with the intention of gathering support for the cause only by George Jesse's renamed Society for the Total Abolition and Utter Suppression of Vivisection. Jesse's society had an income of a few hundred pounds per year, largely the result of his advertisements attacking vivisection and pleading for funds. Since the society had no executive officer but himself, Jesse could spend virtually all of his income on pamphlets and advertisements. His characteristic productions, filled with legions of exclamation points, and expressing a spluttering indignation worthy of Colonel Blimp, appeared for over twenty years in the *Morning Post*, the *Standard*, the *Athenaeum*, and the *Daily Chronicle*. The absurdity of his style made his attacks on the bastions of vivisection little more than laughable to contemporaries and caused the *Times* to refuse his advertisements altogether.[135]

As with other movements, then, the final success or failure of antivivisection depended very much upon editorial attitudes toward it, for such attitudes determined not only which letters appeared in correspondence columns, but also the nature and extent of news coverage and of editorial comment. The extraordinary diffusion of discussion of antivivisection in the last quarter of the nineteenth century would make a precise definition of its reception by various sections of the press an extremely laborious and probably unrewarding task. In practice, outside the great newspapers, the reception given to the movement by a particular periodical seems to have depended very much upon the personal background of the journalist involved, for the movement had relatively little correlation with partisan political or religious postures. Unlike, say, controversy over compulsory vaccination, or education, or temperance legislation, the discussion of vivisection rarely had any distinctive local

[135] A good example of Jesse's handiwork is *"The Times" on Vivisection* (?London, 1881).

flavor or consequences; neither had it any economic ramifications. Hence, a sampling of the literature outside the major newspapers, the professional journals, the prestigious reviews, and movement propaganda reveals repetitive discussion completely derivative from them.

About the treatment of the movement at the hands of the press, however, it may be worthwhile to delineate some general trends. Until early in 1876, the antivivisection movement, such as it was, enjoyed remarkable sympathy from the press (the medical journals always excepted). Later opponents such as the *Times*, the *Saturday Review*, the *Pall Mall Gazette*, and *Punch* were prepared to support the cause in the early seventies. The two major factors behind the ultimate loss of sympathy, which these and other organs of opinion experienced, initially appeared in 1876. They were, first, the overwhelming and unified expressions of antipathy by the medical profession toward Carnarvon's bill and the movement behind it, and, secondly, the excesses of hysterical elements within the movement. The prestige of the medical profession and the frantic tactics of enthusiastic antivivisectionists (*London* referred to "grown persons" taking to "public fainting in the cause of the lower animals")[136] combined progressively to weaken the position of organized antivivisection with the influential press. By the late eighties, Cobbe could say "three-quarters of the press" are "dead against us" and note "the persistent boycotting of our meetings by the leading newspapers."[137] Late in the century, the *Abolitionist* complained that the writers of obituaries in the leading daily and weekly newspapers inevitably failed to mention that "the deceased eminent person . . . was an ardent Anti-Vivisectionist."[138] Others claimed that even when meetings were reported, journalists were biased against the movement.[139]

[136] Quoted in *Home Chronicler* (1878), 214.

[137] F. P. Cobbe, *A Charity and a Controversy* (London, 1889), p. 2 and *In the Long Run* (London, 1887), p. 3.

[138] *Abolitionist* i (1899), 83.

[139] T. A. Howard, *Vivisection. A Reply to a Letter written by Dr. Caton to the Chairman of an Anti-Vivisection Meeting held in the*

Among the factors working against coverage of the antivivisection movement by the press was public distaste for the gruesome experimental details so frequently retailed by speakers on behalf of the movement. Such ghastly images could be conjured up by the mere word "vivisection"—hence the use of the term was avoided by the scientific and medical interest. In an article on "Government by Journalism," W. T. Stead characteristically exaggerated to make the general point about the editorial sense of decorum:

> . . . it is doubtful whether, if an edict were to be issued condemning every [prostitute] to the Lock Hospital to be vivisected at the medical schools for purposes of demonstration, the more decorous of our journals would deem the wrong scandalous enough to justify the insertion of a protest against so monstrous a violation of human rights. The medical journals would of course enthusiastically support it; the *Saturday Review* would empty vials of its sourest ink over the indecent Maenads and shrieking sisters who publicly denounced such an outrage on humanity and womanhood; and the great majority of the papers would avoid the subject as much as possible, in the interests of public morality and public decency.[140]

The failure of antivivisection in the press is manifest when one lists the influential dailies and weeklies definitely antipathetic to it after 1876. Included are the *Times*, the *Standard*, the *Daily Telegraph*, the *Pall Mall Gazette*, and the *Saturday Review*. Opposition also came from *Punch*, the *Illustrated London News*, and the blood-sports organs, *Field*, and *Land and Water*. Apart from Hutton's *Spectator*, only the *Daily Chronicle* and the *Echo* among the dailies and weeklies could be relied upon consistently to back the movement. The *Globe*, the *Morning Post*, and the *Daily*

Y.M.C.A., Liverpool, in April, 1897 (London and Liverpool, 1897), p. 5. Cf. the suspicion of the press in *Animals' Guardian* iv (1893–1894), 54, and *Home Chronicler* (1876), 136.

[140] W. T. Stead, "Government by Journalism," *Contemporary Review* xlix (1886), 670. Note the contiguity of antipathy to the Contagious Diseases Acts with antivivisection.

News covered antivivisectionist activities, but were open to both sides of the question.

With the exception of *Macmillan's Magazine* (Macmillan's also published *Nature*), the intellectual monthlies and quarterlies permitted considerable discussion from both sides. Particularly popular as centers of debate were the *Contemporary Review*, the *Fortnightly Review*, and the *Nineteenth Century*.

The movement achieved a wide and fairly sympathetic hearing in the religious press, although here again defenders of experimental medicine were not absent. Staunchly antivivisectionist were the *English Churchman*, the *Rock*, the *Referee*, the *Irish Ecclesiastical Gazette*, and the *Record*. The *World*, *Our Work*, *Church Bells*, the *Christian*, the *Tablet*, and any number of other journals of varying religious stance gave the issue extensive but generally inconclusive discussion.

Although women played a paramount role in the movement, periodicals addressed to them were by no means unanimously in support of it.[141] Among them, the best friends of the cause were the *Lady's Pictorial* and the feminist *Englishwoman's Review*. Most women's journals ignored the movement.

Circulation data for 1870 for some of the periodicals listed is available.[142] An educated guess would suggest that periodicals clearly unsympathetic to the movement had a combined circulation from three to five times that of the periodicals clearly sympathetic to the movement for most of the last three decades of the century.

Antivivisectionists frequently agonized over their mistreatment at the hands of the press—especially at the hands of the *Times* and of *Punch*—but were quite unable to analyze the source of their difficulties. These difficulties lay not only in their failure to establish the validity of their

[141] See *Verulam Review* iv (1894), 113–118. Cf. *Journal of Science* 3rd series, iii (1881), 732.

[142] A. Ellegard, "The Readership of the Periodical Press in Mid-Victorian Britain II. Directory," *Victorian Periodicals Newsletter*, No. 13 (1971), 3–22.

arguments, but in a concomitant and perhaps antecedent phenomenon: the loss of credibility consequent upon their style of agitation. All the references to the controversy I have found in *Punch* before 1876 were either sympathetic or neutral toward antivivisection.[143] In early 1876, the magazine began to satirize and then to criticize the movement.[144] Its campaign was not modified when a deputation of antivivisectionists sought its support.[145] *Punch* lampooned what it took to be antivivisectionist ignorance and hypocrisy, with cartoons, satiric squibs, and direct criticisms.[146] A *Punch* article of 1878, under the title "Harvey a Humbug!," included the following barbed account of an imaginary meeting:

> On Monday, the First instant, a numerously attended meeting of members of the Anti-Physiological Society was held at St. Luke's Hall. Besides Anti-Vivisectionists, the assemblage included Anti-Tobacco Leaguers, Compulsory Good Templars, Peculiar People, and other enthusiasts of the same intelligent description . . . [According to the chairman] too much had been made of [Harvey's] discovery and all such discoveries. Patients continued to die in spite of them and doctors differed. It was argued that Harvey's . . . discovery enabled surgeons to perform great operations before unthought of, taking up and tying arteries, and so on—things just as well let alone. He didn't believe in doctors, nor surgeons neither—he owned himself one of those whose faith was faith alone (*Cheers.*). Faith against physic, he said; ay, and against surgery too. The faith that sufficed for the cure of a fever would be just as equal to setting a broken bone ("Hear!"). Perhaps mere faith had never yet been tried in that case, but whenever it was, of course it would prove no less effectual than in the other. A living faith would nullify the necessity for operations on live creatures.[147]

[143] See *Punch* lxvi (1874), 28; lxvii (1874), 257; lxviii (1875), 62–63; lxix (1875), 22.

[144] Ibid. lxx (1876), 93, 105, 265.

[145] Ibid. lxxi (1876), 40, 43.

[146] Ibid. lxxi (1876), 1, 41; lxxii (1877), 205; lxxv (1878), 204.

[147] Ibid. lxxiv (1878), 157.

Punch's dislike for the style as well as the substance of the agitation is also evident in the following poetic advice to an antivivisectionist, dedicated to the physician, Sir William Jenner:

Then pule no more about the life
Of dog or cat in crass correction;
The man who gives you back your wife
Was taught, he owns, by vivisection.
We bid the painter's art expand,
Nor curb the pens of men of letters.
Yet sad to think 'tis in this land
Alone that Science walks in fetters.

So let the young men gain the lore
Of Jenner, in a firm reliance
Of those who cry "Excelsior!"
And scale the shining heights of Science.
They work with clear, untiring eyes,
Each day some mighty truth revealing.
Leave sham humanity its lies,
And cry God speed to men of healing![148]

The alteration in attitude toward antivivisection, exemplified above, was, its humor excepted, entirely typical of the movement's experience with a great portion of the press. The unfavorable nature of this experience dictated that the movement itself initiate the vast efflorescence of specifically antivivisectionist pamphlets and periodicals after 1876. In any case, the ragtag collection of miscellaneous lay periodicals that backed the movement in no way measured up to the strength and prestige of those that opposed it. Generally, the influential periodicals were satisfied with the Act of 1876 and unwilling to consider further legislation.

THE ATTACK UPON EXPERIMENTAL MEDICINE

Parallel to the movement's program of public education and political action was its second major tactical front:

[148] Ibid. lxxxvii (1884), 191.

the attack upon experimental medicine itself, an attack against the individuals involved in animal experimentation, the profession that condoned their activity, and the institutions that were thought to support it. Antivivisectionist initiatives in monitoring the literature of experimental medicine and routinely comparing the results with the lists of licensees in the inspector's annual reports produced a certain number of supposed violations of the Act of 1876. The cases were brought to the attention of the Home office, some of them eventually proving to be genuine violations. Such systematic scrutiny of the operation of the Act undoubtedly affected its administration, though not to the point where antivivisectionists could feel any real satisfaction. Not unnaturally, the movement turned its attention to tactics that might produce a direct curtailment of experimental medicine; such a curtailment would be more visible than revisions in administration and more quickly realized than the ultimate goal of legislated total abolition.

Obvious targets for pressure tactics were large hospitals that depended upon charitable donations for a major portion of their operating funds. In early 1876, the Baroness Burdett Coutts was calling for a charity boycott of any hospital that had a vivisector on its staff.[149] The eventual publication of names of licensees in the annual reports of the administration of the Act of 1876 made such a tactic considerably easier to carry out. In 1879 leaflets were sent to supporters of University College, St. Thomas's, and King's College Hospitals, suggesting that subscriptions be withheld because of the presence of licensed vivisectors on the staffs of these hospitals.[150] Other antivivisectionists urged fellow workers to replace their normal donations on the annual Hospital Sunday with a written pledge not to give until vivisection was abolished.[151] Victoria Street maintained a list of hospital charities, showing those that were and were not approved.

[149] *Lancet* i (1876), 138, and *Lady Burdett Coutts on Hospital Funds* (London, n.d.).

[150] *B.M.J.* ii (1879), 356.

[151] *Home Chronicler* (1878), 412–413.

> Not a single physiologist takes out a license for Vivisection or writes an article, or gives a lecture in defence of the practice, but he thereby becomes a marked man; and any institution wherewith he is connected becomes an object of mistrust to an ever enlarging section of that very class of humane and religious persons by whom all public charities are habitually supported.[152]

Money was not to be withheld from hospitals merely because they supported animal experimentation by having licensees on their staffs or by providing facilities for an associated medical school, although these were reasons enough. The most compelling argument of all was based upon the antivivisectionists' belief that "the animal torturer of the morning is too often the theatre-operator of the afternoon"[153] and thus,

> In the interests of the sick poor, every humane person may justly hesitate to give money to support an Institution where the physicians and surgeons *at whose mercy the patients will be,* are upholders of the practice of vivisection or licensed vivisectors themselves.[154]

To spokesmen for the movement, it was clear that ambitious medical men looked upon charity patients as "clinical material," simply so much experimental fodder.[155] The doctors' aims were to advance science, to further their own careers, to educate medical students, but not to cure sick people.

There is evidence that a certain amount of experimental research was prevented or suppressed by hospital author-

[152] *The Hospitals and Vivisection* (London, 1883), p. 4; *Zoophilist* iv (1884), 5, 161–163; *Anti-Vivisectionist* vi (1879), 276.

[153] *Hospitals in Danger* (London, c. 1884). The pamphlet went on, "The vivisector, however great his skill and distinction, is never safe to be trusted with a hospital patient on an operating table with a tray of knives at his side." Cf. *Vivisection at Hospitals. Demand for a Rate-Aided Laboratory* (London, 1897).

[154] *Zoophilist* iv (1884), 5; italics are in the original.

[155] *Hospitals. The Vivisectionist View of Them* (London, n.d.,); E. Bell, *Vivisection and the Hospitals* (London, 1895).

ities who feared the taint of vivisection and its possible consequences. J. W. Ogle complained in his Harveian Oration for 1880 that

> . . . a series of experiments likely to lead to important practical results has lately been arrested in two hospitals with which I am acquainted, from a fear that, with the extravagant clamour of the anti-vivisectionists in the ears of the public, the interests of the hospital might be injured if a license were taken out . . .[156]

Such events were, however, relatively few and far between. The major flaw in the tactic was that for all the ostentatious concern for the interests of the sick poor manifested by the movement, insofar as the tactic succeeded it merely reduced the hospitals' capability of dealing with the ills of charity patients. It was overwhelmingly the case that hospitals stood by the researchers on their staffs (though hospitals rarely supported experimental research financially) at the inevitable cost of antivivisectionist subscriptions. Some leaders of the movement, including Lord Shaftesbury, were well aware of this, and pleaded against the tactic.[157] The lay press generally saw the tactic as proof that antivivisectionists thought more of the sufferings of animals than of human beings.[158] Nevertheless, antivivisectionists continued to agitate against the so-called "vivisecting hospitals" through the eighties and nineties.[159]

Other institutional props of experimental medicine were

[156] J. W. Ogle, *The Harveian Oration, 1880* (London, 1881), p. 54. See also J. Bland-Sutton, *The Story of a Surgeon* (London, 1930), pp. 35–36.

[157] *Home Chronicler* (1878), 204–205; *Animals' Guardian* iii (1892), 47–48; Earl of Shaftesbury to the Countess of Portsmouth, 4 June 1877, Henry E. Huntington Library and Art Gallery, Frances Power Cobbe Papers, CB.124.

[158] See, e.g., *Public Opinion* xxix (1876), 93–94.

[159] See, e.g., *Verulam Review* i (1888–1889), 194–211; vii (1897), 251–256; S. Coleridge, "Some London Hospitals and their Audited Accounts," *Contemporary Review* lxxvii (1900), 343–350; S. Coleridge, *The Diversion of Hospital Funds* (London, 1901).

also attacked. The campaign against the Brown Institution of the University of London, initiated by R. H. Hutton and George Jesse, was maintained by the movement.[160] A second important target was the British Institute for Preventive Medicine, which was modeled upon the Pasteur Institute in Paris.[161] In 1887, a petition by British proponents of experimental medicine (including J. E. Erichsen, then inspector under the Act of 1876) appealed to the Royal College of Surgeons to use a large benefaction recently received to set up a physiological and pathological laboratory at the College. A barrage of antivivisectionist counterpetitions succeeded in eliciting from the Council of the College assurances that the College would not permit vivisectional research.[162] Government support for research, especially support by the Local Government Board for the work of the pathologist Emanuel Klein, was constantly under attack from the movement, which made the most of Klein's foreign antecedents and his statements before the Royal Commission.[163]

Antivivisectionists attempted to counteract the penetration of physiology into university and school curricula. F. P. Cobbe circularized over seven hundred head schoolmasters in England with her pamphlet *Physiology as a Branch of Education.* On the basis of her survey, she pro-

[160] See *Third Annual Report of the British Union for the Abolition of Vivisection.* (Bristol, 1901) pp. 5, 7.

[161] H. J. Reid, *Science on "Ticket of Leave"* (London, 1889); *Why Sign the Memorial to the Home Secretary?* (London, 1891); T. A. Williams, *What I Saw at the Pasteur Institute. An Englishman's Personal Experience* (London, 1894). See also the *Zoophilist* for the late eighties and early nineties.

[162] *The Queen's Jubilee Among the Doctors* (London, 1887); *Will England Bear It?* (London, 1887); *Experiments Forbidden at the College of Surgeons* (London, 1889). The Council's decision undoubtedly owed more to that body's innate conservatism than it did to antivivisectionist sentiment.

[163] J. Colam (ed.), *Further Proceedings Against Vivisection* (London, c. 1879), pp. 21–22; *Zoophilist* vi (1886), 57–58; F. P. Cobbe, *Public Money. An Enquiry Concerning an Item of its Expenditure* (London, 1892); A *Foretaste of the Institute of Experimental Medicine* (London, 1898).

vided information for parents as to schools where children "will be safe from imbibing [the] ardour of research."[164] Movement literature harped upon the superiority of literary and classical over scientific education.[165] Financial support for medical schools, research, and research studentships or scholarships was under continual and occasionally successful pressure.[166] The movement led a boycott of the Society for the Promotion of Christian Knowledge when the S.P.C.K. gave its imprimatur to bacteriologist Percy Frankland's popular work, *Our Secret Friends and Foes.*[167]

The most publicized nineteenth-century confrontation between the movement and the educational advance of experimental medicine followed the election of J. S. Burdon Sanderson as Professor of Physiology at Oxford in 1883. Antivivisectionists both within and without the University worked feverishly to defeat a motion in the University's Convocation to provide financial support for a physiological laboratory. Anyone holding an Oxford degree had the right to vote in Convocation; hence the movement circularized hundreds of potential voters—largely country clergymen—against the motion. Unfortunately for the movement, its attempt to involve nonresident degree-holders in the dispute backfired when resident members came to see the issue as hinging upon something quite different from the morality or immorality of vivisection. Mark Pattison wrote to the Regius Professor of Medicine, Henry Acland,

> I look upon the emancipation of the University from the control of the clergy as a glorious revolution, and one which was necessary to its very existence. It is therefore

[164] "Physiology as a Branch of Education," *Zoophilist* iii (1883–1884), 128–129; *Zoophilist* vi (1886), 121; *Letters of Head School-Masters* (London, 1886).

[165] See, e.g., *Verulam Review* v (1895), 9–11.

[166] See, e.g., *Anti-Vivisectionist* (1878), 268; H. N. Mozley, *The Stealthy Advance of Vivisection. The John Lucas Walker Studentship* (London, 1886); J. H. Clarke, *The British Medical Association As Vivisector* (London, 1896).

[167] *Animals' Guardian* iii (1893), 193; *Verulam Review* iv (1894), 97–103; vii (1897), 283–286.

> most alarming to find claims put forward on behalf of the clergy to control science and learning . . .[168]

Burdon Sanderson remarked of his close 88–85 vote victory in the first and most serious confrontation:

> Everyone here, so far as I know, regards it as an important victory of University versus outsiders, and this is just what is best for us. There are many people who do not like experiments on animals who are nevertheless glad that any attempt to interfere with the internal management of the University should be defeated.[169]

Subsequent agitation to block support for experimental medicine at Oxford was less successful still. By 1891 Frances Power Cobbe was protesting the final outrage: the girls of Somerville College attended lectures by Burdon Sanderson and the biologist E. Ray Lankester.[170]

One of the major tactical failures of the antivivisection movement was its inability to provide institutional alternatives to "vivisecting" hospitals and medical schools. Anti-

[168] M. Pattison to H. W. Acland, 3 March 1884, Bodleian Library, Oxford, Acland MS d.59 f.189.

[169] J. S. Burdon Sanderson to ?G. Burdon Sanderson, 9 February 1885, Sinclair Collection, Woodward Library, University of British Columbia, Letterbook III, pp. 28–29.

[170] For an introduction to the events at Oxford, see *B.M.J.* i (1884), 290; ii (1885), 548–549, 567–568; *Nature* xxxi (1885), 453–454; *Public Opinion* xlv (1884), 163–164; *Spectator* lvii (1884), 180; lviii (1885), 299, 310, 343, 375, 386–387, 471; E. A. Freeman, *A Speech made in the Theatre, Oxford, February 5, 1884, on the vote for the proposed Physiological Laboratory* (Wells, 1884). There is a good deal of material in pamphlets and manuscripts relating to the Oxford controversy in the University's Bodleian and Radcliffe Science Libraries. The material was probably collected by Bodley's Librarian E. B. Nicholson, who, with the historian E. A. Freeman, led the antivivisection party within the University. John Ruskin resigned his Oxford professorship as a protest against funds being spent upon experimental medicine while they were withheld from his Slade School. *Zoophilist* v (1885–1886), 6. For Cobbe's protest about Somerville, see *Spectator* lxv (1891), 944. Physiology instruction to the girls of Girton College, Cambridge, was another target.

vivisectionists gave moral support to an apparently short-lived Society for the Protection of Hospital Patients, which aimed to restore the cure of illness as the leading priority of the large hospitals. As things stood, the society charged, such an objective was strictly secondary to the advancement of individual professional careers and to the education of medical students. Publicity, legal redress, and increased lay control of hospitals were needed to rectify the situation, which left charity patients at the mercy of the doctors.[171] When it came to embodying such principles in an institution of their own—in periodic attempts to found an antivivisection hospital—the movement simply could not enlist a permanent medical staff to give full time and effort to the project. Despite collecting some £2,000, the proposed Shaftesbury Hospital appprently failed for lack of staff in the late eighties.[172] Homeopaths and other schools of medical unorthodoxy received an enthusiastic hearing from antivivisectionists. The movement never succeeded, however, in articulating a research program of its own that proved convincing to any significant proportion of the orthodox profession. The Leigh Browne Trust, founded in the mid-nineties, had "the object of promoting and encouraging original research in Physiology and Biology without experimentation on living animals of a nature to cause pain." One of its first publications was "Dr. Carne Ross's valuable pamphlet on the use of cinnamon in the treatment of cancer and other diseases."[173] The Trust commissioned Benjamin Ward Richardson's *Biological Experimentation*, an exploration of the possibilities for painless research,[174] and, indeed, present day antivivisectionists maintain the hope that advances in tissue culture and other in vitro

[171] "The Truth about the Hospitals," *Verulam Review* vii (1897–1898), 358–361.

[172] See the admittedly hostile account by V. Horsley, "The Morality of 'Vivisection'," *Nineteenth Century* xxxii (1892), 805. Cf. *Verulam Review* ii (1890–1891), 69–80; vii (1897–1898), 22–34; viii (July, 1899), inside front cover.

[173] *Leigh Browne Trust* (London, 1895).

[174] B. W. Richardson, *Biological Experimentation. Its Function and Limits* (London, 1896).

methods may make substantial inroads into current levels of animal experimentation.

Antivivisectionist enthusiasm did not stop at placing the utmost pressure possible upon the institutional supports of experimental medicine. A second tactic involved the attempt to discredit the profession identified with vivisectional research and to harass individual proponents of it. This tactic involved many different methods and was closely related to the attack upon institutions. The substance of the arguments used against the hospitals and the orthodox medical profession will be discussed in the following chapter. The point to be made here is that the movement embarked upon a campaign to break public confidence in the profession as a whole and to invoke social stigma upon the heads of specific advocates or practitioners of experimental medicine. They sought to rupture the bond between patient and doctor, since the profession's personal contacts with the influential sections of the population militated strongly against the success of the antivivisectionist program of public education. They sought to make the social environment of an individual who performed or supported experiments upon living animals so unpleasant as to remove the attractions of such a career.

At first, antivivisectionists tried those methods of information-gathering that had served the animal protection movement so well in its efforts to enforce Martin's Act. The R.S.P.C.A. utilized "the surveillance of detectives" in its attempt to gather information for its report to the Royal Commission.[175] In the same year, 1875, George Jesse advertised in London newspapers a reward of £20 for information leading to the conviction of an experimentalist under Martin's Act.[176] The *Zoophilist* suggested slyly in 1881,

> Could we but get information as to the doings of the laboratories one half as prompt and as accurate as the vivisectionists manage to obtain of the proceedings of

[175] *Vivisection. The Royal Society for the Prevention of Cruelty to Animals and the Royal Commission* (London, 1876), p. v.

[176] *B.M.J.* ii (1875), 25–26.

> some of our committee rooms, we too could surely afford as well as they to "make it worth the while" of our informant.[177]

Although these methods proved fruitless, an array of other techniques used by the movement succeeded in affecting their targets to a certain extent. These techniques included social sanctions against vivisectors, organized boycotts of all medical men who refused to pledge themselves against vivisection, publicizing the alleged misdoings of individual experimenters and their supporters, and—as a result of the publicity—showering personally abusive mail upon supposed offenders.

To Frances Power Cobbe, the imposition of social sanctions against the immoral was a part of the duties of women, to be applied in the case of vivisection as in that of any other variety of sin.[178] The *Zoophilist* sermonized upon the text of Madame Claude Bernard's domestic difficulties:

> The horror of the position of a wife endowed with humane feelings who discovers that her husband is a vivisector, is one which this story may perhaps bring home to some parents who thoughtlessly open their doors to "men of science" of this sanguinary school and allow their daughters to meet them in society and perhaps to touch their stained hands. Let them bear in mind what misery they may be preparing for their children should such acquaintanceship lead to an ill-omened attachment.[179]

Isabel Burton suggested,

> Could not another pledge be set on foot for women, whereby they should agree not to engage themselves, not to marry Vivisectors, and not to invite them to their houses, not to dance with them at parties, and never

[177] The *Zoophilist* was under the impression—so far as I know, incorrect—that there were paid informers within the ranks of antivivisection. *Zoophilist* i (1881), 52.

[178] See F. P. Cobbe, *The Duties of Women. A Course of Lectures* (Boston, 1881). Cf. A. Helps, *Animals and their Masters* (London, 1873), pp. 41–44.

[179] *Zoophilist* iv (1885), 85.

> knowingly to associate with them? I am sure women could stamp out Vivisectors, if they would only be of one mind. If public opinion is but strong enough, in a short time a Vivisector will be unable to show his face anywhere.[180]

The efficacy of sanctions like these is difficult to assess. The loss of the society of such as proposed them would be unlikely to be keenly felt by their targets, but the sentiments expressed undoubtedly were shared, though less enthusiastically, by enough people to affect the social atmosphere perceptibly. The secretary of the International Association for the Total Suppression of Vivisection did not fail to point out that a vivisector in the neighborhood lowers property values![181]

Combined with social sanctions against those most intimately involved in experimental medicine was an attempt to channel patients away from any practitioner who refused to avow his opposition to it. Charlotte Elizabeth had called for such a boycott in the 1840s, and now the International Association moved to compile a central register of antivivisectionist medical men.[182] Advertisements were placed in the *Times* calling for the names of such practitioners, but the profession took the threat implicit in these tactics very lightly indeed.[183] According to Cobbe, such a boycott was intended to succeed less through economic pressure than as another form of social sanction:

> A great physician or surgeon may have his waiting-room daily crammed by a hundred adoring patients, open-mouthed to receive sentence from him of life or death. But yet, if he learn that two or three gentlemen suffering from the complaint which it is his specialty to cure, have openly sought advice elsewhere, because they have been disgusted with his defence of Vivisection, and that three

[180] *Home Chronicler* (1877), 507. A number of ladies wrote letters to the *Home Chronicler* in 1876 suggesting that their sex, at least, should try to do without doctors entirely.

[181] Ibid. (1876), 348.

[182] *Medical Press and Circular* ii (1876), 453; *Home Chronicler* (1876), 104; (1878), 93; *Anti-Vivisectionist* vi (1879), 169.

[183] *Lancet* ii (1878), 204; *B.M.J.* ii (1876), 291.

> or four ladies have ceased to invite him to their dinner parties on the same grounds, the effect (as we know from reliable report) is truly glavanic![184]

In order to provide antivivisectionists with reliable information upon individual experimenters, the Secretary of Victoria Street, Benjamin Bryan, compiled *The Vivisectors' Directory; Being a List of Licensed Vivisectors in the United Kingdom* . . . , published in 1884.[185]

A. P. Childs, a retired surgeon and editor of the *Home Chronicler* and the *Anti-Vivisectionist,* did his part to lower the prestige of his profession by repeatedly publicizing the court appearances of medical men and especially of medical students.

Antivivisectionst opprobrium was frequently directed at specific individuals. Victoria Street routinely challenged any exponent of the experimental method or critic of antivivisection by personal letter.[186] The letter, asking for substantiation of the statements made, was simply an attempt to draw leaders of the profession into controversy and thereby confer status upon the movement. After the late seventies, few medical men obliged, infuriating antivivisectionists such as Charles Adams and George Jesse.

A more direct and public method of attack was through the publication in movement literature of sensationalized descriptions, in laymen's terms, of an individual's experiments or his claims to medical eminence. Among those subjected to such treatment were John Burdon Sanderson, William Rutherford, Samuel Wilks, C. S. Roy, Joseph Lister, Victor Horsley, and David Ferrier. It cannot have been pleasant for the latter to see flyers such as *Professor Ferrier's Experiments on Monkey's Brains* or *Ferrieristic Brain Surgery. A Candid Condemnation* circulating in the

[184] *Zoophilist* iv (1884), 115–116.

[185] B. Bryan (ed.), *The Vivisectors' Directory; Being a List of Licensed Vivisectors in the United Kingdom, Together with the Leading Physiologists in Foreign Laboratories* (London, 1884). See esp. pp. iii–iv. Cf. *List of Medical Men Who Have Declared Themselves Opposed to the Practice of Vivisection* (London, c. 1883).

[186] *Zoophilist* i (1881), 122–123.

streets and parlors of London.[187] When in 1879 the Bishop of Peterborough defended the Act of 1876 during the debate over Lord Truro's total abolition bill, he was roundly assailed by the antivivisectionist press.[188] Such publicity brought to the individual named—whether David Ferrier or the Bishop of Peterborough—a certain amount of anonymous abuse from fanatical members of the movement. This remains true to the present day.[189]

There can be no doubt that the antivivisectionist attempts to make things unpleasant for practitioners of experimental medicine succeeded to the extent that they drew a reaction. One physiologist, H. Newell Martin, published his own pamphlet detailing the *Zoophilist's* misrepresentations and falsehoods against him. He circulated the pamphlet to the patrons and officers of the Victoria Street Society.[190] Medical men complained that they were "objects of suspicion to their neighbours" and of "the current quips at the expense of physiologists."[191] An appeal to the British Medical Association to support experimental medicine was phrased as follows:

> It was a matter of duty and of justice that the Association should stand by those who had performed experiments and who had thereby incurred considerable odium, and perhaps in some instances even damaged their reputation and success in life, in consequence of appearing in public as advocates of this method of investigation.[192]

[187] *Professor Ferrier's Experiments on Monkey's Brains* (London, 1885); *Ferrieristic Brain Surgery. A Candid Condemnation* (London, 1887).

[188] See, e.g., A. De Noë Walker, *The Right Reverend Father in God the Bishop of Peterborough on Vivisection* (2nd ed., Norwich, 1882).

[189] See *Anti-Vivisectionist* vi (1879), 480; J. C. MacDonnell, *Life and Correspondence of William Connor Magee* (London, 1896), ii. 113–117; *B.M.J.* ii (1881), 846; G. Gordon-Taylor and E. W. Walls, *Sir Charles Bell. His Life and Times* (Edinburgh and London, 1958) reprints some modern (1950) samples of this unappealing genre on pp. 234–235.

[190] H. N. Martin, *A Correction of Certain Statements Published in the 'Zoophilist' also a Castigation and an Appeal* (Baltimore, 1885).

[191] *B.M.J.* i (1876), 796; ii (1893), 749.

[192] *Lancet* ii (1892), 326. See also *Nature* xiii (1876), 343; *Journal of Science* new ser., vi (1876), 334.

The movement's final tactical approach involved the attempt to wean the public from its confidence in medical practitioners. Until the mystique of the doctor-patient relationship could be broken, antivivisectionist efforts at public education could have only limited success.[193] As Eleanor Vere Boyle wrote to Frances Power Cobbe, "Every one of these Doctors probably reigns over a little kingdom of patients whom he has perhaps pulled through danger, and who swear by him."[194] The answer to the problem was to impugn medical judgment on matters moral:

> Don't go to your doctor for advice on the subject. This is mainly a great *moral* question—a question of eternal right and wrong. You don't go to your butcher for bread; you don't pay your lawyer for his theology; you don't go to the clergy for instruction in law. So you go to your doctor for medicine, not morals. Therefore don't ask him to prescribe for you on a tremendous *moral* question. You are as capable of judging as he—perhaps better.[195]

The net effect of antivivisectionist pressure on the medical profession was to increase professional unanimity on the necessity of experiments upon living animals. The medical leadership stood almost without exception behind the experimentalists. As a result, antivivisectionists attacked the profession as a whole, thus quickly bringing the mass of practitioners in line behind their leaders on the issue. Some antivivisectionists, including Charles Adams, de-

[193] See, e.g., *Home Chronicler* (1877), 565–567. Cf. M.R.C.S., *Twelve Years' Trial of the Vivisection Act (39–40 Vict., c. 77). Has it stopped the Scientific Torture of Animals in England?* (London, 1889), pp. 3–6.

[194] E. V. Boyle to F. P. Cobbe, 24 January [?1887], Henry E. Huntington Library and Art Gallery, Frances Power Cobbe Papers, CB.34.

[195] T. A. Howard, *Vivisection. A Reply to a Letter Written by Dr. Caton to the Charman of an Anti-Vivisection Meeting held in the Y.M.C.A., Liverpool, in April, 1897* (London and Liverpool, 1897), p. 26. Cf. E. Bell, *Vivisection on the Increase. My Doctor Tells Me that 'There is very little Vivisection now in England'* (London, c. 1896).

plored the movement's injudicious hostility to the whole profession, tarring it for the sins of a mere handful of its members. To them, such a tactic alienated a good portion of the public and, more important, effectively isolated the many potential converts to the cause within the ranks of ordinary practitioners.[196] Leaders of the movement, however, adamantly insisted upon the necessity of the tactic. A passage from the *Zoophilist,* probably written by Frances Power Cobbe, clearly sets out the grounds for their viewpoint:

> We are often advised by kindly-intentioned friends to forbear from citing instances of the blunders of doctors, and counselled rather to conciliate their profession by a profuse employment of . . . honorific epithets . . . But what is it that our good friends would have us do? Here are a number of men who arrogate to themselves almost the characters of impeccability and infallibility; and who, in full conclave solemnly sanctioned the very practice which it is the reason of our existence as a Society to suppress and abolish! . . . Till the medical profession generally recants the doings of the Congress of 1881 *in re* Vivisection, it will be only part of our duty of self-preservation to show cause for believing that it frequently makes very serious mistakes. In setting itself against the moral sense of the nation it is preparing the collapse of its own just influence.[197]

In setting itself against the power and prestige of the medical profession, the antivivisection movement ensured the collapse of *its* would-be influence.

CONCLUSION

By the time the antivivisection movement emerged in the mid-seventies, a rich variety of agitational tactics had been

[196] *Home Chronicler* (1876), 445; (1877), 565–567; *Zoophilist* iv (1884–1885), 187–188; v (1885–1886), 202–203; *Champion* i (1884–1885), 7–9.

[197] *Zoophilist* ii (1882–1883), 23.

developed by other crusades, such as antislavery, temperance, Chartism, and the Anti-Corn Law League. Antivivisection used these tactics imaginatively, aggressively, and indefatigably. Indeed the movement constitutes almost a glossary of nineteenth-century methods for mobilizing and focusing public opinion upon a specific issue. The tactics available had their limitations; such phenomena as the antivivisectionist meeting transformed into a triumph for animal experiment, the lapse into tedious dogmatism by movement literature, and the embarrassment of forged petitions were difficulties by no means confined to antivivisection. But in general, the movement's failures—especially the failure to generate any kind of serious political response between 1876 and the turn of the century—were not primarily tactical ones.

The two basic and interrelated flaws of the antivivisectionist program were disunity among adherents and misjudgment of the temper of the general public. These flaws may in turn be traced to a problem that seems fundamental for all voluntary associations: the tendency for such associations, despite their nominally democratic ethos, to be run by small self-perpetuating coteries of dedicated enthusiasts.[198] This phenomenon of "minority rule" or the "iron law of oligarchy" is exemplified in particularly acute form by Victorian antivivisection. The movement initially served to articulate public concerns that might not otherwise have entered the national political and intellectual domain. By serving as a locus for the expression and affirmation of certain values, antivivisection developed a limited but dedicated following, for whom the movement provided group consciousness and a sense of belonging. This much the historian can glean from a close examination of the literature of the movement. What does not emerge from the existing documentary evidence is the sense that the effective direction of the movement owed anything to anyone outside the very small circle of

[198] Cf. D. L. Sills, "Voluntary Associations," in D. L. Sills (ed.), *International Encyclopedia of the Social Sciences* (New York, 1968), xvi. 362–376.

London personalities who devoted themselves to it. Indeed, Frances Power Cobbe herself exercised an extraordinary dominance over the movement from the mid-seventies to the mid-nineties. There are probably a number of reasons for this. Few people had the interest, abilities, temperament, and time to involve themselves deeply in the movement. The cause seemed to generate relatively few concrete tasks to be performed, at least in comparison to movements such as ragged schools or workhouse visiting. Supporters of the movement might go to annual meetings, but except for the members of the relatively small executive committees, would have little occasion to maintain continuing contact with one another. In any case, however widely spread was support for antivivisection, its formal activity was crucially guided by Frances Power Cobbe and her Victoria Street allies such as Lord Shaftesbury, Lord Chief Justice Coleridge, Cardinal Manning, the Society's Executive Committee, and its permanent staff.

The efficacy of antivivisectionist activity was greatly reduced by conflict within the movement. This conflict did not originate in religious, political, or class stresses, of which there seem to have been rather few. Structural pressures for minority rule were, however, aggravated by what Cobbe's critics in the movement saw as her unseemly and overweening desire to dominate the cause and the public attention it attracted. Cobbe's vendettas with Kingsford and Charles Adams were microcosms of Victoria Street's rivalries with other antivivisection societies. The resulting failure to preserve unity produced inefficiency, fragmentation of effort, and dissipation of energy in internal squabbling. The desire to monopolize initiative was particularly damaging to antivivisection, but by no means unique to it. As Harrison has remarked:

> Philanthropic sectarianism ensured that charities were as inefficient in relieving human misery as the religious bodies in spreading Christianity. They were often more energetic in denouncing alternative remedies for the problems they attacked than in tackling them. Considerations of personal

ambition, class prejudice, and denominational proselytism no doubt lie behind such virulent enmities.[199]

A second important consequence of the concentration of power into relatively few hands was loss of contact with popular sentiment. The leadership of the movement lacked the invigoration of fresh personalities with fresh ideas from the mid-seventies to the mid-nineties. They failed to expand the constituency for antivivisection significantly by harnessing the general public distaste for animal experiment because they saw the issue in black and white terms: whoever is not committed to total abolition is against us. They were completely unwilling to compromise the policy of total abolition, while the greater part of the public was unwilling to go nearly so far. They became isolated from their potential constituency and enmeshed in the mechanics of agitation until agitation became an end in itself. The tone and content of propaganda, the attack upon the medical profession, and the abuse of individuals were all tactics predicated upon a policy unacceptable to the majority, and all helped in alienating public opinion, if we may take the national press as a reliable indicator. The estrangement of the movement from the R.S.P.C.A. and its hubris in launching a frontal assault on the medical profession as a whole symbolize the failures of perception that ensured final ineffectuality. The substance of arguments and principles aside, the style of antivivisectionist agitation—ghastly placards, unscrupulous petitioners, vitriolic pamphlets—undoubtedly put off potential supporters and encouraged public apathy toward the movement.

[199] B. Harrison, "Philanthropy and the Victorians," *Victorian Studies* ix (1966), 365.

9. The Mind of Antivivisection: Medicine

Nothing could be more fatal to a historical understanding of the antivivisection movement than to dismiss it, as some contemporaries were wont to do, as no more than a haven for cranks and misfits. The movement certainly had such within its ranks, but among its spokesmen were highly intelligent and articulate people whose polemical writings can tell us a great deal about Victorian attitudes. Like a distorting mirror, these writings reflect in exaggerated form the sentiments of a great many Victorians of the middle and upper classes, just as the hearing achieved by the movement indicates a general distaste for animal experiment on the part of those same classes. Implicit in the explorations of the assertions and arguments of the movement in Chapters 9, 10, and 11 will be the sense that such arguments are to be taken not only as the holy grail of the converted, but also as signposts for opinions diffused more generally but less explicitly through the society as a whole. Antivivisectionist rhetoric struck resonant chords with a great many people who were unwilling to give formal allegiance to the movement.

In their campaign to end the use of living animals as experimental subjects, antivivisectionists found themselves face to face with the medical profession. The political implications of this confrontation have been considered at some length in previous chapters. What remain to be discussed are the details of the movement's indictment, which covered not only experimental medicine, but also extended to the metropolitan elite of hospital appointees and consultants who promoted it, and finally to the profession as a whole.

Medicine was one of the classic trio of learned professions, which also included law and the church. Unlike the latter, medicine had undergone profound changes during the nineteenth century. If, as a recent study by Peterson demonstrates,[1] it was still during the latter part of the century the least prestigious of the three, medicine nevertheless had grown to a formidable position of institutional strength and social influence by this time. Previous chapters have been sufficient to show this position of relative power and its importance as a resource for the political shelter of experimental research. In order to appreciate better the antivivisectionist attack on medicine, however, it may be worthwhile to look somewhat more closely at the changes

[1] M. J. Peterson, "Kinship, Status, and Social Mobility in the Mid-Victorian Medical Profession," (University of California (Berkeley) Ph. D. thesis, 1972), Chapter 5. Much of the following introductory discussion is based upon this study, and upon F. N. L. Poynter (ed.), *The Evolution of Medical Practice in Britain* (London, 1961); F. N. L. Poynter (ed.), *The Evolution of Medical Education in Britain* (London, 1966); S. W. F. Holloway, "Medical Education in England, 1830–1858: A Sociological Analysis," *History* xlix (1964), 299–324; C. E. Newman, *The Evolution of Medical Education in the Nineteenth Century* (Oxford, 1957); Z. Cope, *The Royal College of Surgeons of England. A History* (London, 1959). On the political relations of the profession, see J. L. Brand, *Doctors and the State* (Baltimore and London, 1966); H. H. Eckstein, *Pressure Group Politics: The Case of the British Medical Association* (London, 1960); E. M. Little, *History of the British Medical Association, 1832–1932* (London, 1932); P. Vaughan, *Doctors' Commons, A Short History of the British Medical Association* (London, 1959); R. Stevens, *Medical Practice in Modern England: The Impact of Specialization and State Medicine* (New Haven, 1966); A. Beck, "The British Medical Council and British Medical Education in the Nineteenth Century," *Bull. Hist. Med.* xxx (1956), 150–162, and "Issues in the Anti-Vaccination Movement in England," *Medical History* iv (1960), 310–321; D. L. Cowen, "Liberty, Laissez-Faire and Licensure in Nineteenth Century Britain," *Bull. Hist. Med.* xliii (1969), 30–40; R. J. Lambert, "A Victorian National Health Service: State Vaccination 1855–1871," *Historical Journal* v (1962), 1–18; and R. M. MacLeod's case studies, "The Edge of Hope: Social Policy and Chronic Alcoholism, 1870–1900," *J. Hist. Med.* xxii (1967), 215–245; "Law, Medicine, and Public Opinion: The Resistance to Compulsory Health Legislation, 1870–1907," *Public Law* (1967), 107–128, 189–211; "The Frustration of State Medicine, 1880–1899," *Medical History* xi (1967), 15–40.

experienced by the profession during the century, as well as at the internal structure of the profession as it came under heaviest pressure from the movement.

By the seventies, medicine in Britain was emerging from a most fundamental transition: from the rather loose association of three "estates" widely varying in size, education, social status, skills, power, and clientele, which prevailed in the early part of the century, to the modern two-tiered profession that, for all the distinctions built around the consultant-general practitioner dichotomy, has comprised a body of individuals of increasingly homogeneous education, skills, and rights to practice, qualified through increasingly consistent procedures.[2] The gulf between the medical and surgical elite of London in the seventies—the Jenners and the Pagets—and the general practitioner of the Midlands (himself no doubt a duly qualified Member of the Royal College of Surgeons and a Licentiate of the Society of Apothecaries or the Royal College of Physicians), though large, does not bear comparison to that of earlier days. For it is difficult, indeed, to see the Oxbridge-educated grandees of the Fellowship of the Royal College of Physicians of London, the gritty surgeons, and the lowly drug-trading apothecaries, as part of a single early nineteenth-century profession of medicine. In comparison to their successors several decades later, they qualified in such different ways (through a maze of no less than nineteen different licensing bodies), ministered to such different people, using such a variety of skills, for such different rewards, that the speed of transformation in the lineage of the profession is startling until it is seen in the context of the equally rapid changes that were taking place generally in nineteenth-century British society. Much of the detail of this transformation, in terms of the politics, prestige, career structure, financial basis, and folkways of the profession, has been unearthed and analyzed by Peterson in her superb and as yet unpublished study. The relevant broader factors may be summarized as the political and

[2] M. J. Peterson, "The Mid-Victorian Medical Profession," Chapter 1.

social changes associated with industrial growth and urbanization and the intellectual changes associated with the vigor of the medical sciences in Europe, both clinical and basic.

The Victorian period saw a vast increase in the administrative activity of the state, activity that included the intervention of government inspection and regulation into a wide array of hitherto relatively unimpeded jurisdictions. Principal among these new activities, for our purposes, was the attention paid to the protection of the public with respect to the provision of services and the preservation of health and sanitation. Thus, public health legislation and regulation were constantly before Parliament and in the press during the middle and latter part of the century, and though the profession regularly complained that it was insufficiently consulted, there can be little doubt that this activity increased the public's sense of its right to health, and, hence, increased pressure on the services of the profession, both at Whitehall and in British society generally. More people expected better medical care than ever before. Further legislative attention was paid to medicine during the fifties, when persistent reform activity revolving around qualification requirements for the registration of medical practitioners finally resulted in the Medical Reform Act of 1858. While Peterson shows that some of the changes often attributed to this Act were in fact the result of practical demands that caused change well before 1858,[3] the Medical Reform Act may be seen as symbolic of the transition from three estates to a much more homogeneous single profession, albeit one comprehending important distinctions of status. The Act confirmed the power of the Royal Colleges and the medical elite over the governance of the profession by creating a General Medical Council dominated by them, while it gave impetus to the process already in motion of breaking down the radical distinctions of education, function, and rights that had long existed among the three estates of physician, surgeon, and apothecary. The establishment of a somewhat more consistent set of

[3] Ibid., pp. 30–58.

pathways to a registration that conferred a single, clear-cut set of rights to the practice of medicine and surgery in Britain may initially have been seen by the tiny elite of the Fellowship of the Royal Colleges as an intolerable adulteration of their status and encroachment on their rights. What it meant in the long run for the rest of the populace was an increasing confidence in the training and skills of the local general practitioner, who was no longer a genus entirely apart from the medical eminences of London or Edinburgh. The Act of 1858 and the associated political experience acquired by the medical leadership substantially altered the position of the profession in national political life, for it provided a base from which the leadership could claim to speak for the whole profession, however inaccurate such a claim may have been with respect to certain issues around which intraprofessional political controversy continued to rage.

These questions of political power and social perceptions were intimately related to advances in medicine and changes in the organization of medical care and medical education, seen against the background of increasing public sensitivity to the efficacy and quality of health services. For instance, the discoveries of vaccination, anesthesia, and antisepsis had important consequences. They changed popular attitudes toward a profession wielding such mysterious but powerful tools, and they reordered the mobilization of resources for training and the provision of care involving them. Such clinical innovations and the significance of such sciences as anatomy and physiology dictated a progressive formalization of medical education, including systematic direct observation and experience in an institutional setting. First, the apprenticeship system, which was standard for surgeons and apothecaries, was combined with the private schools of anatomy of the early part of the century. During the forties, experience in the hospitals, newly emerged as medical schools, became more or less standard, even for the Oxbridge-educated physicians, and apprenticeship died out entirely. Well before the medical Reform Act of 1858, hospitals were bringing together the hitherto separate estates of physician and surgeon and providing a common experience for future medical practitioners.[4] The position of hospitals in the system of providing health care itself had changed radically as a result of increasing demand and therapeutic innovation.[5] By the seventies and eighties hospitals were emerging as virtually the only source of medical care for the seriously ill, rich and poor alike. Previously, they had been the refuge of the penniless and the helplessly ill; no one who could afford to pay for medical care would utilize hospital facilities. Now, with the medical profession taking the lead in the foundation and expansion of more hospitals for more people, such institutions came to be seen as a vital national resource close to the interests of people from all parts of the social spectrum. The policies and practices of hospitals, even those related solely to charity patients, came to be of interest to the public as a whole. In hospitals, patients with similar cases could be grouped together to facilitate efficient treatment, instruction of students, and clinical research. Hospital posts became eagerly sought after as indicators of professional status; such positions were not only challenging medically, but were lucrative as well, for the middle and upper classes would patronize the private practices of men with hospital appointments, on the assumption that such appointments denoted superior medical ability.[6] Thus the coveted hospital appointment formed the basis for a new distinction within the profession: that between the elite metropolitan consultant and the general practitioner. The London consultants numerically constituted less than two percent of the profession, but they effectively dominated it through an interlocking web of hospital appointments, medical school chairs, and Fellowships and offices in the Royal Colleges.[7]

[4] Ibid., pp. 26–30 and Chapter 2.

[5] The best general account is B. Abel-Smith, *The Hospitals 1800–1948* (London, 1964).

[6] See Abel-Smith, *The Hospitals,* pp. 16–31 and Peterson, "The Mid-Victorian Medical Profession," Chapter 4. As Peterson makes clear, the assumption was not inevitably correct.

[7] See Peterson, "The Mid-Victorian Medical Profession," pp. 196–197, and Chapter 4.

It was certain of the leading members of the London medical elite who were promoting experimental medicine during the seventies and eighties and whose political experience and connections were vital to its survival in the face of antivivisectionist pressure. They saw experimental physiology, pharmacology, and pathology as the most rapid route to a new, more powerful array of therapeutic tools. Further, experimental medicine was a rigorous, systematic, esoteric, scientific discipline that could serve as the intellectual touchstone for training and qualification in a profession whose guardianship of its own body of specialized knowledge and skills was a fundamental part of its raison d'etre. Initiation into the mysteries of the basic medical sciences would provide a cachet and an insight for the genuine professional far different from that of the self-educated quack or even that of the older medical man who qualified through apprenticeship and ward-tramping. The experimental approach would avoid both the false system-building and the therapeutic nihilism that, its promoters believed, had damaged public confidence in medicine. These views were taking an increasing hold among the consultant-professor-Fellows of the metropolis, as they were promulgated by leaders such as James Paget, William Jenner, and William Bowman, and by experimentalists such as Michael Foster, John Burdon Sanderson, and Thomas Lauder Brunton. While on many issues, notably those involving the structure of fees, patterns of patient care, and prosecution of quackery, the vast majority of general practitioners were by no means at one with the closely knit London fraternity who governed the profession,[8] there was, as we have seen, a remarkable closing of ranks in response to the external threat of antivivisection. This occurred despite the fact that few of the medical men who signed petitions and attended British Medical Association protest meetings in 1876 would have performed or even have seen vivisections during the course of their medical education. Likewise, the leadership of the Association for the Advance-

[8] Ibid., pp. 351–388.

ment of Medicine by Research called for general support from the rank and file (in the form of subscriptions rather than public agitation) in 1882, and it did not even feel the necessity to democratize the organization, as advocated by Ernest Hart, in order to ensure such support. When laymen appeared to menace the autonomy of medicine, the mass membership of the B.M.A. and the professional aristocracy were quite prepared to bury their differences in a show of strength. It is hard to imagine the surgeons and apothecaries of the twenties and thirties rallying in similar fashion behind the then Fellows of the Royal College of Physicians.

To summarize, the antivivisectionist attack on experimental medicine, which ultimately included the medical profession as a whole and its institutions, came fairly late in a period of significant change for the profession. Insofar as this change involved an overall increase in the power, homogeneity, and unity of the profession with respect to the national political arena, the movement was to find its frontal attack on the profession frustrated. This was particularly true to the extent that the critique drove the medical leadership as a body to the public support of a politically and numerically weak handful of experimentalists, and in turn, to the extent that it drove the general practitioners to a (less than instinctive) support of the leadership on this particular issue. On the other hand, some aspects of change, such as the new elitism of the consultant hospital appointees, the rise of the hospitals, and the educational and institutional ambitions of experimental medicine, were by no means universally acceptable to the profession or, indeed, to the public at large. The antivivisectionist critique may be thought of as the attempt to exacerbate these schisms, as the sotto voce grumbles of the disaffected and disenchanted writ large.

This chapter is devoted to an exploration of the movement's attack upon experimental medicine, the medical profession, and its institutions. As such, it is a mosaic, an attempt to form a composite of the widely varying and sometimes contradictory opinions of antivivisectionists, opinions that resist classification or simple characterization. Let us begin with the question of animal experiment and then move on to the broader attack on the profession and its leadership.

ANIMAL EXPERIMENT

Antivivisectionists gave innumerable justifications for their concern with the issue of animal experiment. They viewed it as a central moral dilemma for the society at large and at one time or another linked its resolution with that of almost every other social and ethical problem extant. At the most basic level, however, the preoccupation with vivisection allegedly grew out of what the movement took to be a grievous outrage to the educated classes' franchise upon moral leadership. Experiments upon living animals were performed not only by individuals whose social position carried with it the noblesse oblige to set a good example; such experiments had also the explicit sanction of the state:

> With what consistency, or hope of success, can we pursue our warfare against the ruffianism of the street or the alley, or the other manifold forms of cruelty engendered of luxury or vice, when our remonstrance may be met by the plea that, do what the perpetrators may, they cannot begin to compete with the licensed horrors of the laboratories?[9]

The movement viewed its mission as a crusade against the professionalization and institutionalization of experimental medicine, which was seen as a trend, of foreign origin, expanding rapidly in Britain. At the crudest level,

[9] [E. Maitland], *"The Woman" and the Age: A Letter Addressed to the Right Hon. W. E. Gladstone, M. P., First Lord of the Treasury . . .* (London, 1881), p. 16. See also D. Mushet, *The Wrongs of the Animal World* (London, 1839), p. 192; *Spectator* xlix (1876), 555, 673; *Zoophilist* i (1881–1882), 10; *The Debate on Mr. Reid's Bill in the House of Commons, April 4, 1883* (London, 1883), p. 11; B. Wilberforce, "Women, Clergymen, and Doctors," *New Review* viii (1893), 93.

xenophobia could be appealed to by references to "foreign demons of vivisection" and the practice could be stigmatized as "not congenial to the soil; it comes from abroad. . . ."[10] However, while *Leisure Hour* asked, "surely there must also be many men of cultivated and humane minds in the medical profession, who would side rather with the old English than with the new French school of physiology?,"[11] the movement as a whole attempted to erase national distinctions, and thereby taint British experimentalists with the sins of their continental counterparts. Antivivisectionists emphasized the solidarity of British and continental researchers: Britons attended overseas schools, they discussed and praised continental experimeters in their textbooks, and they subscribed to memorials for such as Claude Bernard. How then could they claim that British experimental medicine was more humane than the French or German variety?[12] Furthermore, experimental medicine was represented as widespread, systematically pursued, and growing at an alarming rate. Enthusiasts were apt to speak in terms of "hundreds of thousands" of animals subjected to experiment. The Victoria Street Society, on the other hand, compiled considerable data on the actual extent of research that might involve experimentation upon living animals.[13] In Europe in the early eighties, Victoria Street found 143

[10] W. Lescher, *The Practice of Vivisection* (Stroud, 1894), p. 21. See also [M. Wemyss], *Licensed Cruelty* (Gloucester, c. 1886); J. Moffatt, *The French Exhibition of Horrors: A Sermon on the Sin of Torturing Animals* (Toronto, 1879).

[11] *Leisure Hour* (1876), 555.

[12] See, e.g., F. P. Cobbe (ed.), *Bernard's Martyrs. A Comment upon Claude Bernard's Leçons de Physiologie Operatoire* (London, 1879), esp. p. iv; A. Kingsford, *Unscientific Science* (Edinburgh, 1883), pp. 9–10; C. A. Gordon, *The Vivisection Controversy in Parliament . . .* (London, 1888), pp. 11–12. Cf. one physiologist's defense, G. F. Yeo, "The Practice of Vivisection in England," *Fortnightly Review* xxxvii (1882), 352–368.

[13] See, e.g., B. Bryan (ed.), *The Vivisectors' Directory; Being a List of Licensed Vivisectors in the United Kingdom, Together with the Leading Physiologists in Foreign Laboratories* (London, 1884); G. M. Rhodes, *The Nine Circles of the Hell of the Innocent. Described from the Reports of the Presiding Spirits* (London, 1892).

"vivisecting laboratories" with 241 "vivisectors."[14] The Society believed the experiments performed to be largely repetitious exercises for training or testing:

> The greater number are, so to speak, *stock* experiments. They are gone over by each new recruit in the army of science who takes up the study of the organs concerned, and may be likened more properly to the scales and exercises of the musical practitioner, than to the purposeful operations of the surgeon.[15]

Indeed, "It is a fact that most experiments now carried out have been done before—there is, in fact, little or nothing absolutely new to be done."[16] To antivivisectionists, the really ominous aspect of this widespread, redundant cruelty was that its ostensible purposes were receiving the kind of institutional recognition that opened up career possibilities. Vivisection was not the infrequent court of last resort for the dedicated medical man: "It is on the contrary a *profession*, a regular and independent business; to which

[14] Included in the calculations were laboratories of physiology, anatomy, pharmacology, bacteriology, pathology, histology, teratology, etc. The summary:

Country	*Laboratories*	*Vivisectors*
France	39	85
Britain*	32	45
Italy	32	51
Germany	20	29
Netherlands	5	8
Switzerland	5	6
Austria	4	7
Sweden	4	6
Denmark	1	3
Norway	1	1

* Includes Ireland.

Source: *Zoophilist* ii (1882–1883), 173.

[15] F. P. Cobbe, *Light in Dark Places* (London, n.d.), p. 4.

[16] B. Bryan, *The Inspectors' Report and Vivisectors' Returns for the Year 1897* (London, 1898), p. 4. See also E. Bell, *Vivisection on the Increase. My Doctor Tells Me That "There is very little Vivisection now in England"* (London, c. 1896).

men devote themselves with ardour and ambition. . . ."[17] The traditional humanity of the English medical profession,

> . . . is being altered, under the pretext of supplementing medical knowledge, by a DISTINCT CLASS *exclusively engaged* in physiological speculations which are to give scientific certainty to medical practice, or at least, as the next excuse, to promote abstract science.[18]

Many physiologists were not medical men at all, but self-interested scientific *entrepreneurs:*

> It was not the medical profession calling out for scientific aid, but the biologists, infected by foreign fashions, making themselves a necessity, and erecting a *distinct profession* over the heads of the medical faculty, with secrets which nobody but the initiated can understand . . .[19]

It was this expansionist momentum, and a lively sense of its consequences upon the continent, that spurred R. H. Hutton, as a Royal Commissioner, as a member of the Senate of the University of London, and as editor of the *Spectator,* in his battle against "scientific cruelty."[20] Hutton epitomized antivivisectionist sentiment when he wrote,

> You cannot by any possibility inaugurate a new and highly distinguished profession of persons whose business it is known to be to inflict on animals any amount of suffering requisite for the special purpose of benefitting men, without giving a new impulse to the selfishness of men in every other grade of life . . .[21]

[17] F. P. Cobbe, *Bernard's Martyrs. A Comment on Claude Bernard's Leçons de Physiologie Operatoire* (London, 1879), pp. x–xi. Italics here and below, are the original author's.

[18] B. Grant, in J. Macaulay, B. Grant, and A. Wall, *Vivisection, Scientifically and Ethically Considered in Prize Essays* (London, 1881), p. 105. Italics in the original.

[19] Ibid., pp. 108–109. Italics in the original.

[20] See, e.g., *Spectator* lx (1887), 279.

[21] R. H. Hutton, "The Biologists on Vivisection," *Nineteenth Century* xi (1882), 34.

In their attempts to convince the public of the cruelty and unjustifiability of experiments upon living animals, antivivisectionists had a choice between two basic lines of argument. In general terms, they could argue either from the immorality of the method or from the inutility of the results. While in practice the two approaches were frequently combined, the distinctive characteristics of each are worth examining for what they reveal about the antivivisectionist mentality.

The argument that, whatever its results, a method of research that causes pain to sentient creatures was ipso facto immoral had as an important corollary that antivivisectionists need not meet proponents of experimental medicine upon the latter's home ground. That is, if the utility of the method was of no consequence to its legitimacy, the movement could and did argue that "Right and wrong can never be a medical question," but "a question for the country at large, and not for the doctors, to settle."[22] This theme, that the conscientious citizen was as competent to decide the morality of the method as any learned scientist or doctor, was reiterated again and again in the literature of the movement. Lord Chief Justice Coleridge asked in the early eighties:

> Why should a venerable osteologist, a world famed naturalist, or a couple of most illustrious physicians, be any better judges than a man of average education, and average fairness, when the question is what is the limit (it being I think certain that there *is* one) between lawful and unlawful knowledge, and lawful and unlawful means of gaining it . . . ?[23]

[22] H. N. Bernard, *Vivisection: How should Christians Regard it?* (London, c. 1885), p. 4; J. H. Clarke, *The Plea for Vivisection—What it means* (London, 1887).

[23] *The Lord Chief Justice of England on Vivisection* (London, c. 1882), p. 15. See also *A Few Words on Vivisection, By a Member of the Birmingham Speculative Club* (London and Edinburgh, 1875), p. 4; C. A. Gordon, *The Vivisection Controversy in Parliament* . . . (London, 1888), pp. 45–50; W. S. Lilly, *An Independent Opinion on Vivisection* (London, 1885), pp. 2–3; W. Lescher, *The*

Such assertions were, of course, part and parcel of the movement's attempt to reduce the influence of medical men over their patients with regard to the issue in question. The argument also partook, as did the movement as a whole, of growing public suspicion of the arrogance of the scientific establishment:

> If society be right in refusing to recognise the infallibility of a purely ecclesiastical caste in matters affecting the public conscience,—as, for instance, in respect of religious persecution—it is equally right in refusing to admit the assumption of infallibility on the part of a caste exclusively scientific and materialist, in matters similarly affecting the public conscience.[24]

The argument from the immorality of the vivisectional method, then, had the advantage of avoiding direct confrontation with the technical complexities involved in a discussion of the utility of experiments upon living animals. It was also a clear and simple battle cry with which to appeal to the public. Frances Power Cobbe had antecedent philosophical basis for preferring such a position. She saw herself as a proponent of intuitive ethics and denied the applicability of inductive evidence to moral judgment.[25] Whatever the medical benefit of animal experiments, it was too dearly bought at the price of national complicity

---

Practice of Vivisection (Stroud, 1894), p. 11; R. H. Hutton, "The Anti-Vivisectionist Agitation," *Contemporary Review* xliii (1883), 512–513.

[24] A. Kingsford, *Unscientific Science* (Edinburgh, 1883), p. 37. Cf. E. Maitland, "Vivisection," *Examiner* (1 July 1876), 738–739.

[25] "Secondly, speaking as an Intuitive Moralist—one who holds that moral laws are independent of Utility and that they are Useful because they are Right not Right because they are Useful, I decline for my own part to entertain the question of the possible usefulness of vivisection." F. P. Cobbe, *Meeting at Stoke Bishop* (London, 1883), p. 4. See also F. P. Cobbe, *Life of Frances Power Cobbe* (Boston, 1894), i. 101, ii. 418ff. For general background to the Utilitarian-Intuitionist debate over ethics, see J. B. Schneewind, "Moral Problems and Moral Philosophy in the Victorian Period," *Victorian Studies* ix (1965), 29–46.

in a brutal and demoralizing practice. "You may buy knowledge at the cost of sin," but that cost was too great.[26] As the Report of the Committee of the Victoria Street Society put it in 1878,

> Better it is in the supreme interests of man as well as brute that all the vaunted benefits to be won by Vivisection (were they ten-fold greater than they are ever likely to be) should remain unattained to the end of time rather than that by familiarity with the impassive pitilessness of the physiologist our race should lose those sentiments of mercy and sympathy which are of more worth than a million facts of science.[27]

Of what precisely did the moral price of vivisection consist? In the first place, it should be noted that antivivisectionists were convinced that a very large majority of experiments upon living animals involved severe and often prolonged suffering for the subject. Anesthetics were regarded as "the Will-o'-the-wisps of Science misleading sincere but imperfectly informed Anti-Vivisectionists and luring them into the bogs and quagmires of physiological deception off the straight, hard road of Abolition."[28] It was claimed that anesthetics were rarely used, and when they were, they were ineffective. Hence, said George Hoggan, "I am inclined to look upon anesthetics as the greatest curse to vivisectible animals."[29] Instead of anesthetics curare was

[26] *Spectator* liv (1881), 527. See also R. Quiddam, "Vivisection," *St. James's Magazine* xxxviii (1876), 286; R. H. Hutton, "The Biologists on Vivisection," *Nineteenth Century* xi (1882), 30; V. Lee [V. Paget], "Vivisection: An Evolutionist to Evolutionists," *Contemporary Review* xli (1882), 796–797, 811; A. Kingsford, *Unscientific Science* (Edinburgh, 1883), pp. 31–32.

[27] *Transactions of the Victoria Street Society for the Protection of Animals from Vivisection During the Half-Year Ending Dec. 31st, 1879* (London, 1880), p. 19.

[28] *Anaesthetics and Vivisection* (London, 1887), p. 1.

[29] G. Hoggan, *Anaesthetics and the Lower Animals* (London, 1875), p. 2. Cf. *Do We Exaggerate?* (London, 1889). More than anyone else, Hoggan was responsible for antivivisectionist cynicism about anesthetics. See the correspondence on the subject in *Spectator* lv (1882), 184, 263–264, 292–293, 326–327, 358–359, 392–393.

often used, which simply paralyzed the animal without anesthetizing it.[30] Curare captured the imagination of the movement, inspiring with particular horror many who came to think of it as actually intensifying pain.

Given, then, the inevitable suffering involved in virtually all vivisection, what was the system of experimental medicine other than a base pursuit of mere bodily health by means of selfish cruelty to sentient living creatures? It was not only that causing pain to an animal for reasons other than its own ultimate welfare was wrong, there was also the hardening and demoralization of the human beings who were party to such experiments. What of the impressionable young students, especially the would-be doctors? Lewis Carroll looked grimly forward to ". . . a day when successive generations of students, trained from their earliest years to the repression of all human sympathies, shall have developed a new and more hideous Frankenstein—a soulless being to whom science shall be all in all."[31] This theme of the inevitable degradation inherent in vivisection was reiterated constantly in the movement's literature.[32]

Defenses of experimental medicine were far fewer in number than the antivivisectionist broadsides, and were largely confined to addresses before medical and scientific bodies or to articles in the intellectual reviews. Except during the period when the A.A.M.R. was founded, leaders of the profession felt that there was little to be gained by a public joining of issues with the movement. After all, how could the public judge the validity of technical points of science and medicine? This aloof posture had reached such an extent that in 1892 the *Times*, no friend of antivivisection, commented that "Men of science who take up an attitude of lofty disdain towards popular agita-

[30] See, e.g., *Anaesthetics and Curare in Vivisection* (London, c. 1882).

[31] L. Carroll, "Some Popular Fallacies About Vivisection," *Fortnightly Review* xxiii (1875), 854. For more on Carroll's views, see S. D. Collingwood, *The Life and Letters of Lewis Carroll* (New York, 1891), pp. 165–171, 187–191.

[32] See, e.g., F. P. Cobbe, *The Higher Expediency* (London, 1882).

tions and refuse even to show the futility of the assertions of the agitators may some day find themselves unexpectedly overwhelmed by a sudden rush of the sentimentalists."[33] As we have seen, the profession regarded itself as the guardian of a sacrosanct body of esoteric knowledge, for whose validity it took sole responsibility. The public that was to benefit from that knowledge could not expect to exercise any arbitrative function regarding it.

Those proponents of experimental medicine who did publish articles justified experiments upon living animals almost exclusively on the grounds of practical utility. Living creatures suffered a great deal of pain in agriculture and sport; why, then, should not a certain minimal amount of animal suffering be sanctioned for the pursuit of knowledge that would be valuable to society for years to come? Generally speaking, professional spokesmen ignored the antivivisectionist's insistence upon judging the question solely on the grounds of the intrinsic morality of the method.[34] When they did notice the argument, they invariably denied the possibility of assessing an action without reference to its results: "While for the purpose of dialectics it may be possible to separate morality and utility, it is impossible to perform this operation when dealing with mundane affairs and questions of daily living and dying."[35] As long as the utilitarian consequences weigh in the matter, "It is hopeless for any lady, however eminent in philanthropy, however famed in literature—or, indeed, for any class of women or men, excepting trained physiologists, to form an opinion of the slightest value on this point."[36]

[33] Quoted in *Public Opinion* lxii (1892), 544.

[34] See, e.g., S. Wilks, "Vivisection: Its Pains and Its Uses," *Nineteenth Century* x (1881), 936–948, who claimed that the inutility of vivisection was the crux of the movement's case. See the remarks of Edmund Gurney in "A Chapter in the Ethics of Pain," *Fortnightly Review* xxxvi (1881), 778–796, and "An Epilogue on Vivisection," *Cornhill Magazine* xlv (1882), 191–199.

[35] V. Horsley, "The Morality of Vivisection," *Nineteenth Century* xxxii (1892), 804–811.

[36] J. Hutchinson, "On Cruelty to Animals," *Fortnightly Review* xxvi (1876), 309.

Some supporters of the experimental approach, notably scientists in *Nature* and the *Journal of Science*, resented the insistence upon the practical utility of vivisection and claimed that it could be justified simply on the grounds that it advanced man's knowledge.[37] Whatever their private sentiments, however, most defenders of vivisection were content to represent the question solely in terms of medical benefits. They recognized in such an argument a means of forcing antivivisectionists to attack experimental medicine on grounds where expertise would tell most effectively. Even more important, they realized that many laymen were prepared to countenance vivisection in the interests of health although they were quite unsympathetic to the plea that it advanced scientific knowledge.

Many antivivisectionists also perceived the public's interest in its own health and sensed that Victoria Street's pride in the solely moral nature of its appeals to the public, through the mid-eighties, was unjustified and misconceived. Inevitably, it was the maverick Charles Adams who voiced these reservations,

> . . . the world at large . . . holds . . . that the mere fact of Vivisection being, as its practisers assert it to be, profitable to humanity, renders it *ipso facto* morally right. . . . it is upon this axiom that the chief, indeed the only effective, defence of the practise is based.
>
> With those who hold by this axiom, and they form, and we fear always will form, the overwhelming majority of mankind, the ethical argument essentially fails. And this being so, the exceeding importance, to the battle we have to fight, of that minor premiss of the vivisector's syllogism which consists in the assertion that vivisection is of profit to mankind becomes at once only too apparent. Between its meek acceptance and its bold rejection lies the whole difference between energetic attack and almost desperate defence.[38]

[37] *Nature* xiv (1876), 65; *Journal of Science* vii (1885), 599. See also *British and Foreign Medico-Chirurgical Review* lx (1877), 146–148.

[38] *Verulam Review* iii (1892–1893), 205–206. Cf. *Zoophilist* i (1881), 190–191, probably written while Adams was still editor.

In the end, both sides were to find the public more impressed by the promise of concrete benefits in health and therapeutics than by the intuitive illegitimacy of vivisection or the intrinsic worth of scientific knowledge.

Before discussing the movement's attack upon claims for the medical utility of experiments upon living animals, it may be worthwhile to examine antivivisectionist attitudes toward bodily health, since these lie at the root of their unwillingness even to recognize such claims. To most antivivisectionists, the argument that vivisection was justified because of its medical value was nothing more than a blatant appeal to man's most selfish, cowardly, and elemental instincts of self preservation. An animalistic obsession with bodily health could be the only basis for entertaining such an argument, for disease was the divinely ordained consequence of sin and folly, and was to be borne as such. Experimental medicine based upon the plea of practical utility was simply "slighting a sovereign Providence, by searching desperately in the midst of bloodshed and torments for the balm of human life."[39] It meant that man was unwilling to bear his allotted share of pain in the world: "there can be no offence more shocking and no act more dastardly than this of trying to shift the natural punishment of our own sins and vices and stupidities, on to the shoulders of those who are powerless to resist us."[40] The ubiquity of this kind of argument in antivivisectionist controversy reveals to us a segment of Victorian opinion for whom medical advance, regardless of source, appeared as a tampering with the order of things, an unwarranted intrusion of rather distasteful medical doctrine into public consciousness. For many such people, experimental medicine was a fundamentally misdirected effort, grovelling in the entrails of animals when the path to progress lay in the prevention of disease, through sanitary measures

[39] D. Mushet, *The Wrongs of the Animal World* (London, 1839), p. 213. See also pp. 200–207.

[40] M. Caird, *A Sentimental View of Vivisection* (London, c. 1893), p. 28.

and, most pressingly, through moral improvement.[41] The role of suffering as a means to grace was an important part of the theodicy associated with classical natural theology, which, as we shall see in the following chapter, was a central tenet in the antivivisectionist creed. To an unsympathetic medical observer, "it was these same people or others of a like kind who, when medical science invented a means of alleviating the 'tortures' of child-birth, denounced the invention as an impious attempt to save women from the punishment rightly inflicted on them by an offended deity."[42]

To the popular novelist Ouida, medical knowledge was responsible for a "culture of cowardice" in which people

> . . . are taught perpetually to endeavour to defend themselves from the proximity of death by the most minute cares and the most elaborate precautions; they are to pass their whole existence in a stench of disinfectants; they are to see deadly organisms in everything they touch; they are to suspect injury to themselves in every breeze which blows; they are to shrink in fear of contamination from the rosy lips of a little child, and flee from the good-natured gambols of a merry dog; the pleasant odour of a freshly turned furrow to them speaks of poisonous exhalations and mephitic vapours; the prick of a pin may mean tetanus, and the humming of a blue-bottle fly can only preface an inoculation of carbon.[43]

According to Ouida, "The increase of dread and nervousness in the world in general is due to physiologists . . . ,"

[41] Such opinion was, of course, by no means unknown within medical circles. For an excellent discussion of "the hostility of certain sanitarians to animal experimentation," see L. G. Stevenson, "Science Down the Drain," *Bulletin of the History of Medicine* xxix (1955), 1–26. Cf. A. Wall in J. Macaulay, B. Grant, and A. Wall, *Vivisection, Scientifically and Ethically Considered in Prize Essays* (London, 1881), pp. 288–291.

[42] R. Harvey, *The Pasteur Institute and Vivisection* (Calcutta, 1895), p. 19.

[43] E. Lee, *Ouida: A Memoir* (New York, 1914), p. 325.

who teach "The blind and brutal passion of self-preservation," which is "destructive of moral character."[44]

Not only, then, did many antivivisectionists regard the spiritual and moral aspects of experiments on living animals as prior and far higher considerations than any practical and merely physical benefits that might accrue from them,[45] some members of the cause showed a decided disgust for any considerations of bodily welfare and an animus against doctrines that to them constricted and choked the spontaneity of life. For others, medical research seemed an enterprise in blatant contradiction to a deeply felt belief in a kind of natural theology of disease. Physical afflictions were divine in origin, and attempts to circumvent them were a futile impertinence bordering on blasphemy.

Frances Power Cobbe christened "the elevation of Bodily Health into the summum bonum" as "Hygeiolatry." Hygeiolatry had as its necessary consequence "Doctor's Doctrine": "that any practice which, in the opinion of experts, conduced to bodily Health or tends to the cure of disease, becomes ipso facto, morally lawful and right."[46] It was in this light that Cobbe's particular crusade could be seen in its broadest moral perspective:

> To contend against Vivisection is, then, not against any exceptional or transitory evil, but against those besetting sins of the age of which it is the outcome,—selfishness and cowardice, and the pitilessness characteristic of cow-

[44] Ouida [L. de la Ramée], *The New Priesthood* (London, 1893), pp. 30, 32.

[45] See also *Spectator* lii (1879), 1011; G. R. Jesse, *"The Times" on Vivisection* (?London, 1881); A. Kingsford, *Unscientific Science* (Edinburgh, 1883), pp. 31–32; J. H. Clarke, *The Plea for Vivisection—What it means* (London, 1887), pp. 6–7; R. Wagner, "An Impossible Science," *Animals' Guardian* i (1891), 76–78; *Sir R. T. Reid, Ex-Attorney-General, on Vivisection* (London, 1892); E. Carpenter and E. Maitland, *Vivisection* (London, 1893), pp. 14–17, 20–25; *Quotations from Great Thinkers* (London, c. 1895), p. 10.

[46] F. P. Cobbe, "The Scientific Spirit of the Age," *Contemporary Review* liv (1888), 135; "Hygeiolatry," *The Peak in Darien* (Boston, 1882), p. 81; *Doctor's Doctrine. A Correspondence in the "Leicester Daily Post"* (London, 1889).

> ards; overestimate of the Body as compared to the Soul; overestimate of Knowledge as compared to Love.[47]

Given this kind of view of the ethical significance of opposing experiments on living animals, the movement's ambivalent attitude toward a critique of the utility of vivisection becomes easier to understand. Ultimately, however, almost all of the movement's leaders recognized that practical medical and scientific arguments against claims for the benefits of vivisection represented one indispensable path toward the conversion of a public quite unmoved by simple appeals to the intrinsic immorality of the method. An ever-increasing proportion of antivivisection literature consisted of attacks upon the supposed triumphs of experimental medicine. Proponents of experimental medicine had instanced a growing number and variety of scientific and especially medical advances based upon vivisectional research: Harvey's discovery of the circulation of the blood, Aselli's discovery of the lacteals, John Hunter's development of techniques for tying arteries in cases of aneurism, Edward Jenner's introduction of vaccination for smallpox, the testing and development of anesthetics, and so on. As fast as such examples were put forward, they would be demolished by the spokesmen for antivivisection, who employed any number of approaches in debunking such "spurious" claims. They could deny the medical value of the knowledge gained; they could deny the scientific validity of the conclusions drawn. A regular objection to animal experiment was that the pain and suffering resulting from it caused the kind of physiological disturbance that invalidated any possible results. It was frequently argued that experiments upon living animals were not absolutely necessary as a basis for discoveries alleged to have come from such experiments: anatomical research or clinical observation or some other alternative means would have led to the same conclusions. A most sophisticated statement of this point of view came from positivist medical men

[47] F. P. Cobbe, *The Significance of Vivisection* (London, 1891), p. 3.

J. H. Bridges and Richard Congreve.[48] Debates over the historical evidence pro and con vivisection resulted in a huge but largely sterile literature.[49] Proponents of experiment were content to appeal occasionally to the reading public, for the most part without attempting to engage in specific controversy with antivivisectionists. From such an engagement they had nothing to gain, and they refused to be drawn. On the other hand, in the mid-seventies, they lacked a single, clinching, readily understandable example of a widespread medical practice based inevitably and unambiguously upon knowledge gained from experiments upon living animals. Without such an example, it was entirely possible for an earnest layman to read the literature on both sides of the question and—especially if he were unwilling or unable to assess the relative professional stature of the spokesmen for each side—emerge with the sense that the central question was still very much open.

An example may serve to illustrate both the importance and the complexities of the relationship between laboratory advances and the medical implementation of techniques based in part upon them, a problem that, in its broadest sense of the link between science and technological innovation, continues to plague students of the subject. In Chapter 7, the importance of the prosecution of David Ferrier in 1881 to the establishment of the A.A.M.R. in the following year was discussed. The choice of Ferrier as a subject of prosecution was dictated by his apparent vulnerability in not possessing a license, but it had a profound effect on the scientific and medical community because his work on the localization of the motor and sensory functions within the cerebral cortex of the brain was regarded as highly significant. In 1860, nonvivisectional work in the

[48] R. Congreve and J. H. Bridges, "Vivisection," *Fortnightly Review* xxiii (1875), 435–437; J. H. Bridges, "Harvey and Vivisection," *Fortnightly Review* xxvi (1876), 1–17.

[49] Three typical specimens of this literature are W. G. Gimson, *Vivisections and Painful Experiments on Living Animals: Their Unjustifiability* (London, 1879); C. A. Gordon, *Remarks on "Experimental" Pharmacology* (?London, c. 1883); C. Adams, *The Coward Science* (London, 1882).

clinic and the dissecting room had enabled Paul Broca to localize the speech center in the cerebrum and, thus, to challenge the doctrine of cerebral equipotentiality, the idea that any part of the cortex can subserve the collective functions of the cerebrum as well as any other similarly sized part. This doctrine continued to dominate neurological thought into the seventies, although the demonstration of localized electrical excitability in the cerebra of living dogs by Fritsch and Hitzig in 1870 was a major turning point toward localization. From the early seventies to the late eighties, Ferrier extended and confirmed the findings of Broca and Fritisch and Hitzig with a meticulously executed series of ablation and stimulation experiments upon the exposed cerebral hemispheres of animals such as dogs, cats, rabbits, guinea pigs, and monkeys. By ablating a specific area of the cerebrum, and observing subsequent paralysis or sensory failure in the living animal, Ferrier could develop a series of designations of specific areas in the brain that subserved specific motor or sensory functions. Electrical stimulation of the cerebrum allowed similar correlations to be made. The experiments had to be painstakingly carried out and reported, in order to carry conviction in what became a very hotly contested area of medical debate. The epic confrontation between localization and equipotentiality took place between their leading champions, Ferrier and Friedrich Goltz, at the International Medical Congress in 1881. It was this confrontation, at which Ferrier's experimental work carried the day, that received the wide publicity that stimulated Ferrier's prosecution. Because of the work of Ferrier and others, "by the close of the century the main cortical centres for motor functions and the various sensory modalities in mammals were established."[50]

What of the practical consequences of the scientific doctrine of cerebral localization? Tumors and foreign objects embedded in the brain had long presented intractable prob-

[50] R. M. Young, *Mind, Brain and Adaptation in the Nineteenth Century* (Oxford, 1970), p. 240. This discussion is based principally upon pp. 224–245.

lems for clinicians. Motor and sensory symptoms could be described, but any pain in the cranium was usually dispersed widely, preventing sufficient diagnosis for surgical intervention at a precise point in the skull. The doctrine of cerebral localization provided a theoretical basis for inference from symptoms to locus of tumor, but the experimental data supporting it were derived from research on animals rather than man. Thus the doctrine gave a mandate to medical men to begin a vast enterprise of correlation of patients' symptoms during life with the results of postmortem dissections of their brains. This enterprise, combined with the data already available from animals, slowly became such as to permit diagnosis quite adequate for the purposes of neurosurgery. Parallel advances in practical techniques for surgery in an extraordinarily delicate area of the body emerged, such that modern neurosurgery can give substantial hope where none existed in the seventies.

The history summarized here would, of course, provide little challenge to the antivivisectionist critic seeking to disprove the value of experiments upon living animals in therapeutic advance. He would simply point to Broca's nonexperimental identification of the speech center, to the extensive work in neurological theorizing, clinical observation, and postmortem, and claim that these activities alone could have lead, or perhaps did lead, to neurosurgery as we know it today. Antivivisectionist literature on the medical utility of experiments on living animals comprises exhaustive reviews of similar issues, filled with tortuous and ingenious reasoning to the same effect. The layman is left with the usual conflict of testimony. Perhaps it would be appropriate to give Ferrier the last word, recognizing that it would scarcely be accepted by our antivivisectionist spokesman. Ferrier admitted the problems of extending results obtained from experiments with animals to man, but he went on,

> Notwithstanding these difficulties and discrepancies, many of which will be found, on careful examination, to be more apparent than real, experiments on animals under

> conditions selected and varied at the will of the experimenter, are alone capable of furnishing precise data for sound inductions as to the functions of the brain and its various parts; the experiments performed for us by nature, in the form of diseased conditions, being rarely limited, or free from such complications as render analysis and the discovery of cause and effect extremely difficult, and in many cases practically impossible.[51]

It is generally felt that it was the accomplishments of bacteriology and immunology during the last two decades of the nineteenth century that provided the example required by spokesmen for experimental medicine.[52] There could be no denying that the discoveries involved were based on experiments on living animals. The consequences for medical practice eventually affected the lives of a large number of people and received unequivocal endorsement from the medical establishment and the government. Hence, the rise to fame and institutional prominence of bacteriology and immunology during the eighties was particularly bitterly opposed by the movement, which perceived the threat clearly. Antivivisectionist attacks upon Louis Pasteur and his method for the cure of rabies flooded forth until the issue seemed almost a monomania. The imminent demise of Pasteur's system was forecast from the mid-eighties until the late nineties, when the matter was suddenly dropped. The antivivisection movement had all but said that the success or failure of immunological medicine constituted a crucial test case; it could scarcely face the fact of its success.

In addition to a substantive discussion of the merits of experiments upon living animals as a means of improving medical therapeutics, antivivisectionists also engaged in a concerted campaign to convince the public that defenses of vivisection on the grounds of utility were an insincere

[51] Quoted in ibid., p. 238.

[52] Cf. R. H. Shryock, "Freedom and Interference in Medicine," *Annals of the American Academy of Political Science* cc (1938), 36, and "Public Relations of the Medical Profession," *Annals of Medical History* n.s. ii (1930), 308–309.

sham. The real motive of vivisectors, it was said, was not improvement of the public's health, but the indulgence of a passionate and morbid curiosity. At the expense of untold animal suffering, experimenters were covertly in pursuit of their own selfish and dilettantish objectives: the advance of mere abstract science and the achievement of fame. According to one antivivisectionist medical man,

> Vivisectors, finding that public opinion is opposed to the heartless plea of the mere advancement of science, have lately invoked the aid of the medical profession to shield them from popular reprobation. It is the policy of the biologists to mix up their cause with that of medical progress.[53]

In their reckless competition for knowledge, researchers privately

> . . . laugh at the idea of these experiments being made for discovering new means for preventing and curing disease, and assert that they do these foul and evil deeds purely and simply for the advancement of science—surely science run mad. I believe that in many instances these atrocities are committed for the purpose of gaining a little spurious notoriety as the exponents of advanced biology.[54]

Critics in this vein had a certain amount of proof from the writings of experimentalists themselves—especially continental physiologists—to support their contention.[55] Fur-

[53] J. Macaulay, *The Claims of Vivisection* (London, 1882), p. 4.

[54] *Fourth Annual Conference of the Scottish Society for the Total Suppression of Vivisection, Held on 26th May, 1883* (Edinburgh, 1883), p. 12. See also *Have Pity* (London, c. 1882), p. 4 and W. H. Llewelyn, *Vivisection: Shall it be Regulated or Suppressed?* (London, 1880), pp. 4–5; E. Berdoe, *The Real and Pretended Aims of Vivisection* (London, 1901). In the last named pamphlet, written in 1890, Berdoe complained, ". . . I detest the hypocrisy of pretending to benefit humanity so that permission may be obtained to do atrociously cruel acts in the interests of merely abstract science."

[55] See Note 37 above, and S. Wilks, "Vivisection: Its Pains and Its Uses," *Nineteenth Century* x (1881), 945–947.

thermore, was not an increasing proportion of experimentalists without any medical qualification whatever?[56] Regular scrutiny of the research literature of experimental medicine revealed vast differences between its ostensible nature, as presented in the lay press by its proponents, and the unvarnished truth as it manifested itself in professional journals.[57] The sterility of the method, and the sole concern of researchers with their reputations, was manifested by their continual controversies and mutual contradictions. Doctors differed; how then could there be any certainty or progress? Medical and scientific disagreement was a fruitful source of public suspicion, to be magnified and capitalized upon at every opportunity.[58] Other antivivisectionists went so far as to claim that the main purpose of experimental medicine was to view animals in pain: ". . . the whole interest and excitement of a physiological experiment on a living animal, both to operators and spectators, is necessarily dependent on closely watching its contortions on the rack, which their scientific training enables and binds them to appreciate minutely."[59]

Besides such supposedly empirical evidence of the nonmedical motives of experimentalists, there were a priori scientific and religious arguments against contentions that vivisection could aid the advance of medical practice. There were fallacies intrinsic to the method, such as the suffering of the animal, the use of anesthetics, the state of health of the animal, and its diet, which allegedly rendered any

[56] B. Bryan, *Legalized Vivisection in 1893* (London, 1894); B. Grant in J. Macaulay, B. Grant, and A. Wall, *Vivisection, Scientifically and Ethically Considered in Prize Essays* (London, 1881), pp. 99–117.

[57] F. P. Cobbe, "Vivisection and its Two-Faced Advocates," *Contemporary Review*, xli (1882), 610–626.

[58] *Verulam Review* iii (1892–1893), 295; cf. R. H. Shryock, "Public Relations of the Medical Profession," *Annals of Medical History*, n.s. ii (1930), 314–315.

[59] H. N. Oxenham, "Moral and Religious Estimate of Vivisection," *Gentleman's Magazine* ccxliii (1878), 731; see also P. C. Hill, *Man's Dominion over the Lower Animals not Unlimited* (London, c. 1885), p. 4 and E. Gurney, "A Chapter in the Ethics of Pain," *Fortnightly Review* xxxvi (1881), 780–781.

general conclusions drawn from experiments on living animals highly questionable. How much more suspect then was the attempt to extrapolate such conclusions over the vast species differences between the lower animals and man! And how could one duplicate states of natural diseases by artificially induced ones?[60]

Anna Kingsford put the question of species error in a particularly interesting context when she wrote:

> . . . it is precisely the subtle but enormous differences existing between the manifestations and character of the nervous system as we see them in man and as we see them in other animals, which distinguishes the former from the latter, and which endows vivisectors with the legal right they now possess to inflict on anthropoid apes injuries and mutilations which, if they inflicted the same on men, would be held to render the perpetrators guilty of crime. When, therefore, it is understood that this occult nervous differentiation is capable of constituting a distinction so vast, how is it possible to suppose that the study of biological function in the beast is capable of explaining satisfactorily the mysteries of human life?[61]

It should be noted that many scientific arguments against the validity of conclusions drawn on the basis of vivisectional research had appeared and continued to appear in discussions in the professional journals. They were raised by medical men—who often had little sympathy for organized antivivisection—as difficulties in the interpretation of specific results and even as fatal flaws in the method

[60] *A Few Observations on Vivisection* (London, 1875), p. 4; A. Wall in J. Macaulay, B. Grant, and A. Wall, *Vivisection, Scientifically and Ethically Considered in Prize Essays* (London, 1881), pp. 217–228; A. Kingsford, "The Uselessness of Vivisection," *Nineteenth Century* xi (1882), 171–183; C. A. Gordon, *Remarks on "Experimental" Pharmacology* (?London, c. 1883); C. A. Gordon, *The Vivisection Controversy in Parliament . . .* (London, 1888), esp. pp. 32–45; E. Blackwell, *Erroneous Method in Medical Education* (?London, 1891), Cf. W. Ballantine, *Some Experiences of a Barrister's Life* (New York, 1882), pp. 17–18.

[61] A. Kingsford, *Unscientific Science* (Edinburgh, 1883), p. 13.

itself. The discussion of the value of experiments upon living animals that took place in Parisian medical circles in the sixties provides the best example. Disagreement within the profession over the validity of the method, the therapeutic significance of its results, and the place of experimental medicine in medical education provided an important polemical foothold for antivivisectionists. These debates seemed to show a breakdown in professional solidarity on the issue and supplied spokesmen for the movement with many of their "scientific" arguments.

The second, and for many antivivisectionists more binding, argument against the possibility of benefits to humanity arising from vivisection was religious in nature. The argument from the divine design of the universe was predicated upon the world view of natural theology and a vague preference for the naive inductivism (perhaps better termed, in this context, "observationalism") associated with it. The Bishop of Durham put the view quite aptly in a sermon at Westminster Abbey in 1889:

> If the world were the work of an Evil Power, or if it were the result of a chance intervention of force and matter, it would be at least possible that we might gain results physically beneficial to ourselves by the unsparing sacrifice of lower lives. But if He who made us made all other creatures also . . . then I find it absolutely inconceivable that He should have so arranged the avenues of knowledge that we can attain to truths which it is His will that we should master only through the unutterable agonies of beings which trust in us.[62]

Voicing similar convictions, R. T. Reid, parliamentary total abolitionist, added, "I cannot convince myself that secrets impenetrable to steady and diligent observation can be extorted from nature by means of unnatural experiments."[63] Among those who shared such views were Lord Shaftes-

[62] *The Late Bishop of Durham (Wescott) on Cruelty* (London, ?1912).

[63] *Sir R. T. Reid, Ex-Attorney-General, on Vivisection* (London, 1892), p. 4.

bury and Frances Power Cobbe.[64] Arguments like these as a rationale for hostility toward experimental medicine provide a link to antiscientific sentiment also expressed by members of the movement. As we shall see, such sentiment was based in part upon the belief that the scientists of the seventies and eighties had sacrificed the humility of the natural theological viewpoint for an overweening self-importance that failed to take account of the divinely appointed limitations to science.

Antivivisectionist pessimism about the value of vivisection reached its logical conclusion with Sir George Duckett, a member of the committee of George Jesse's Society for the Abolition of Vivisection, who in declining to appear before the Royal Commission in 1875 gave the following capsule summary of his opinions:

> All that I could say would be, what the major part of the kingdom would say,—that the practice of vivisection, an abomination introduced from the continent, is horrid and monstrous, and goes hand in hand with Atheism. Medical science has arrived probably at its extreme limits, and has little to learn, and nothing can be gained by *repetition* of experiments on living animals.[65]

HOSPITALS AND THE MEDICAL PROFESSION

After arguments outlining the immorality and inutility of vivisection per se, the movement expanded the scope of its target to include hospitals, the medical profession, and the scientific establishment. In attacking these institutions, they linked their cause with issues that were troubling important segments of society, segments that had no direct

[64] Lord Shaftesbury, *The Total Prohibition of Vivisection* (London, 1879); *Transactions of the Victoria Street Society for the Protection of Animals from Vivisection During the Half-Year Ending, Dec. 31st, 1879* (London, 1880); F. P. Cobbe, *A Controversy in a Nutshell* (London, 1889), pp. 1–2; *Bishop Barry and Canon Wilberforce on Vivisection* (London, 1903).

[65] "Report of the Royal Commission on the Practice of Subjecting Live Animals to Experiments for Scientific Purposes," *Parl. Papers 1876* xli (C.-1397), Appendix ii, p. 603.

stake in the question of vivisection. One can sense from the way in which antivivisectionist opinions were picked up and discussed in parts of the press that the movement's concerns, though expressed rather immoderately, were nevertheless symptomatic of deep and widespread public interest.

One such area of social concern, in which the movement participated intensely, focused upon the quality of patient care in hospitals. The crux of the antivivisectionist indictment, which achieved a wide hearing, lay in the charge that hospitals had been transformed from benevolent and merciful institutions for the cure of the sick into instrumentalities for the self-interested advance of medical careers.

Spokesmen for the movement had always insisted that arguments for vivisection on the grounds of adding to knowledge of human disease logically extended to demand for human subjects. Ouida based her portrayal of antivivisection as in part a question of self defense upon this point. She feared especially for those who could not protect themselves: "What single argument for the scientific torture of animals would not apply equally to the right of science to seize and use for vivisection the pauper, the idiot, the maniac, the halt, the blind, the hopelessly diseased amongst men?"[66] While the *Medical Times and Gazette* responded that the contention had the same logical status as the argument that eating the flesh of animals implied the justification of cannibalism,[67] antivivisectionist scrutiny detected both intent to commit and actual cases of, "human vivisection."

[66] Ouida, "The Rights of Animals," *Animals' Guardian* i (1890), 13; *The New Priesthood* (London, 1893), pp. 61–67, 72–75; E. Lee, *Ouida: A Memoir* (New York, 1914), pp. 323–324. See also *Home Chronicler* (1878), 391; *Anti-Vivisectionist* vii (1880), 53; *Reply by John Bowie, Esq., M.D., To Professor Rutherford on Vivisection* (Edinburgh, c. 1880), p. 7; M. Thornhill, *The Morality of Vivisection* (London, 1885), pp. 8–9; E. Berdoe, *The Vivisector in the Lunatic Asylum* (London, 1897); E. Berdoe, *A Catechism of Vivisection. The Whole Controversy Argued in all its Details* (London, 1903), pp. 126–134.

[67] *Medical Times and Gazette* i (1885), 18.

"Human vivisection" was defined as,

> . . . the practice of subjecting human beings, men, women, and children, who are patients in public charitable institutions, hospitals, or asylums, to experiments involving pain, distress, mutilation, disease, or danger to life, for no object connected with their individual benefit, but for scientific purposes.[68]

Statements by proponents of experimental medicine that the prohibition of experiments on living animals would drive doctors to experiment upon patients[69] were taken by the movement as evidence of medical willingness to perform such experiments. The concrete examples of experiments upon human beings that were publicized by the movement occurred largely in France and Germany. Typical of such examples were tissue graft experiments, in which tissue from cancerous or tubercular lesions was implanted in an uninfected part of the original patient's body. Sometimes, such tissue would be transferred into the bodies of other, fatally ill, patients. Other experiments were bacteriological or galvanic in nature.[70] Most of the specific cases of "human vivisection" culled from British research literature involved experimental pharmacology.[71] The most publicized British case, centering upon London physicians Sidney Ringer and William Murrell, is worth examining because it reveals some of the complexities of the question of experimentation upon human beings and its interpretation by the antivivisectionist press.

[68] The definition is from *Illustrations of Human Vivisection*, published somewhere in the United States in 1906.

[69] See, e.g., *Public Opinion* xxxvi (1879), 66; *Medical Press and Circular* i (1883), 344–345.

[70] See, e.g., *Zoophilist* i (1881), 229; vi (1886), 25–26; Brayfytte, "Our Continental Models," *Animals' Guardian* i (1891), 126–127; F. P. Cobbe, *Cancer Experiments on Human Beings* (London, c. 1892); *Animals' Guardian* iv (1893–1894), 185; *Medical Experiments on Human Beings* (London, 1893); B. Bryan (ed.), *Anti-Vivisection Evidences* . . . (London, 1892), pp. 112–118.

[71] *The Experimental Physiologist in the Hospital* (London, c. 1880); *Verulam Review* vi (1896–1897), 273–284.

In 1883, Murrell, Lecturer in Materia Medica and Therapeutics at the Westminster Hospital, administered doses of nitrite of sodium to a number of hospital outpatients. The dosages given were calculated from previously published results, which, as it happened, had been based upon experiments using extremely impure specimens of the drug. Since Murrell's supply of nitrite of sodium was unadulterated, the dosages he administered were far too strong. The eighteen outpatients involved uniformly suffered great discomfort. Subsequent experiments by Ringer on frogs and cats conclusively demonstrated the toxic character of the drug. In November 1883, the *Lancet* published Ringer and Murrell's hastily prepared account of their experiments. Their paper effectively described the toxic effects of nitrite of sodium and corrected the inaccurate recommended dosage extant in the literature. What it failed to do was to specify that the clinical experiments by Murrell preceded those by Ringer in the laboratory, that the clinical subjects were all suffering from similar complaints for which drugs chemically similar to nitrite of sodium had previously proved efficacious, and that the effects of the drug upon the patients were not as severe as expressions used in the paper—many of them taken verbatim from the patients' own comments—seemed to indicate. Ernest Bell, a stalwart medical antivivisectionist, wrote a letter to the *Standard* (no friend of the movement) exposing to the lay public what he took to be Ringer and Murrell's deliberate and irresponsible experimentation upon human subjects. To the movement and to a good portion of the press, including the *Standard*, the *Spectator*, and the *Medical Times and Gazette*, Ringer and Murrell had apparently followed up conclusive laboratory experiments by recklessly causing extreme suffering to patients who could not possibly benefit by the proceedings. Furthermore, editors wondered, had the patients been informed of the experiments to be carried out? Surely their permission should have been obtained. The matter closed after explanatory communications from the principals and issue of a report of the Censor's Board of the Royal College of Physicians, exonerating Ringer and

Murrell of anything more serious than lack of care in writing their paper. By this time, however, the affair had been thoroughly ventilated in both the lay and professional press. To antivivisectionists, Ringer and Murrell's misdoings were a striking confirmation of expectations. The movement's clear propaganda victory greatly aided its campaign to shake public confidence in the hospitals.[72]

The complex of functions that hospitals subserved and the variety of motives that impelled medical staffs of hospitals became the object of deep suspicion on the part of the antivivisection movement. To antivivisectionists, the functions of hospitals as centers of medical education left patients, especially poor patients, to suffer indignities at the incompetent hands of medical students. Furthermore, the prestige of an appointment to the staff of a hospital or the faculty of a medical school was seen as a constant temptation to unscrupulous experimentation.

An altruistic protest against the mistreatment of charity patients became another stick with which to beat vivisection and its proponents. Such patients were poked and prodded by hordes of eager students:

> We wonder how many of those who contribute on Hospital Sunday are aware that they are providing funds for the procuring of delicate-looking young girls to be stripped before a class of medical students, simply to enforce a lecturer's assertion that scars cannot be seen until clothing has been removed.[73]

Not only were charity patients repeatedly and unnecessarily examined by students, but their treatment might be delayed

[72] The best single source for the Ringer-Murrell controversy is *Zoophilist* iii (1833), 221–226, where many of the principal documents are extracted. See also "Report of the Censor's Board" (Annals of the Royal College of Physicians of London, 31 January, 1884), pp. 336–338. I am grateful to Jeanne Peterson for bringing this last reference to my attention, and to the Royal College of Physicians of London for permitting me to examine it. Typical of antivivisectionist use of the matter is G. R. Jesse, *Experiments on Patients by Two Hospital Physicians* (London, 1885).

[73] *Verulam Review* ii (1890–1891), 74. See also "The Medical Profession and its Morality," *Modern Review* ii (1881), 311.

in the interests of clinical study, new drugs or methods of surgery might be tried out on them, or they might be treated by mere students in the process of training. Reckless and unnecessary surgery, the result of a vivisectionist mentality, might easily fall to their lot. All in all, charity patients were simply serving as guinea pigs for their social betters:

> For in its relation to Science and her operations, the working class occupies, and is beginning slowly but surely to realize that it occupies, a position perilously more near to that of the vivisector's subject than to that of the patient of the fashionable physician. The "rising democracy" is beginning to learn that the true material of the experimental scientist is not the laboratory dog but the hospital patient. And that whatever gains may be achieved by the process will be achieved, not for its benefit, but at its expense.[74]

No wonder the poor, thought of as they were as so much "corpore vilia" or "clinical material," were terror stricken at the thought of having to enter a hospital.[75]

The definitive statement of this indictment of the hospitals was the work of Edward Berdoe, a member of the executive committee of the Victoria Street Society. Berdoe was a medical man with a practice in the East End of London. In addition to his antivivisectionism, he was also opposed to the Contagious Diseases Acts and compulsory vaccination, and, completing a consistent pattern of heresy, had leanings toward homeopathy. In 1887, Berdoe published, under the pseudonym Aesculapius Scalpel, a novelistic account of the adventures of a medical crusader not unlike himself entitled *St. Bernard's: The Romance of a Medical Student. St. Bernard's* detailed in fictional form

[74] Ibid., 223. Cf. iii (1892–1893), 68–71, and *A Burning Question* (London, n.d.).

[75] G. R. Jesse, *Extracts from and Notes upon the Report of the Royal Commission on Vivisection, Refuting its Conclusions* (London, 1876), ii. 27n; *Experiments on Patients by Two Hospital Physicians* (London, 1885), p. 28; B. Abel-Smith, *The Hospitals 1800–1948* (London, 1964), p. 105; M. J. Peterson, "Kinship, Status and Social Mobility in the Mid-Victorian Medical Profession" (University of California (Berkeley) Ph.D. thesis, 1972), pp. 229–230.

all of the abuses mentioned above, and more besides. It caused a sensation in medical circles, and was widely reviewed and discussed. The first edition sold out within a year, and the second edition was accompanied by *Dying Scientifically: A Key to St. Bernard's.* The latter was a slender volume in which Berdoe listed the iniquities he had fictionalized in *St. Bernard's* and substantiated their existence with specific references to the medical literature, especially to the *B.M.J.*[76]

Berdoe undoubtedly succeeded in his intention to stimulate wide interest in the issues he raised in so unorthodox a manner:

> It was not possible to draw public attention to these abuses by any other method than that of writing a story, as interesting as it might be, embracing all the facts. A treatise on hospital management would have fallen stillborn from the press. The abuses complained of in St. Bernard's have been ventilated over and over again in the medical and lay papers, and nothing has been done to rectify them. I could only expect the most violent denunciation from medical critics. . . I have read the history of many reforms, and I well knew the sweet ways of my professional brethren and their methods of dealing with their opponents.[77]

It was to rectify just such evils as those depicted by Berdoe that the Society for the Production of Hospital

[76] A. Scalpel [E. Berdoe], *St. Bernard's: The Romance of a Medical Student* (1st ed., London, 1887; 2nd ed., 1888), and *Dying Scientifically: A Key to St. Bernard's* (London, 1888). Attribution of authorship from J. Kennedy, W. A. Smith, and A. F. Johnson (eds.), *Halkett and Laing's Dictionary of Anonymous and Pseudonymous English Literature* (Edinburgh and London, 1929), v. 161. Reading Berdoe's work, one is drawn irresistibly to a comparison with M. H. Pappworth's *Human Guinea Pigs* (London, 1967).

[77] *Dying Scientifically: A Key to St. Bernard's* (London, 1888), p. 5. According to Berdoe, he wrote the book pseudonymously because otherwise his criticisms might have been taken to apply only to the hospital at which he had trained, whereas they applied equally to most other hospitals; see p. 12.

Patients was originated in 1897. Various remedies were proposed to alter the situation: accurate, comprehensive records kept open to the public, lay control of hospitals, creation of a hospital avowedly for the purpose of experimentation on patients, and legal recourse for abused patients.[78]

In their earnest search to unearth scandals, antivivisectionists often failed to distinguish among bona fide experiments, new procedures being used in the hopes of benefiting the patient, examinations for purposes of medical instruction, and "exploratory" operations or procedures solely for diagnostic purposes.[79] A good example was the highly publicized dispute over the Chelsea Hospital for Women during 1894–1896. In this case, a high mortality rate of around forty-five per cent in "exploratory" abdominal surgery for diagnostic purposes was parlayed by the movement into a widespread impression that the Chelsea medical staff was experimenting upon the unfortunate women under its care. While the mortality rate certainly reflected badly upon the competence of the staff, inquiry failed to uncover any experimental motivation for the operations performed. Nevertheless, antivivisectionists and sympathetic organs of the press spoke of "experiment on women at the Chelsea Hospital" and succeeded in associating the general concern aroused with their own objectives: abolition of experiments upon living animals and defeat of the proposal for the British Institute of Preventive Medicine, to be located in Chelsea.[80]

In sum, the antivivisectionist case against the hospitals was predicated upon the belief that medical enthusiasm for knowledge and notoriety left no room for the humane and merciful treatment of the diseased and the injured. The medical student received from his teachers "gallons

[78] *Verulam Review* vi (1896–1897), 352–362; vii (1897–1898), 53–59. See also ii (1890–1891), 248–249.

[79] See, e.g., Ouida, *The New Priesthood* (London, 1893), pp. 45–48.

[80] See P. R. O., Home Office Papers, Series 45/9879/B15780. See also the *Daily Chronicle* for the period in question, esp. 15 May 1894.

of scientific sack to a miserable halfpennyworth of therapeutic bread" and was "sent into the world, degreed and diplomaed, a sworn devotee of science."[81] The way to professional advancement no longer lay in curing the sick:

> But now quite another programme presents itself as that of modern medical advisers. The pursuit of Physiological and Therapeutic Science has become the foremost object, and the cure of patients has fallen into the background. It would be a bold thing to affirm now that when a patient sends for one of the new school of doctors the only, or even the uppermost idea in the doctor's mind is "how best I may cure him." Much more likely that idea is: "Here is a case to be diagnosed-tabulated-concluded by a successful *post-morten,* and then published in the medical journals."[82]

Modern medicine was a medicine "without patients,"

> Your patient is a mistake altogether. A vulgar concrete; a stupid unscientific excresence, certain to disturb the fair proportions of the pure abstract scientific argument.[83]

An assessment of the extent to which criticism of the hospitals by antivivisectionists and others was justified would require a great deal of additional research. It should be noted that the hospital system was the subject of considerable debate, both within and without the profession; this debate turned upon a variety of issues, most of which had nothing whatever to do with experiments upon patients.[84] The profession as a whole has rarely shown a sense of public relations that might mitigate the hostility and suspicion with which it was, and is, regarded in various

[81] *Zoophilist* ii (1882–1883), 71.
[82] Ibid. iii (1883–1884), 210.
[83] Ibid. ii (1882–1883), 174. See also *Scientific Medicine* (London, c. 1882) and *The Scientist at the Bedside* (London, 1882).
[84] B. Abel-Smith, *The Hospitals 1800–1948* (London, 1964).

quarters.[85] Consider the *B.M.J.'s* frank pessimism: "The enlightenment of the public, as to the true nature of hospital work, is a matter of questionable feasibility."[86] In any case, whatever the basis in fact for their contentions, antivivisectionists were not unprovoked. Two examples should suffice. The first is a paragraph from an eminent London physician's defense of Ringer and Murrell, published in the *Standard.*

> I think we, as medical men, should not attempt to conceal from the public the debt of gratitude they owe to the "corpora vilia"—for such there are, and will be, as long as the healing art exists and progresses. So far from there being a reason why moral and pecuniary support should be refused to hospitals on the ground that the inmates are made use of otherwise than for treatment, there is ground why more and more should be given to them, in order to compensate by every possible comfort for the discomforts necessarily entailed by the education of succeeding generations of medical men, and the improvements in our methods of coping with disease.

The writer went on to deplore the futility of "hysterical agitation" by "sentimentalists" attempting to modify laws of nature, "one of the plainest of which is that the few must suffer for the many."[87] This blunt declaration was widely reprinted in the movement's literature, as was the final clause of the *Medical Press and Circular's* argument against the demands of total abolitionists:

> . . . if experiments on animals are prohibited, then human beings must necessarily become the subjects of the roughest possible treatment. Refinements of research under these circumstances, moreover, would be simply impossible, and every medical man would perforce become, whether he wished it or not, a vivisector of his patients; for it cannot

[85] R. H. Shryock, "Public Relations of the Medical Profession," *Annals of Medical History* n.s. ii (1930), 308–339.

[86] *B.M.J.* i (1886), 837.

[87] Letter of A. De Watteville, *Standard* (24 November, 1882).

> be gainsaid that the attendant's duties to his profession are a higher claim than any advanced by the men and women who come to him for cure and relief.[88]

Here was perfect grist for the antivivisectionist mill.[89]

One reason why the movement's criticism of the scientific—or "scientistic"—cast of modern hospital practice achieved such a wide hearing was that it coincided with a larger discussion taking place within the profession over alternative views of the future of medicine. Although many shades of opinion were represented in the discussion, the two major schools of thought involved those who thought that medical practice must be reconstituted on a strictly scientific basis and those who keenly resented the idea that medicine could be anything but an art: "All attempts to wed the art and the science of medicine have proved failures."[90] During the last twenty years of the nineteenth century and the first few of the twentieth, profound tensions within the profession crystallized around this polarization of viewpoint.

Spokesmen for the methods of "scientific medicine"—some of them not medical men at all, but scientists in the strictest sense—saw themselves as progressives who spoke for an approach that could advance medicine incalculably faster than mere clinical research on fortuitously provided hospital patients. To insure maximum progress, pathology, physiology, pharmacology, and the rest had to

[88] *Medical Press and Circular* i (1883), 344–345. See also M. Mackenzie, "Is Medicine a Progressive Science," *Fortnightly Review* xxxix (1886), 852, where the use of condemned criminals for painless experiments is advocated.

[89] Experiments upon human beings continue to be a matter of debate. There is now a large literature upon the subject, of which M. H. Pappworth, *Human Guinea Pigs* (London, 1967), "Ethical Aspects of Experimentation with Human Subjects," *Daedalus* xcviii, no. 2 (Spring, 1969), B. Barber et al., *Research on Human Subjects: Problems of Social Control in Medical Experimentation* (New York, 1973), and J. Katz (ed.), *Experimentation with Human Beings* (New York, 1972), are important examples.

[90] *Scientific Medicine* (London, c. 1882), p. 3.

be rebuilt upon the knowledge provided by experiment. Such a program represented an activist approach, to settle

> . . . whether medicine should wait upon time and circumstances, upon the accidents of life, upon the habits, or even whims of society and fashion; or whether, it should form circumstances to its needs, turn accident to good purpose, and wrest from Nature that which she freely gave to him who asked, but which she resolutely withheld from the listless bystander; whether, in short, medicine should remain and be forever relegated to the limbo of observation, or whether it should become an experimental science.[91]

The major difficulty in the way of experimental medicine lay in the lack of institutional and financial support for its proponents, coupled with the apparent "incompatibility between research and practice." It was clear that ". . . original work does not *pay*, and the reputation of being a discoverer is often disadvantageous to the practitioner" and "the pursuit of truth for its own sake in the present arrangement of things requires first of all the possession of independent means."[92] Hence, a concerted campaign during the last thirty years of the century pressed medical schools, universities, and government to commit resources to experimental research.[93]

[91] *B.M.J.* i (1882), 477. See also *Nature* xxiv (1881), 329–332; *Lancet* i (1882), 530.

[92] M. Mackenzie, "Is Medicine a Progressive Science," *Fortnightly Review* xxxix (1886), 851.

[93] The campaign for the endowment of science and basic medical research was far more complex and far-reaching than can be indicated here. For some information and bibliography on the pleas put forth on behalf of experimental medicine, see R. D. French, "Some Problems and Sources in the Foundations of Modern Physiology in Great Britain," *History of Science* x (1971), 28–55. The fundamental reference point for British proponents of the experimental approach was, of course, the laboratory system of France and Germany. See also R. M. MacLeod, "Resources of Science in Victorian England: The Endowment of Science Movement 1868–1900," in P. Mathias (ed.), *Science and Society 1600–1900* (London, 1972), pp. 111–166.

In reaction to this campaign, many traditionally-oriented members of the profession became resentful and, indeed, hostile to the demands put forward by a small but vocal group. They saw themselves as preserving the humanitarian values of medicine, which were about to be jettisoned in the wake of the experimental approach. In their view, medical research without patients was a delusion and a sham. On the contrary, they believed that one major means for advance lay in clinical research using strictly inductive and observational methods.[94] As one of them instructed his students in the late eighties:

> When you enter my wards your first duty is to forget all your physiology. Physiology is an experimental science—and a very good thing no doubt in its proper place. Medicine is not a science but an empirical art.[95]

To such clinicians, experiments upon animals were dangerous distractions from the true paths of medical advance: clinical casework, the postmortem and gross pathology, pathological chemistry, microscopic anatomy, and other patient-centered research would provide data upon which to base inductions. They strongly suspected the new drugs developed by experimental pharmacologists such as Lauder Brunton, and they often felt that Joseph Lister's antiseptic spray or David Ferrier's research on cerebral localization were simply temptations to reckless surgery. Clinical cases alone could present all the variety and complexity required by research work. By no means all of the advocates of clini-

[94] For an early, crude example of the case for induction and observation, see the following works by George Macilwain, a veteran antivivisectionist: *Medicine and Surgery One Inductive Science . . .* (London, 1838); and *Remarks on Vivisection, and on Certain Allegations as to its Utility and Necessity, in the Study and Application of Physiology* (London, 1847).

[95] K. D. Keele, *The Evolution of Clinical Methods in Medicine* (London, 1963), p. 104. See pp. 103–107, and cf. R. H. Shryock, *American Medical Research: Past and Present* (New York, 1947), pp. 68–73.

cal research or opponents of the experimental approach were sympathetic to the movement. Medical antivivisectionists, however, inevitably considered themselves part of this larger school within the profession.[96]

A final and very significant element in the creed of some of the conservatives who opposed the experimental approach was their faith in the advance of techniques in sanitation. At this point, with emphasis on the prevention rather than on the cure of disease, the means emerged by which animal experimentation was to be rendered obsolete—even granting the extravagant claims made for the latter. Stevenson, in his article subtitled "On the Hostility of Certain Sanitarians to Animal Experimentation, Bacteriology, and Immunology," has clearly portrayed the cluster of issues at the heart of the "sanitarian syndrome."[97] In particular, such a viewpoint rejected not only animal experimentation, but also the germ theory of disease based upon it. Thus, sanitarians opposed vivisection, compulsory vaccination, the Contagious Diseases Acts and rigid orthodoxy in matters of drugs and dosage.[98] Hygienic research was ethically unimpeachable in itself, and it taught that the one sure means to the prevention of disease was through the improvement of morality. Disease was the divinely sanctioned wage of sin and folly. As one antivivisectionist medical man put it,

[96] J. Macaulay, B. Grant, A. Wall, *Vivisection, Scientifically and Ethically Considered in Prize Essays* (London, 1881); A. Kingsford, "The Uselessness of Vivisection," *Nineteenth Century* xi (1882), 171–183; L. Tait, *The Uselessness of Vivisection as a Method of Scientific Research* (London, 1882); *Dr. Bell Taylor on Vivisection* (London, c. 1885); C. A. Gordon, "The *Cui Bono* of Experiments on the Brains of Animals," *B.M.J.* ii (1886), 1240; *A Bold Stand for Truth . . .* (London, 1888); C. A. Gordon, *The Vivisection Controversy in Parliament . . .* (London, 1888); E. Berdoe, *Sir Spencer Wells on Vivisection and the National Health* (London, 1890); C. Bell Taylor, *For Pity's Sake* (London, 1893); B. W. Richardson, *Biological Experimentation. Its Function and Limits* (London, 1896); E. Blackwell, *Scientific Method in Biology* (London, 1898).

[97] L. G. Stevenson, "Science Down the Drain," *Bulletin of the History of Medicine* xxix (1955), 1–26.

[98] Stevenson does not deal with these last two issues in his article.

> When man is in harmony with his environment he has physical health, but he can never have that socially, or on a large scale until he is first brought into mental harmony with the laws of his being and the righteous purpose of his Creator. The medicine of the future must largely consist therefore, in trying to bring about this holy adjustment . . .[99]

In contrast to the filth and gore purveyed by immunological and physiological doctrines, the laws of hygiene taught simple cleanliness and moral purity, the natural rather than the artificial.[100] To the three medical men identified by Stevenson as espousing these views—Benjamin Ward Richardson (in his later years), George Wilson, and W. G. Collins—can be added the bulk of those medical practitioners who were prepared to be identified with the antivivisection movement.

The medical antivivisectionists were a mere handful in number, considerably less than one per cent of the approximately 22,000 medical practitioners registered in Britain in 1883. It has proved possible to trace some 78 qualified and registered medical practitioners who had committed themselves publicly to the antivivisectionist position by 1883. Sixty-six of these were among the *Medical Men Opposed to Vivisection* listed in a Victoria Street pamphlet, while the remainder were the authors of books or articles supporting the movement.[101] They were traced principally through the *Medical Register* and the *Medical Directory*. This sample is not exhaustive, for it does not include individuals who were medically qualified but unregistered,

[99] *Animals' Guardian* iv (1893–1894), 98.

[100] B. W. Richardson, *Biological Experimentation. Its Function and Limits* (London, 1896), pp. 124–130; A. Wall, in J. Macaulay, B. Grant, and A. Wall, *Vivisection, Scientifically and Ethically Considered in Prize Essays*, pp. 288–289; E. Blackwell, *Scientific Method in Biology* (London, 1898), pp. 64–68.

[101] *Medical Men Opposed to Vivisection* (London, 1883). This list, of course, included a number of the most prolific authors, such as Charles Bell Taylor, Edward Berdoe, A. P. Childs, John H. Clarke, W. Gimson Gimson, C. A. Gordon, George Hoggan, and Lawson Tait.

such as Anna Kingsford, nor a number of authors or individuals listed in the Victoria Street pamphlet, who claimed professional status but could not be traced in the sources listed. The geographic characteristics further emphasize the incompleteness of the sample, for they indicate that it is biased toward doctors resident in areas where the movement was most active. London, where the movement was strongest, was the home of thirty-two per cent of the sample of medical antivivisectionists, although only about twenty per cent of all doctors lived there. Furthermore, of the approximately sixty-eight per cent of the sample who were provincial practitioners, nearly half, or thirty-three per cent overall, were resident in five provincial hotbeds of antivivisectionist activity: Bristol and Clifton, Bath, Brighton, Torquay, and the Isle of Wight. Practitioners from the north and the midlands are conspicuously underrepresented in the sample. Public allegiance to antivivisection among the medical profession was strongest where the movement was strongest. It is hard to imagine that the preexisting potential for antivivisectionist sentiment among the profession was geographically linked, except that one might expect it to be stronger in the provinces than in the metropolis, where the professional elite promoting experimental medicine held sway.[102] One can thus conclude that the laymen at the heart of the movement succeeded in eliciting some support from members of the profession with whom they came in most direct contact, but they failed in the attempt to develop fully the potential support in this most important constituency.

The most telling characteristic of those in the sample is age: the average medical man in the sample had, by 1883, been practising his profession for as long as thirty

[102] Indeed, among Victoria Street's members were "a vast number more medical friends of all ranks in the provinces than in London," *Zoophilist* iv (1885), 149–150. It is not clear whether this means that Victoria Street exceeded the 80/20 provincial/metropolitan distribution for the medical profession as a whole, or simply reflected it. In any case the sample indicates that the relative strength of the movement among doctors was greater in London.

Table III. Year of first medical qualification of a sample of antivivisectionist medical men (1883)

Year	*Number*
1826–1835	12
1836–1845	12
1846–1855	24
1856–1865	10
1866–1875	14
1876–1883	6

years! The data in Table III, which show this clearly, are radically at odds with the age structure in a profession whose total numbers increased some fifty per cent between 1855 and 1875.[103] Victoria Street numbered among its medical members a disproportionate number of retired doctors.[104] Antivivisectionists attributed this support among older members of the profession to the sagacity of age and to its comparative insulation from the pressure of professional sanctions. They looked forward to increasing support as younger members of the profession perceived the failures of experimental medicine and became sufficiently established to express their sentiments publicly. In retrospect, however, we can see that the age structure of the sample clearly indicates that antivivisectionist opinions within medicine were largely the province of men no longer in touch with the mainstream of the profession.

Although there were twenty-seven M.D.s among members of the sample,[105] and about twelve per cent of the sample were Fellows of the Royal Colleges (this contrasts

[103] M. J. Peterson, "Kinship, Status, and Social Mobility in the Mid-Victorian Medical Profession," (University of California (Berkeley) Ph.D. thesis, 1972), p. 425.

[104] *Zoophilist* iv (1885), 149–150.

[105] Ten from St. Andrews, eight from Edinburgh, three from Glasgow, two from Aberdeen, and one each from Oxford, Dublin, Jena, Zurich, and Erlangen.

with about six and one half per cent for the profession as a whole in 1876[106]), the medical antivivisectionists cannot be said to have achieved much professional renown. Principal among the few who did were Charles Bell Taylor, an ophthalmologist, Henry Monro, a prominent member of the Royal College of Physicians of London and a leading medical psychologist, and Lawson Tait, the Birmingham abdominal surgeon whose contributions to his specialty and to aseptic methods of surgery were the subject of great controversy. Rather, however, than the pursuit of professional status, many medical antivivisectionists seem to have chosen to express not only an antipathy toward the experimental approach, but also toward the metropolitan medical establishment—"a certain clique of hyper-scientific men"[107]—who advocated it. The demands of this "clique" were characterized as follows,

> *A complete autocracy* was claimed by the chief ministers at the shrine of physiology, and by some few others who worshipped before the dripping altar. . . . *All* that the *public has to do* with the matter is to find institutions in which the practising physiologist may carry on operations, which require appliances that are beyond his private means; and *all that the legislature is bound to do* in the matter *is, to protect the physiologist* in his practices, to the extent dictated by his scientific purposes—according to his own conscience, of which it is presumption for "outsiders" to form a moral judgment.[108]

Antivivisectionist literature written by members of the profession is filled with similar expressions of alienation from the supporters of experimental medicine and their backers in the professional elite, despite the power of the latter

[106] M. J. Peterson, "The Mid-Victorian Medical Profession," p. 221.

[107] E. Berdoe, *An Address on the Attitude of the Christian Church Towards Vivisection* . . . (London, 1897), p. 12, and *The Healing Art and the Claims of Vivisection* (London, 1890), p. 7.

[108] B. Grant, in J. Macaulay, B. Grant, and A. Wall, *Vivisection, Scientifically and Ethically Considered in Prize Essays* (London, 1881), pp. 201–202. Italics in the original.

to wield professional sanctions against the recalcitrant. Certainly a sense of being discriminated against—as a provincial practitioner (Charles Bell Taylor and Lawson Tait), as a woman (Anna Kingsford, Frances Hoggan, and Elizabeth Blackwell)—is discernible in many of the active medical antivivisectionists. A number of the remainder such as John H. Clarke or Edward Berdoe, were either homeopaths or had definite sympathy for homeopathy.

Although the movement was prepared to give various heterodox schools of medicine—notably homeopathy—a respectful hearing,[109] its attention remained focused upon the orthodox body of the profession, which harbored vivisection and as a whole stood behind its leaders on the issue. "So long as they throw the cloak of their authority over Vivisectors, so long it is our business to make holes in that somewhat well-worn mantle," wrote the *Zoophilist*.[110] Thus the movement developed an indictment, not only of the "hyperscientific" leadership, but also of the rank and file, and of its would-be members, the students.

Medical students, as a class, had always been subject to a certain amount of odium for their riotous behavior and continual clashes with the law.[111] In 1867, a young practitioner wrote a book-length vindication entitled *Medical Students of the Period. A few words in defense of those much maligned people. . . .*[112] A standard response to the accusation of unruliness was that young men in difficulty with the authorities often pretended to be medical students in the hope that they would be treated indulgently.[113]

[109] See A. Ruffer, "The Morality of 'Vivisection,'" *Nineteenth Century* xxxii (1892), 813.

[110] *Zoophilist* iv (1885), 149.

[111] W. Gilbert, "The London Medical Schools," *Fortnightly Review* xxxi (1879), 47–48.

[112] R. Temple Wright, *Medical Students of the Period. A few words in defence of those much maligned people; with digressions on various topics of interest connected with medical science* (Edinburgh and London, 1867).

[113] R. B. Carter, "The London Medical Schools," *Contemporary Review* xxxiv (1878–1879), 586; W. H. S., "Medical Students at Work. Sketches at St. Thomas's Hospital," *Graphic* ii (1886), 361.

Whatever the justice of the reputation, antivivisectionists were not slow to take advantage of it—especially after medical students at Edinburgh and London broke up meetings organized and paid for by the movement. The *Home Chronicler* and the *Anti-Vivisectionist* never failed to publicize the misdemeanors of medical students.[114] The terms in which medical students were anathematized may be indicated with a pair of examples. The first is extracted from a novel by the Rev. R. St. John Tyrwhitt,

> In Oxford new studies were producing new students. Earnest and flabby young men in spectacles came up professing biology and nothing else . . .
> . . . they looked forward to vacations of unlimited vivisection in Rome and Paris. When they get possession of the medical profession, I hope we may all die of regular, well understood old complaints, possessing no features of interest, for we shall certainly get taken the rough way round in the cause of science if we venture on novel symptoms in our dread progress.[115]

A second example is taken from the *Daily Chronicle's* account of the new (vivisecting) school of medical men:

> The new school consists of . . . enthusiasts who have only just passed from the stage at which young men go forth from the hospitals on football or boat-race nights to parade the West End in gangs, knock foot passengers off the pavement and then, in the interests of what they call sport, destroy the glasses of some more or less innocent proprietor of a West-End drinking-bar; and, having returned to their Bayswater or Bloomsbury lodging in the early morning and tried to sleep off the effects of bad whisky and worse cigars, go forth to gloat over men older than themselves destroying human lives in the interests of science.[116]

[114] See, e.g., *Anti-Vivisectionist* ix (1882), 32, 72.
[115] Quoted in *Animal World* xii (1881), 45.
[116] Quoted in *B.M.J.* i (1894), 1143–1144.

In their attempts to mobilize sentiment against the profession proper, antivivisectionists appealed to a wide variety of psychosocial interests and prejudices, including snobbery, pietism, prudery, antielitism, financial self-interest, anality, obsession with individual liberty, and feminism. During this process, medical men saw their way of life, their duties, and their view of the world scrutinized and, in general, maligned in a way probably unparalleled before or since. I do not wish to suggest that the critique developed was unique to antivivisection or to the period in question. Many features of it appeared long before the movement emerged and many remain as matters of active public concern to the present day. It is clear, however, that the antivivisectionist attack on the profession—combined with the contemporaneous attacks by agitators against the Contagious Diseases Acts and compulsory vaccination—constituted some of the most sustained and intense social pressure that medical men had ever experienced.

The principal document of the antivivisectionist campaign was a lengthy article in the *Modern Review* for 1881 entitled "The Medical Profession and its Morality." Although unsigned, the article was widely (and correctly) attributed to Frances Power Cobbe. A great many of the themes used to stimulate suspicion and distrust of the profession were concisely expressed by Cobbe in this piece of journalism.[117]

During the nineteenth century the status and social position of the great bulk of the medical profession had been steadily consolidated, while the political activity of its leadership grew substantially. Between 1850 and 1883, thirty-six British doctors were knighted and sixteen were granted baronetcies.[118] According to Cobbe, literature "represents the heroines of at least half the novels of the last decade as passionately adoring their doctors." This accession to influence and public approval on the part of the profession

[117] [F. P. Cobbe], "The Medical Profession and its Morality," *Modern Review* ii (1881), 296–328.

[118] R. H. Shryock, "Public Relations of the Medical Profession in Great Britain and the United States: 1600–1870," *Annals of Medical History*, n.s. ii (1930), 324.

did not go unchallenged. Cobbe noted primly that "a lady's medical advisor is the last person with whom it is natural or desirable that she should associate the notion of love-making."[119] She and others deplored the sacrosanctity of the profession, its aggressive arrogation to itself of decisions about national welfare that should be the responsibility of all.[120] The profession, it was charged, cloaked the narrow and selfish pursuit of occupational interests with a spurious humanitarianism.[121] Medical men practiced the most collusive "trades-unionism." They refused to indict fellow practitioners for even the most egregious errors. In their greed, they overcharged, overprescribed, overtreated, undertreated, and generally failed to honor their responsibilities to a trusting clientele.[122] They were "doubly treacherous to women" in their failure to countenance female medical practitioners, their failure to combat the excesses of female fashions (in the form of tight-lacing and skimpy evening wear), and their espousal of the Contagious Diseases Acts.[123]

According to antivivisection spokesmen, the history of medicine was "largely a history of human folly," its remedies either "exceedingly Costly," "very Painful or very Disgusting."[124] The practice of medicine was predicated upon the most selfish and basest motives: the desire for bodily health, the attempt to escape the just deserts of immoral and insanitary life.[125] Indeed, Edward Maitland thought that medical science, by identifying itself with vivisection, constituted "a deliberate conspiracy to demonise the race"

[119] Cobbe, op. cit., 297.

[120] See, e.g., *Anti-Vivisectionist* ix (1882), 31–32; Ouida, *The New Priesthood* (London, 1893), pp. 10–14, 23–24.

[121] *Zoophilist* iii (1883–1884), 187; *Professionalism in Morals* (London, 1889).

[122] Cobbe, op. cit., 311–319.

[123] Ibid., 321–325.

[124] W. S. Lilly, *An Independent Opinion on Vivisection* (London, 1885), p. 2; [F. P. Cobbe], "Sacrificial Medicine," *Cornhill Magazine* xxxii (1875), 427–438.

[125] F. P. Cobbe, "Hygeiolatry," in *The Peak in Darien* (Boston, 1882), pp. 81–91.

by "the reconstruction of society on ethics which can only be described as those of hell."[126] He felt that "The vivisector is in the position of one who poisons the well-spring to which the whole community resorts."[127] Not only was the rationale of medicine grossly self-centered, but medical men themselves were materialists and agnostics—traits scarcely desirable in those ministering to the sick and the dying, who must be in positions of peculiar spiritual vulnerability.[128] Practitioners needed a living faith to counteract the "downward pressure" of days of "disease-mongering":

> There is, of course, a great and ever-present temptation to a physician to view things from the material or (as our fathers would have called it) the carnal side; to think always of the influence of the body on the mind rather than of the mind on the body; to place the interests of health in the van and those of duty in the rear; to study physiological rather than psychological phenomena; nay, to centre attention on the morbid phases of both bodily and mental conditions rather than on the normal and healthful ones of *mens sana in corpore sano.*[129]

Even granting the spiritual bankruptcy of the profession, however, one was scandalized to learn that many medical men blatantly recommended immorality to their male patients as a means of health.[130]

Given the disgusting details of bodily dysfunction, which were the medical man's daily lot, it was no wonder that no one of real sensibility would undertake the study of medicine. A slavish devotion to externals atrophied the

[126] E. Maitland in E. Carpenter and E. Maitland, *Vivisection* (London, 1893), p. 25, and E. Maitland, *Anna Kingsford. Her Life Letters Diary and Work* (London, 1896), i. 303.

[127] E. Maitland, "Vivisection," *Examiner* (1876), 738–739.

[128] [F. P. Cobbe], "The Medical Profession and its Morality," *Modern Review* ii (1881), 308–309; E. Carpenter in E. Carpenter and E. Maitland, *Vivisection* (London, 1893), p. 13.

[129] Cobbe, *Modern Review* ii (1881), 309.

[130] Editorial postscript, *Modern Review* ii (1881), 327–328. Cf. P. T. Cominos, "Late Victorian Sexual Respectability and the Social System," *International Review of Social History* viii (1963), 44–45.

higher mental and spiritual faculties and repelled minds of true quality from the profession. In fact, "it appears that the majority of British doctors are either the sons of men of the secondary professional classes or of tradesmen, and in some cases (especially in Scotland) of intelligent artisans." The British profession was, then, "a *parvenu* profession, with all the merits and the defects of the class."[131] The coarseness of mind and "gross and unrefined tastes" to which medical science appealed made "the personal associations to be endured . . . scarcely less repulsive than the methods to be employed."[132]

Antivivisectionists further attacked the supposedly lofty motives of the medical profession with the imputation to its scientific idols of objectives no more honorable than financial gain and personal notoriety. It was the *Verulam Review*—notable elsewhere for its willingness to extend its good offices to the rank and file of the profession—that took the lead in these accusations of base self-promotion. The following shaft was typical of many:

> As "scientists" M. Pasteur and Dr. Koch are, no doubt, very much on a par. But the latter would seem to lack something of that keen insight into the mainsprings of human gullibility which forms so important an element in the great art of commercial success, and of which the former is so distinguished an example.[133]

Charles Adams, editor of the *Verulam Review*, quite outdid himself in satirizing the vicissitudes of immunological research, conducted as it was in the glare of public attention not entirely shunned by the research workers themselves. Adams traced the typical pathway of the "New Science

[131] *Modern Review* ii (1881), 299–304.

[132] [E. Maitland], *"The Woman" and the Age . . .* (London, 1881), p. 11; E. Maitland, "The Doctors and the Vivisection Bill," *Examiner* (1876), 684; A. Kingsford, "The Uselessness of Vivisection," *Nineteenth Century* xi (1882), 181.

[133] *Verulam Review* v (1895), 18; see also ii (1890–1891), 150–157, 253–268; iii (1892–1893), 210–214; vi (1896–1897), 145–150, 165–187.

Cure" from the initial optimism and wide publicity of "the Exultant Period," through various stages to "the Reactive Period" when "Hyperexia of Hope passes more or less swiftly into the Cold Fit of Disgust."[134]

While the antivivisection movement made a serious tactical blunder in presuming to take on the whole of the medical profession, its indictment of the profession certainly had sufficient merit to be widely noticed. Grass roots antimedical sentiment, particularly on the part of women, emerged again and again in the correspondence columns of the movement's periodicals.[135] Such sentiment was not created by antivivisectionist propaganda alone; it predated the organized movement and was mobilized on behalf of the movement. Antivivisectionists linked their cause to antimedical sentiment, which was rooted in a great variety of religious, social, and economic factors. MacLeod's studies of public confidence in medical doctrines indicate that "tendencies toward Shavian distrust of the medical profession, which were to become overt in the Edwardian period, were being kindled well before 1900."[136] One could now suggest that G. B. Shaw's early twentieth-century critique of the medical profession was neither original nor prophetic. On the contrary, Shaw is best seen as merely the most visible figure in a long tradition of critics of physicians and physic.[137]

SUMMARY

During the Victorian period, popular perceptions of medicine changed as radically as did the profession itself. Health care became less and less a privilege of the rich and more and more the right of every citizen. This brought about new institutional structures for providing care and

[134] Ibid., vi (1896–1897), 160–162.

[135] For a particularly good example, see G. Douglas, "War to the Knife," *Home Chronicler* (1878), 12–13.

[136] R. M. MacLeod, "Medico-legal Issues in Victorian Medical Care," *Medical History* x (1966), 48.

[137] R. Boxill, *Shaw and the Doctors* (New York, 1969) describes Shaw's viewpoint very well, but Boxill's attempt to place it in historical context is grossly misconceived and inadequately researched.

educating doctors, a new and deeper involvement of government with medicine and public health, and an increasing willingness on the part of the general public to speak out on medical issues that concerned them. Clearly the changes involved here are extremely complex, and looking at them as manifested in the antivivisection debate is looking at only one aspect of a broader picture. While this particular aspect involved a whole gamut of controversial problems, much of the critique of the movement may be characterized as part of a single line of argument.

That line of argument centred on the rejection of so-called "scientific medicine," the veneration of a personal, humane style of medicine directed toward the relief of the sufferings of individual patients, and the promotion of a naive sanitarianism, in which a confusion of dirt, disease, and sin was juxtaposed with a similar confusion of cleanliness, health, and moral restraint. The experimental approach and the belief that the advancement of knowledge was fundamental to the professional obligation seemed to antivivisectionists a sinister and threatening attempt to disjoin material phenomena of bodily function from the moral and religious foundations of which they were the outward manifestation. Medical science was discounting religious faith, and medical scientists were infiltrating and dominating the profession. The bedside manner was being crushed under a new orthodoxy of collective treatment in hospitals with hideous, artificial, "science cures." The medical elite was seen to be riding roughshod over the preferences of the profession's clientele by the use of their status and statutory power to advance experimental medicine.

In trying to pursue these concerns beyond simply articulating and disseminating them, the movement ran into the brick wall of professional solidarity and prestige. No matter how skillfully the antivivisectionists attempted to capitalize on the general practitioners' resentment of consultants, of hospitals, which were seen to be competing for patients, and of experimental medicine, they did not succeed in driving a wedge between the rank and file and the leadership. No matter how persistently they publicized scandals in the hospitals and tried to organize for the protection of hospital patients, they failed to arouse the public sufficiently to carry their plans to any effect.

There are, perhaps, two main reasons for this failure. Most significant in an immediate sense was the simple fact that the public was prepared to endorse vivisection if it conduced to human health and was prepared to believe medical assurances that this was the case. Successes such as anesthesia had brought the profession a credibility within the community too strong for antivivisectionists to counter. It would seem that utilitarian arguments continue to carry the day.[138] In a broader perspective, the antivivisectionist failure was merely one of the first in a long series resulting from the persistent attempts of laymen to exercise leverage over the modern profession. In this sense, the movement may be seen as part of a very broad tradition of protest against the hegemony of power and expertise in industrial society.

[138] *Animal Experimentation* (Report No. 39, National Opinion Research Center, University of Chicago, 1949).

10. The Mind of Antivivisection: Science and Religion

The paucity of source material that might provide access to a statistical sample of antivivisectionists and, thus, to some assessment of their personal characteristics is as serious a problem in assessing the religious composition of the movement as it is in the areas of political opinions and reform commitments of friends of the cause. Hence the religious background of the movement must be explored less on the basis of generalizations from firm data regarding denominational allegiances than by inference from the religious views expressed in antivivisectionist literature. Such views are especially significant for our topic in that they were frequently articulated with reference to that vexed Victorian problem, the so-called "conflict between science and religion." However inaccurate this rubric to describe the actual intellectual situation, it captures nicely the bitter contention of the debate. To continue the suspect analogy, the conflict was a very confused war, in which antivivisection may be seen as a comparatively well defined battle.

There are some basic observations that can be made about the movement's religious composition. It was not a sectarian movement; it was rather, as Stevenson has said, ecumenical.[1] The question had no theological bearings that made of it a sectarian issue—with perhaps one exception—and, indeed, my reading of the literature confirms Stevenson's conclusion that even attempts to find simple biblical authority for antivivisection foundered.[2] The single excep-

[1] L. G. Stevenson, "Religious Elements in the Background of the British Anti-Vivisection Movement," *Yale Journal of Biology and Medicine* xxix (1956), 125–157.

[2] Ibid., 134–137.

tion arose in connection with the teaching of the Catholic church that it is a theological error to hold that man has any duty toward animals; he has, however, a duty to God and to himself not to treat animals wantonly or cruelly.[3] This issue was almost entirely academic in Britain because of Cardinal Manning's very public allegiance to the movement. Its ramifications may, on the other hand, have been crucial for the fate of the movement in Catholic countries, where antivivisectionist agitation conspicuously failed.[4]

Contemporary impressions of the religious support for antivivisection are fragmentary and contradictory. Antivivisectionists agreed only upon the disgraceful failure of the various denominations to commit themselves to the movement in their corporate capacities. As for individual supporters, the established church, and especially the high church, was conspicuous by the absence of its members from the ranks of the movement.[5] Nevertheless, eminent churchmen were to be found on both sides of the question—and the same is true for virtually all of the great variety of religious opinions represented within a movement that consciously tolerated such diversity.[6] Among Unitarians, for example, antivivisectionists Frances Power Cobbe and James Martineau were counterbalanced by partisans of experimental medicine such as W. B. Carpenter. Dissenters gave important support to the movement, but Jonathan Hutchinson, a surgeon and an outspoken defender of vivisection, was a Quaker, and many of Britain's physiologists and pharmacologists came from noncomformist backgrounds. Even if we can give a cautious assent to Stevenson's contention that "Evangelicals and those of similar faith and sympathy, occupied almost all of the chief positions in the anti-vivisection societies,"[7]—one thinks instantly of

[3] See, e.g., G. Tyrrell, "Jesuit Zoophily: A Reply," *Contemporary Review* lxviii (1895), 708–715; "A Catholic View of Vivisection," *Public Opinion* lxv (1894), 641.

[4] Stevenson, op. cit., 128–132, 134.

[5] See *Spectator* lii (1879), 1042–1043; *Zoophilist* i (1881), 103.

[6] *Home Chronicler* (1877), 793–794; (1878), 331.

[7] Stevenson, op. cit., 155.

Shaftesbury and of Evangelical antiintellectualism[8]—it is clear this hardly settles the issue of religious background.

Parenthetically, it should be noted that the broad religious spectrum of antivivisection seems to have caused little sectarian strife within the movement; it was the Victoria Street Society, after all, that brought Shaftesbury and Manning together for the first time. Yet the movement, and Frances Power Cobbe in particular, never hesitated to employ theological odium against those who presumed to withhold support from antivivisectionist dogma.[9] This tendency became exacerbated after the movement's failure to win over the support of "the Humane Jews of England." Charles Adams persistently referred to Ernest Hart, editor of the *British Medical Journal,* as E. Abraham Hart, and he once remarked of Hart's journalistic accomplishments:

> To have pushed such a paper [the *B.M.J.*] into such a position is a feat of which any press-man might well be proud, a feat which would hardly have been within the compass of any who had not in his veins the blood of that pre-eminently pushing race, his connection with which Mr. Hart would seem to be so curiously anxious to ignore.[10]

Antivivisection and anti-Semitism were closely allied in Germany, and in reviewing a German tract of that tenor, the *Zoophilist* did not scruple to refer to "the uncongenial influences of Judaism and Materialism."[11] Such remarks were not unnoticed; in 1894, one Morris Rubens published a pamphlet that, with anguish and indignation, attacked Cobbe and the Victoria Street Society.[12] According to

[8] W. E. Houghton, "Victorian Anti-Intellectualism," *Journal of the History of Ideas* xiii (1952), 302.

[9] See, e.g., Cobbe's disputes with Catholic theologians. F. P. Cobbe, "The Ethics of Zoophily: A Reply," *Contemporary Review* lxviii (1895), 497–508.

[10] *Verulam Review* v (1895), 113.

[11] *Zoophilist* i (1881), 27–28.

[12] M. Rubens, *Anti-Vivisection Exposed, Including a Disclosure of the Recent Attempt to Introduce Anti-Semitism into England* (Part 1, Bombay, 1894).

Rubens, Cobbe and Lawson Tait were attempting to stop the progress of science "by means of the envenomed weapon of Anti-Semitism." The Jews of England were urged to follow the example of their Chief Rabbi and refuse to countenance the movement. It is clear, however, that the association of anti-Semitism with antivivisection was not the product of bona fide theological issues, but rather of the particular mentality of certain spokesmen for the movement. There were Jews who actively supported the cause, according to no less an authority than Frances Power Cobbe.[13]

If the antivivisection movement resists any simple or clear-cut characterization of its sectarian composition, this is not to say that religion was irrelevant to antivivisection sentiment. It was not people who professed a particular formalized creed who would necessarily embrace antivivisectionism, but those, of a number of different communions, whose psychic and intellectual investment in certain issues was greatest. One such issue was a belief in a future life for animals, which will be discussed in Chapter 11. But the most important set of issues revolved around the place of science and scientists in Victorian culture.

Victorian attitudes toward science were very complex, and scholars are only now beginning to study them. Those attitudes involved the interplay of a large number of factors such as the antiintellectualist practicality of a bourgeois-entrepreneurial society,[14] the cult of the irrational,[15] the adoption of "science" as a watchword by militant socialism, the expanding role of technically trained people in state administration,[16] the influence of positivism, and the evolu-

[13] *Zoophilist* i (1881), 103.

[14] W. Houghton, "Victorian Anti-Intellectualism," *Journal of the History of Ideas* xiii (1952), 291–301.

[15] A case in point relevant to antivivisection is Edward Carpenter. See S. Pierson, "Edward Carpenter, Prophet of a Socialist Millenium," *Victorian Studies* xiii (1970), 308–309.

[16] See, e.g., R. M. MacLeod, "The Alkali Acts Administration, 1863–84: The Emergence of the Civil Scientist," *Victorian Studies* ix (1965), 85–112; "Government and Resource Conservation: The Salmon Acts Administration, 1860–1886," *Journal of British Studies* vii (1968), 114–150.

tion of educational thought,[17] among others. What was fundamentally at stake in the debate over the place of science, however, was the primacy of spiritual values and their importance as a basis for discussing the social and intellectual, as well as the explicitly religious, concerns of Victorian society.

Certainly this was true of antivivisectionist attitudes toward science. It is true that some prominent antivivisectionists feared what they saw as the technological consequences of science. Ouida, the popular novelist, in an article, "The Rights of Animals," noted that it was science one had to thank for electricity, "employed at infinite risk, which everyday in one hemisphere or the other slays its victims by an excrutiating death."[18] Stephen Coleridge, a leading figure in the Victoria Street Society at the turn of the century, traced to science such evils as mass production, the degeneration of craft traditions, and the Black Country.[19] But Coleridge also indicted science for the degradation of art and literature and the death of religion, and for most antivivisectionists it was the spiritual rather than the material threat that loomed largest.

While the infliction of pain upon animals was, prima facie, the crux of the movement's concern, many antivivisectionists were simply antiscience, a fact that did not go unnoticed by the movement's critics.[20] For these antiscien-

[17] Lack of scientific education was often cited as a major reason for antivivisectionist sentiment. See, e.g., *Nature* ix (1874), 177; [L. Stephen], "Thoughts of an Outsider: The Ethics of Vivisection," *Cornhill Magazine* xxxiii (1876), 477; *Medical Times and Gazette* i (1882), 146.

[18] *Animals' Guardian* i (1890), 14.

[19] S. Coleridge, *Memories* (London, 1913), pp. 228–239.

[20] See, e.g., J. Hutchinson, "On Cruelty to Animals," *Fortnightly Review* xxvi (1876), 316; S. Wilks, "Vivisection: Its Pains and Its Uses," *Nineteenth Century* x (1881), 938; W. W. Gull, "The Ethics of Vivisection," *Nineteenth Century* xi (1882), 456; J. C. Morison, "Scientific *Versus* Bucolic Vivisection," *Fortnightly Review* xliii (1885), 252; C. S. Myers, "Is Vivisection Justifiable?," *International Journal of Ethics* xiv (1903–1904), 322; P. H. Pye-Smith, *Address to the Department of Anatomy and Physiology of the British Association* (London, c. 1880), p. 9; *Medical Press and Circular* i (1876), 531; *B.M.J.* i (1882), 124.

tists, antivivisection served as a practical issue on which to focus their antipathy toward science, a special case of the general question, "Will science not recognize its moral limits?"[21] The basis of action of the Church Anti-Vivisection League was "the conviction that Morality and Religion have a right to impose and maintain limits upon 'Research' and 'Experiment' in the domain of Nature."[22] Both the tone and content of antivivisectionist literature make it quite clear that the movement involved a generalized hostility to science based upon much more than the use of living animals in experiments by a very few British scientists. George Gore, an electrochemist who was one of the few nonbiological scientists to defend experiments upon living animals, put his finger on the phenomenon. People who oppose science, he wrote, do so because of "the dread that the power which the new knowledge imparts may be used to overthrow their beliefs and sentiments, and thus diminish their happiness." Such resistance, a "retarding and regulating" influence, insured "sufficient stability in human affairs" and prevented "a state of chaos, caused by incessant change and the too frequent introduction of new ideas."[23]

Just what were the "beliefs and sentiments" supposedly in peril? What was the supposed "conflict between science and religion"? Answers to these questions must center upon an examination of the emergence of scientific naturalism, the sciences upon which it seemed to be based, and their joint impact upon natural and revealed theology. Here we may draw a distinction between the methodological presuppositions or assumptions of science, which formed the basis of scientific naturalism, and science as a body of knowledge resulting from the application of the methodology. Although there are difficulties in reconciling literally certain

[21] See, e.g., *Bishop Mackarness (Late of Oxford) and Prof. Ruskin on Vivisection* (London, c. 1885); *Dean of Llandaff on Vivisection* (London, 1894).

[22] *First Report of the Church Anti-Vivisection League . . .* (London, 1890), p. 3.

[23] G. Gore, *The Utility and Morality of Vivisection* (London, 1884), pp. 4, 28–30.

parts of science as a body of knowledge with certain parts of Genesis and Joshua, the far more serious conflict arises between the scientific *assumptions* and the theological *approaches.* This is one important sense in which the "conflict between science and religion" is a misnomer. Equally important is the fact that the Victorians regularly failed to make this distinction, perhaps because the spectacular nineteenth-century successes of science, especially evolutionary science, gave a tremendous impetus to the promotion of a program of scientific naturalism that amounted to an attempt to replace the assumptions of theology with the assumptions of science. This confusion is one key to understanding the Victorian debate. Furthermore, it should be pointed out that this chapter is concerned not with the first statements of the various intellectual positions to be discussed—for the potential differences had existed since the scientific revolution—but with the point at which these ideas and their interrelations came to command a high degree of attention from the intelligentsia and the literate public in general.

Of course, it was not only the pressure from science that raised religious and theological controversy to such an intensity during the Victorian period. Rigorous historical criticism of the Bible and the Philosophic Radical and positivist movements were at least as important as science, for all were part of the broad sweep of naturalistic explanation of man and the universe that unsettled faith through the mid-century period.[24] In addition, essentially political questions surrounding the institutions of religion and their relations to society were also crucial instigators of doubt.[25] It is as well to remember, in the following consideration

[24] L. E. Elliott-Binns, *English Thought, 1860–1900. The Theological Aspect* (London, 1956), pp. 27–31; O. Chadwick, *The Victorian Church* (London, 1966, 1970) ii. 1–35; R. M. Young, "The Impact of Darwin on Conventional Thought," in A. Symondson (ed.), *The Victorian Crisis of Faith* (London, 1970), pp. 13–35.

[25] S. Budd, "The Loss of Faith: Reasons for Unbelief among Members of the Secular Movement in England, 1850–1950," *Past and Present* xxxvi (1967), 106–125.

of the tensions between scientific naturalism and religion, that this was merely part of a more general theological and institutional upheaval that affected virtually all of the communions.

Where science emerges as a widespread social pursuit, the historian frequently finds that it acquires some sort of justificatory rationale, some linkage to the societal context, whether it be, say, practical utility, national prestige, or intrinsic cultural benefit. For a lengthy period of time, British science found its rationale in the religious enterprise of natural theology, and this tendency reached a zenith during the early years of the nineteenth century. Natural theology, in its a priori form, has a lineage dating back at least to Plato. Empirical natural theology originated with Thomas Aquinas and flowered in parallel with the modern natural science that emerged during the seventeenth century. Empirical natural theology may be defined as the systematic treatment of what human reason may learn of God and His attributes from the study of the universe, man, and nature. Natural theology in early nineteenth-century England sought to establish the existence, completeness, and perfection of the Deity through inductive inference from the evidences of purpose and design in the physical universe. Do living organisms exhibit fine structures that interact precisely for the accomplishment of a certain end? Then surely such marvels of adaptation provide a reasonable inductive basis for the assertion of the existence of a divine designer, and their manifest characteristics provide a key to Him. For natural theology, then, the pursuit of science was an act of piety, and every discovery furnished additional evidence of the wisdom and goodness of God. Thus the investigation of physical phenomena in the context of natural theology confirms, supplements, and interprets the image of God, which He has chosen to open to man through the special revelation of the Bible. Reason and empiricism open another avenue to knowledge of God, besides the sacred texts.[26]

[26] On natural theology, see J. Hastings (ed.), *Encyclopedia of Religion and Ethics* (Edinburgh and New York, 1921), xii. 220, 298;

Varieties of natural theological themes had found their way into the writings of English scientists and divines since at least as early as the seventeenth century, but the definitive statement was William Paley's spectacularly popular *Natural Theology, or the Evidences of the Existence and Attributes of the Deity Collected from the Appearances of Nature,* published in 1802. This book, and a series of elaborations on Paley called the Bridgewater Treatises, published in the thirties, had an enormous and lengthy vogue such that the arguments of natural theology became part of the intellectual heritage of successive generations of the literate classes. Paley's book begins with the famous analogy of our inference of a watchmaker when we discover a watch and the necessary inference of a divine designer when we observe the intricate contrivances manifested in natural objects. It goes on to explore at length the utility of the finely integrated structure and function of the various organs of living things, iterating case after case with unflagging enthusiasm, and drawing the appropriate inferences about the attributes of their Creator.

For Paley, and for our purposes here, the most crucial attribute of the Deity was His beneficence and goodness. Thus Paley was concerned to minimize the pain and suffering in the world, to emphasize the harmony of nature, its joys and pleasures, in his justification to man of God's creation. Each organism has its just share of happiness, both nature and society are in perfect providential balance. Where Paley's theodicy recognized the problem of evil, the instrumental value of evil was used to defuse the issue

---

J. M. Baldwin (ed.), *Dictionary of Philosophy and Psychology* (Gloucester, Mass., 1960 (1901)), ii. 137; M. H. Carré, "Physicotheology," in P. Edwards (ed.), *The Encyclopedia of Philosophy* (New York and London, 1967), vi. 300–305; J. Hick, "Revelation," in ibid., vii. 189–190; F. Ferré, "Editor's Introduction," in W. Paley, *Natural Theology. Selections* (Indianapolis, 1963), pp. xi–xxxii; R. M. Young, "Natural Theology, Victorian Periodicals, and the Fragmentation of the Common Context," *Victorian Studies,* in press. I am grateful to Tom Settle for his advice and criticism, here and elsewhere in this chapter.

of God's responsibility for it: suffering and disease were seen as a retributive accounting for sin, and hence to be valued as a stimulus to moral rectitude and a means to grace.

Paley was merely one spokesman for a set of views that was spread widely through British society. The Bridgewater Treatises further popularized the argument from design and emphasized the action of divine providence in natural phenomena. Paley's theodicy, predicated as it was on the principle of human freedom and responsibility based on the doctrine of the Fall, likewise demanded an active, intervening God exercising special providence to maintain the balance of suffering and sin. Hart shows it to have been widely shared in the Established Church through the mid-century period, and she draws a germane conclusion:

> Many examples could be found amongst the sermons and pamphlets produced by anglican clergymen to show that it was very common, specially in the middle years of the century (1830–70), to hold publicly that poverty was ordained by God, that afflictions, even cholera and the plague, were good for people, that the poor should simply trust in God and take a long term view of their miseries. Disease of all kinds was sent by God as a punishment and to teach people to be less sinful. . . . Such doctrines can hardly have assisted the efforts which were being made in many quarters to study disease and public health in a scientific way.[27]

Here, then, is a first point at which our very general exploration of Victorian religious opinion throws light directly on antivivisectionist attitudes. The attitudes in question are those regarding medicine as well as science. If the enterprise of experimental medicine is legitimate, if the causes of diseases are strictly material and accessible

[27] J. Hart, "Nineteenth Century Social Reform: A Tory Interpretation of History," *Past and Present* xxxi (1965), 55. Cf. J. Z. Baruch, "The Relation between Sin and Disease in the Old Testament," *Janus* li (1964), 295–302; J. Hastings (ed.), "Disease and Medicine," *Encyclopedia of Religion and Ethics* (Edinburgh and New York, 1911), vi. 755–757.

to the methods of research, what becomes of Paley's theodicy? Framing the question thus not only impugned divine beneficence and omnipotence, it destroyed the principle of human responsibility. Antivivisectionists could not accept the erasure of moral decision and divine retribution explicit in the systematic attack on the material causes of disease through experimental medicine: "there can be no offence more shocking and no act more dastardly than this of trying to shift the natural punishment of our own sins and vices and stupidities, on to the shoulders of those who are powerless to resist us."[28] Thus, the wholesale rejection of medical research by some members of the movement, noted in the previous chapter. For such individuals, the only avenue of advance seemed through moral and sanitary improvement.[29]

There was, of course, an important group of Victorian believers for whom natural theology was essentially superfluous. Such Christians had no need for evidence, choosing simply justification by faith and the literal truth of the Bible. Principal among them were the Evangelicals, who by the mid-century period had diffused from the Wesleyan origins to revivify both nonconformity and the established church. Evangelical influence was enormous and its role as the "moral cement" of Victorian society was perhaps the single most important force in shaping the age. The Evangelicals were tireless, conscientious, pious, intense, and enthusiastic. Not for them the sophistications of theology, when good works under the eyes of the Lord remained unfinished. Exclusive supernaturalism and confident millennarianism, faith in the Good Book, and a vision of the hand of God intervening everywhere—these would suffice. The revelationism and pietism of the Evangelicals, though less complicated than the natural theological tradition, were to be no less significant in determining Victorian responses to the rise of scientific naturalism.

[28] M. Caird, *A Sentimental View of Vivisection* (London, c. 1893), p. 28.

[29] See L. G. Stevenson, "Science Down the Drain," *Bulletin of the History of Medicine* xxix (1955), 1–26.

The dual traditions of revealed and natural theology were by no means as mutually exclusive as the crude sketches above would indicate. They inevitably coexisted to a greater or lesser extent, especially in the minds of the many Victorian intellectuals who retained traces of Evangelical upbringings. The point is that until the mid-century period, one tradition actively encouraged science, while the other was no worse than indifferent. What happened between about 1840 and about 1880 was that the cumulative influence of an alternative tradition, naturalism, became so great that it could no longer be ignored or absorbed by Victorian theology. The naturalistic approach is one that "limits itself to what is natural or normal in its explanations, as against appeal to what transcends nature as a whole, or is in any way supernatural or mystical."[30] The attempt to explain physical, biological, and human phenomena without reference to supernatural causes emerged to general view in Britain from a disparate variety of historical, philosophical, and scientific inquiries at some point in the late eighteenth century. Among the researches involved were historical investigations of the sacred texts, associationist psychology, utilitarian social theory, and others.[31] Insofar, however, as science, narrowly conceived, was concerned, no explicit program of scientific naturalism could be said to have developed before the middle of the nineteenth century. It is this scientific naturalism, a part of the broader naturalistic alternative that seemed so threatening to traditional theology, that is most germane to antivivisectionist sentiment expressed in the last three decades of the century.

One could scarcely say that the implications of science had not troubled theological interpretations before mid-century. The grim Malthusian scenario of struggle, starvation, and death, for example, seemed a macabre counter-

[30] J. M. Baldwin (ed.), "Naturalism," *Dictionary of Philosophy and Psychology* (Gloucester, Mass., 1960 (1901)), ii. 137–138.

[31] R. M. Young, "The Impact of Darwin on Conventional Thought," in A. Symondson (ed.), *The Victorian Crisis of Faith* (London, 1970), pp. 13–35; "Natural Theology, Victorian Periodicals, and the Fragmentation of the Common Context," *Victorian Studies,* in press.

point to Paley's sunny theodicy until the natural theologians recognized and enunciated the "comforting generalization" that "the Malthusian principle was a means for periodically reestablishing the harmony of nature."[32] Geological research and the biblical history of the earth had likewise raised persistent problems for one another since the late eighteenth century.[33] But it was the triumph of uniformitarian geology, the emergence of evolutionary biology, and the program of scientific naturalism based by their publicists upon them, that by the fifties and sixties seemed to pose definitive and unequivocal challenges to theology. Charles Lyell's insistence upon using only currently observable geological forces to explain the development of the earth subverted belief in divine intervention and cast doubt upon the chronology, the miracles, and the catastrophes of the biblical account. Charles Darwin's statement of the evolutionary development of species, including the development of man and his mind, denied the doctrine of special creations and substituted a new naturalistic teleology for Paley's divine contrivances. Others expanded and articulated this scientific work and some made it the underpinning for "an entirely non-Christian interpretation of man and nature."[34] The broadening scope of the application of natural law to ever more phenomena, including the mind of man, seemed to many Victorians a heretical encroachment upon divine prerogative that destroyed spiritual influence on material events. In particular, Darwin's theory raised to the fore the relationship of man and animals in a way that seemed to leave little room for a unique spiritual nature for man.

The best account of scientific naturalism is Turner's superb and as yet unpublished study, "Between Science and Religion." Following James Ward, Turner describes

[32] R. M. Young, "Malthus and the Evolutionists: The Common Context of Biological and Social Theory," *Past and Present* xliii (1969), 114–116; and "The Impact of Darwin," ibid.

[33] C. C. Gillispie, *Genesis and Geology* (Cambridge, Mass., 1951).

[34] F. M. Turner, "Between Science and Religion: The Reaction to Scientific Naturalism in Late Victorian England," (Yale Univ. Ph.D. thesis, 1971), p. 13.

the basic assumptions of scientific naturalism as, (i) the idea of nature as a mechanism, (ii) the theory of evolution (and atomic theory and the law of conservation of energy) as the working of the mechanism, and (iii) psychophysical parallelism or conscious automatism, "according to which theory mental phenomena occasionally accompany but never determine the movements and interactions of the material world." The world-view of scientific naturalism "separates Nature from God, subordinates Spirit to Matter, and sets up unchangeable law as supreme."[35] The accomplishments of uniformitarian geology, evolutionary biology, and associationist psychology were the basis for the attempt to extend the naturalistic approach to a dominant cultural role: "naturalistic publicists sought to expand the influence of scientific ideas for the purpose of secularizing society rather than for the goal of advancing science internally."[36] Principal among these publicists were John Tyndall, Thomas Henry Huxley, Francis Galton, and Herbert Spencer. Turner adds a number of other spokesmen, including E. Ray Lankester and George Henry Lewes, both of whom performed and defended experiments on living animals.

In exploring the reaction to scientific naturalism, and its relationship to antivivisectionist attitudes, there are a number of possible themes for examination. One that comes to mind immediately is naturalism's emphasis upon treating man as an animal like any other and its denial of any unique spirituality for man.[37] The antivivisectionist response, as Chapter 11 will show, was to deny or downplay the *physical* relationship of man and animals and to extend to at least the higher animals some vestige of the *spiritual* endowment that, it insisted, characterized man. Here, it will be convenient to consider the impact of science and scientific naturalism, first, upon Evangelicalism and Fundamentalism, and, second, upon natural theology. Finally, the collective response of the Victorian literary and theological intelligentsia to the cultural aspirations of scientific

[35] Ibid., p. 14.

[36] Ibid., p. 16.

[37] Ibid., pp. 30–33; R. M. Young, "The Impact of Darwin," op. cit.

naturalism will be discussed. In each case, the attempt will be made to place antivivisectionist sentiment toward science in the context of these larger events.

For those who placed complete faith in the literal truth of the bible, the doctrines of uniformitarianism and evolutionism were an anathema, and the program of scientific naturalism was unthinkable. This blanket rejection was characteristic of Fundamentalists and the wide variety of less well educated believers, as, indeed, remains true to a remarkable extent to the present day. The reaction of Evangelicals generally was little more sophisticated. Evangelicalism had always been more or less dismissive of science, regarding it as a "vanity."[38] When science began to impugn scripture, subvert the immediacy of divine intervention and special providence, and jeopardize the moral meaning of life itself by describing it solely within the confines of the naturalistic approach, this indifference turned to active hostility. Annan has stressed the importance of this reaction (for our purposes here, we may read "naturalism" for "positivism"):

> . . . of the dozen other factors that one could mention, which led to the conflict of science and religion, none is as important as the rise of Evangelicalism. The movement . . . scorned the value of evidences and proofs and wagered all on the conviction of faith. The question was no longer, "How do we believe"?, but "Do you believe?" . . . But this same simplicity rendered it terribly vulnerable to the new weapons in the positivist armoury; and it is not, I think, an exaggeration to see Victorian theology in retrospect as a tireless, and at times almost desperate, attempt to overcome the appalling weaknesses which this simple faith presented to positivist criticism.[39]

[38] D. C. Somervell, *English Thought in the Nineteenth Century* (London, 1929), pp. 103–104.

[39] N. Annan, "Science, Religion, and the Critical Mind," in P. Appleman, L. Madden, and M. Wolff (eds.), *1859: Entering an Age of Crisis* (Bloomington, Ind., 1959), p. 37. On naturalism and positivism, see J. M. Baldwin (ed.), "Naturalism," *Dictionary of Philosophy and Psychology* (Gloucester, Mass., 1960 (1901)), ii. 137–138.

Exclusive reliance upon revelation as the source of historical and physical truth and the basis of faith meant inevitable and interminable difficulties with science.

Evangelical influence, then, was as central to Victorian attitudes toward science as it was to attitudes toward so many other issues. How important a role did it play in antivivisectionist sentiment? Stevenson's thoughtful and erudite account of the theological background to antivivisection emphasizes the awakening of conscience in the Evangelical revival of the previous century as critical to the ultimate emergence of the movement.[40] There can be no doubt that antivivisection owed a conscious debt to earlier attempts to translate organized indignation into political and social action, attempts that consistently involved Evangelicals and Dissenters. But closer links can be discerned. As Stevenson suggests, many antivivisectionist leaders were Evangelicals, or had been raised as such. Certainly the most widely read Evangelical periodical, the *Record,* was firmly behind the movement. Again, there were important parallels of style and substance: "The least attractive features of flamboyant evangelicalism at its worst—its unrestrained indulgence in tear-baths of emotion, its love of unreason, even its belief in the constant operation of 'special providence'—these are also the recurrent features of antivivisectionism."[41] Most important for present purposes, antivivisectionist literature shows a hostility toward science entirely congruent to that of many Evangelicals.

At the root of this hostility was the fear that science was the agency of darker forces, and, in its articulation, spokesmen for the movement addressed themselves to exactly the issues sketched above as fundamental to the broader debate: the primacy of spiritual values, the legitimate scope of scientific inquiry, and man's place in nature. The passion for knowledge was "an inferior part of our being, a part which we may share with demons in the

[40] L. G. Stevenson, "Religious Elements in the Background of the British Anti-Vivisection Movement," *Yale Journal of Biology and Medicine* xxix (1956), 154.

[41] Ibid., 155.

Pit."[42] Indeed, base curiosity was man's first temptation; his fall was traceable to the fruit of knowledge. Modern science was an idol, an evil Moloch, exalting physical discovery to "the supreme place where Christianity would place the spirit of love."[43] It recognized no limit to research, delving into matters no mortal mind could hope to comprehend: "Men want to know what it is that makes up the vast mystery of life; they want to solve everything, to test, probe, analyse, and tell what the soul consists in, and so forth."[44]

One did not need to seek far for the basis of such an impudently sacrilegious program of research. Modern scientists were in the grip of a rampant naturalism.

> Science has declared that man is but a beast which perishes; the superiority once claimed by humanity as having been made in the likeness of the Deity, cannot be put forward by a world which has long been taught by science, to see itself as a mere accidental congregation of atoms.[45]

The aim of scientists was nothing less than "the fall of all religions" and the erection of a philosophy of naturalism in their place, with science governing mankind.[46] With such an ideal, one could scarcely be surprised at the vicious and selfish excesses that emanated from the scientific camp. These were the men who hoped "to found the Religion of the Future, and to leave the impress of their minds

[42] *Letter from a Lady Student of Vivisection* (London, 1883), p. 6.

[43] *Colonel Osborn on Christianity and Modern Science* (London, c. 1889).

[44] *Report of the Annual Meeting of the Victoria Street Society for Protection of Animals from Vivisection* (London, 1881), p. 13. See also *Zoophilist* ii (1882–1883), 70.

[45] Ouida, *The New Priesthood* (London, 1893), p. 74. See also *Anti-Vivisectionist* vi (1879), 183–184; M. Thornhill, *The Clergy and Vivisection* (London, c. 1883), pp. 17–18, 60–62; *Animals' Guardian* iii (1892–1893), 178.

[46] *Home Chronicler* (1877), 682–684; (1878), 21–22; *Anti-Vivisectionist* vi (1879), 104–105; H. N. Oxenham, "Moral and Religious Estimate of Vivisection," *Gentleman's Magazine* ccxliii (1878), 715 730, 732, 736.

upon their age."[47] The credo of science was no more than the crudest expression of "might makes right":

'Tis but the same dead creed,
Preaching the naked triumph of the strong;
And for this Goddess Science, hard and stern,
We shall not let her priests torment and burn:
We fought the priests before and not in vain;
And as we fought before, so will we fight again.[48]

For those believers who were prepared to entertain natural theology as a complement to revealed theology, and human reason as interpreting scripture in the light of the evidence of the divine creation, the response to scientific naturalism was bound to be much more complex. In a series of provocative articles, Young has explored the question extensively, and his analysis underlies much of the approach of this chapter.[49] According to Young, natural theologians and those who thought as they did made two or three different kinds of responses to the challenges posed by scientific naturalism. Many aspects of natural theology could be retrieved by identifying God with the uniform operation of natural law, and effecting a progressive separation of science and theology by weakening the claims of natural theology to illuminate revelation by science. While it could be argued that evolutionary law provided a basis

[47] F. P. Cobbe, *Bernard's Martyrs. A Comment on Claude Bernard's Leçons de Physiologie Opératoire* (London, 1879), p. xvi. See her *Moral Aspects of Vivisection* (4th ed., London, 1882), pp. 9–11.

[48] L. Morris, A *"Song of Two Worlds"* (London, n.d.).

[49] R. M. Young, "Malthus and the Evolutionists: The Common Context of Biological and Social Theory," *Past and Present* xliii (1969), 109–145; "The Impact of Darwin on Conventional Thought," in A. Symondson (ed.), *The Victorian Crisis of Faith* (London, 1970), pp. 13–35; "The Historiographic and Ideological Contexts of the Nineteenth Century Debate on Man's Place in Nature," in M. Teich and R. M. Young (eds.), *Changing Perspectives in the History of Science. Essays in Honour of Joseph Needham* (London, 1973); "Natural Theology, Victorian Periodicals, and the Fragmentation of the Common Context," *Victorian Studies*, in press.

for a deeper understanding of purpose and progress in the universe, the classical argument from design became irrelevant. The change was fundamental, and involved

> . . . the abandonment of the traditional claims of natural theology by emasculating its theodicy. It was no longer a strong enough doctrine to serve a unifying function in the intellectual culture of the period. If the justification of the ways of God to man is that God works by general laws and/or that His intention was that moral and physical laws should be separate, then this would seem to guarantee God's indifference to a careless man of piety who, while praying fervently, stepped over a cliff and perished while obeying the laws of gravity. Of course, this view of God's relationship to nature does not preclude a very personal God serving essentially psychological functions of inspiration and consolation, but it was necessary to confine claims for His efficacy to the afterlife, since science, scientific historiography, and statistics eliminated the claims of Divine creation and the efficacy of prayer.[50]

Hence natural theology no longer provided a rationale by which the pursuit of science could be related to broader cultural goals. On the contrary, the tradition that had legitimized science during the eighteenth and early nineteenth centuries, found itself convulsed by the very successes of its erstwhile handmaiden. As Young puts it, by 1880 theology had become no more than one point of view in a conflict. The change was seen by many who had supported science in the context provided by natural theology as something very much on the order of a betrayal, and their feelings toward science were consequently particularly bitter. Turner captures the phenomenon perfectly:

> The Christian reaction to scientific naturalism is readily understandable. Naturalistic writers claimed that men could live honest, righteous, meaningful lives without God, without faith in Christ's redemptive mission, without the bible, without the church, and without the clergy. That

[50] R. M. Young, "Natural Theology," ibid.

> such claims emanated from members of the profession which only a few years before had aided the clergy in discerning the glory of God in creation occasioned even more resentment on the part of Christians.[51]

The particular resentment of those disillusioned partisans of natural theology provides a striking linkage between antivivisectionst attitudes toward science and the dilution of natural theology at the hands of "the New Nature" of scientific naturalism. For two of antivivisection's most vociferous spokesmen, the Rev. F. O. Morris and Frances Power Cobbe, were just such disillusioned partisans. An examination of their writings shows the bitter perceptions of apostasy with which they greeted evolutionism and its publicists. A demonstration of such opinions in two of the movement's leading figures justifies the assumption that many other antivivisectionists may have shared their views. Case studies of Morris and Cobbe enable one to make an additional subsidiary point about the relationship between antivivisection and broader public sentiment hostile to science or to medicine. The point is that a given person might have been either attracted to the movement through his sharing of one or both of these preexisting broader public sentiments, or he might have been led to share them through an initial concern with the infliction of pain upon animals by experimental medicine.

It seems most likely that the Rev. F. O. Morris was led to antivivisection through his violent antievolutionism. Morris (1810–1893) was rector of Nunburnholme, Yorkshire and a country naturalist of great productivity. He was the author of *A History of British Birds, A History of British Butterflies*, and numerous papers given before the British Association for the Advancement of Science. The emergence of Darwinian theory, with its blatant contradiction of Scripture, and the ascendance within the British Association of the likes of Huxley and Tyndall filled Morris with rage;

[51] F. M. Turner, "Between Science and Religion: The Reaction to Scientific Naturalism in Late Victorian England," (Yale Univ. Ph.D. thesis, 1971), p. 298.

his was the rage of the pious amateur who sees the foundations of his faith shattered by a once benign science now transmogrified at the hands of infidelity and professionalism. In pamphlets and papers he vitriolically attacked the "Darwinian clique" at the British Association and their science of "preposterous and idle vagaries."[52] He became an enthusiastic antivivisectionist and a bitter critic of the R.S.P.C.A.'s inaction on the issue.[53] The contiguity of his opinion of Darwinian theory and his opinion of vivisection is nicely captured in the ponderous title of one of his many tracts: *The Demands of Darwinism on Credulity: A Homethrust at the wretched and shallow Infidelity of the day, and its Twin Sister the Cowardly Cruelty of the Experimenters on Living Animals. Dedicated by Permission to the Right Honourable the Common Sense of the People of England.*[54] He emphasized, "There is a very close connection between the monstrous absurdities of Darwinism and the cowardly cruelties of the perpetrators of Experiments on Living Animals." What else but such cruelty could one expect from men who believed in the survival of the fittest and the doctrine that animals are "conscious automata"?[55] Thus, for at least one member of the movement, there was a direct logical connection between antivivisection and the supposed scientific foundations of the naturalistic program.

In contrast to Morris, Frances Power Cobbe seems to have been brought to her general indictment of science through her original intense involvement in antivivisection. Indeed, she was genuinely respectful of science in 1863, at which point she had just encountered vivisection, but not yet become deeply concerned with fighting it. In an essay of that year on "The Rights of Man and the Claims

[52] See, e.g., F. O. Morris, *All the Articles of the Darwin Faith* (London, 1875).

[53] *Anti-Vivisectionist* vi (1879), 542–543.

[54] (London, c. 1890).

[55] Ibid., pp. 61–62. See also F. O. Morris, *The Curse of Cruelty* (London, 1886), and *Experiments on Living Animals* (no place or date, c. 1888).

of Brutes," she expressed herself in the reverent terms of natural theology:

> "Science" is a great and sacred word. When we are called on to consider its "interests" we are considering the cause of that truth which is one of the three great portals whereby man may enter the temple of God. Physical science, the knowledge of God's material creation, is in its highest sense a holy thing—the revelation of God's power, wisdom, love through the universe of inorganic matter and organic life. The love of truth for its own sake, irrespective of the utility of its applications, has here one of its noblest fields; and no love of the beautiful by the artist, nor of the good by the philanthropist can surpass it in sanctity, *or claim, on moral grounds, a larger liberty.*

Cobbe went on to conclude that "man has a right to take animal life for the purposes of science as he would take it for food, or security, or health," provided that there was no needless infliction of pain. Utilitarian justifications for experimenting upon animals were strictly supererogatory.[56] During the sixties, Cobbe had known a number of members of the London scientific elite, but such relationships ended in 1876.[57] Her closest friend among them was Charles Lyell, who died in 1875. To Cobbe, writing in the early nineties, Lyell was "the man of science as he was of old; devout, and yet entirely freethinking in the true sense." She added,

> But to my memory he will always be something more than *an* eminent man of science. He was the type of *what such men ought to be;* with the simplicity, humility, and gentleness which should be characteristic of the true student of Nature. Of the priest-like arrogance of some representatives of the modern scientific spirit he had not a taint.[58]

[56] F. P. Cobbe, "The Rights of Man and the Claims of Brutes," *Fraser's Magazine* lxviii (1863), 586–682, esp. 593. Italics are mine.

[57] F. P. Cobbe, *Life of Frances Power Cobbe* (Boston and New York, 1894), ii. 440–449.

[58] Ibid., ii. 404, 408–409. Italics in the original.

Between 1863 and the 1880s, Cobbe's attitude toward science changed radically, a change that emerged directly from her antivivisectionist activity, but was expressed in language characteristic of precisely what we might expect from one whose enthusiasm and expectations of science had been cruelly dashed by scientific naturalism. Thus, in 1888 she wrote in an essay criticizing attempts to derive ethical precepts from the knowledge of nature:

> Science has been frequently called the "Handmaid of Religion," and when young and simple she frequently fulfilled that function. Grown old and arrogant, however, she has consigned her mistress to an asylum for imbeciles, while—like other Abigails—she borrows her cap and speaks from her chair.[59]

Precisely when, for Cobbe, did science grow "old and arrogant"? Why, with just that generation of scientists succeeding Lyell's whose work finally subverted the enterprise of natural theology:

> . . . the moralist may discern the production of a certain specific type of hardness, and arrogance which has developed itself in the ranks of Science, ever since science has ceased to be to its followers [what it was to the Keplers and Newtons, and Herschels and Lyells of an earlier time] the highway to religion.[60]

Clearly Cobbe, like Morris, was a classic example of the enthusiast for natural theology doubly bitter over the scientific betrayal.

Although antivivisectionist hostility toward science has been interpreted to this point as part and parcel of the larger clash of ideas between scientific naturalism and the

[59] F. P. Cobbe, "The Lord was not in the Earthquake," *Contemporary Review* liii (1888), 72.

[60] F. P. Cobbe, *The Study of Physiology as a Branch of Education* (London, 1883), p. 4. Cf. the quote from Joseph Hooker in J. Hannah, "A Few More Words on the Relation of the Clergy to Science," *Contemporary Review* ix (1868), 399.

traditions of natural and revealed theology, it was also part of a struggle for cultural leadership between the agressive spokesmen for "the New Nature" and the hitherto dominant literary and theological intelligentsia. Turner, in his study of scientific naturalism, emphasizes the degree to which its proponents were motivated by the desire to improve their own social status and intellectual influence relative to that of the clergy. "Scientific and religious ideas represented merely the weapons" in a conflict "between *some* scientifically oriented men and *some* religiously oriented men" over "which ideas and which men would direct the future of English culture."[61] As Victorian critics of scientific naturalism saw it, certain scientists were getting above themselves, and pronouncing upon social, intellectual, and even religious issues hitherto scarcely deemed to be within the domain of science.

Indeed, the impudent self assurance with which the increasingly powerful scientific fraternity arbitrated some of the major questions of the day moved the critics, with antivivisectionists prominent among them, to formulate an extended analogy between modern science and the Inquisition, between the new and the old priesthood.[62] The new priesthood displayed all the narrow fanaticism, the hypocritical morality, and the intolerance of criticism that had characterized its predecessors. As Ouida expressed it in 1893, "No art, no learning, no achievement dares to say as the physiologist says of his science, that the vulgar must not look on it, nor the multitudes pronounce on it."[63] Scientists pronounced upon public affairs with ill-disguised superiority and hauteur, yet the endless fragmentation and

[61] F. M. Turner, "Between Science and Religion: The Reaction to Scientific Naturalism in Late Victorian England," (Yale Univ. Ph.D. thesis, 1971), p. 37. See also pp. 11, 33–43, 216–218.

[62] Ibid., pp. 39–40.

[63] Ouida, *The New Priesthood* (London, 1893), pp. 6–7. See also E. Maitland, "Vivisection," *Examiner* (1876), 738–739; "The Cui Bono of Vivisection," *Social Notes* i (1878), 7–8; *Debate on Mr. Reid's Bill in the House of Commons, April 4, 1883* (London, 1883), p. 43; C. A. Gordon, *The Vivisection Controversy in Parliament* . . . (London, 1888), p. 46.

specialization of modern science scarcely qualified them for such a role.[64] Like doctors, spokesmen for science presented a united front to exploit public credulity:

> The public is . . . unwilling to believe that a man famed for some one scientific attainment, can be himself mistaken, and thus likely to mislead them; while his colleagues seldom permit themselves to expose, before a popular audience, the nonsense solemnly baled out to them by one of themselves.[65]

Public respect for scientists infuriated Frances Power Cobbe, who alluded venomously to "the science-mongery which is one of the chief modern substitutes for religion":

> . . . to thousands of worthy people it is enough to say that Science teaches this or that, or that the interests of Science require such and such a sacrifice, to cause them to bow their heads, as pious men of old did at the message of a Prophet: "It is SCIENCE! Let it do what seemeth it good."[66]

In their remorseless drive for knowledge—what H. G. Wells called "the strange, colourless delight of these intellectual desires"—scientists went futilely on, never solving problems, merely formulating new ones.[67] Not unexpectedly, their failure to recognize the precedence of moral over intellectual, of spiritual over physical, meant that

[64] C. Adams, *The Coward Science* (London, 1882), pp. 21–22; J. Adam, "Paul Bert's Science in Politics. A Personal Reminiscence," *Contemporary Review* li (1887), 32–33; *Verulam Review* iii (1892–1893), 134–138; E. Maitland in E. Carpenter and E. Maitland, *Vivisection* (London, 1893), pp. 19–20.

[65] A. de Noë Walker, *The British Association for the Advancement of Science and Vivisection* (London, 1882), p. 5.

[66] F. P. Cobbe, *Meeting at Stoke Bishop* (London, 1883), p. 2; "The Scientific Spirit of the Age," *Contemporary Review* liv (1888), 126.

[67] H. G. Wells, *The Island of Dr. Moreau. A Possibility* (New York, 1921 (1896)), pp. 136–137; Brayfytte, "Experimental Physiology. What it is, and What it asks," *Animals' Guardian* ii (1892), 39–42.

> . . . we see the brute instinct of crime, hardly lulled to sleep in the human breast by the religious and moral education of centuries, coming to life again under the aegis of the infallibility of science.[68]

Given this perception of the inevitable tendency of modern science, antivivisectionists were amazed and distressed at the clergy's failure to condemn en masse the practice of experimenting upon living animals. Frances Power Cobbe taunted churchmen with their lack of moral courage; according to her, they wanted peace with science, peace at any price.[69] She went on,

> It would truly seem that blindness has fallen on the eyes of the clergy that they do not perceive that the Science which they shrink from even seeming to oppose or discredit, is the same godless, inhuman Science which teaches that there is no Father in Heaven but only an "unknown and unknowable" Author of a world of struggle and pain, ungilded by a single ray of hope beyond the grave. Can they recognize that this Science—whom they would propitiate at the price of condoning the sin of cruelty—is their own irreconcilable enemy? Do they not foresee that if such Science swell a little higher and sweep over England, it will swallow up their Churches and their Faith . . . ?[70]

The incursion of science into educational curricula naturally appeared to the movement as a particularly serious threat. It could not be denied that in his worship of the intellect, "the scientist too often lacks the ordinary feelings

[68] J. Adam, "Paul Bert's Science in Politics. A Personal Reminiscence," *Contemporary Review* li (1887), 34; *Zoophilist* iii (1883–1884), 56; F. W. Newman, "On Cruelty," *Fraser's Magazine,* n.s. xiii (1876), 533–536.

[69] F. P. C., "The Clergy and Vivisection," *Spectator* lvii (1884), 517.

[70] F. P. Cobbe, *The Churches and Moral Questions* (London, 1889), p. 7. Cf. *Prebendary Grier on the Lichfield Dioscesan Conference and Vivisection* (London, 1892), p. 2: "The fact, too, that we [the clergy] have made mistakes in regard to science in the past renders us very slow to engage in any course which may betray us into similar blunders now."

of human nature."[71] He almost always displayed a lack of sentiment, a pitiless and callous contempt for the humble and the weak, and a total loss of the esthetic sense.[72] The price to be paid for educating the student in science would be a shallow materialism,

> The material (or as our forefathers would have called it, the *carnal*) fact will be uppermost in his mind, and the spiritual meaning thereof more or less out of sight. He will view his mother's tears—not as expressions of her sorrow—but as solutions of muriates and carbonates of soda, and of phosphates of lime; and he will reflect that they were caused—not by his heartlessness—but by cerebral pressure on her lachrymal glands. When she dies, he will "peep and botanize" on her grave . . .[73]

Science taught nothing but a crude and slashing analysis of the physical world, amounting—in the words of a famous public school headmaster—to "a superior grocery assistant's work." Without "a single element of higher mental training," it could hardly be thought useful "to prepare men for life and its struggles."[74]

In sum, the antivivisectionist critique of science was an articulation and crystallization of fears and hostilities shared in varying degree by important segments of opinion of the middle and upper classes. Such a critique evidenced, in particularly naked fashion, a bitter sense on the part of religious thinkers and literary intellectuals that science was becoming the leading institution in late Victorian society.[75] To counteract this trend and to retain their own

[71] *Animal World* xi (1880), 34.

[72] F. P. Cobbe, "The Scientific Spirit of the Age," *Contemporary Review* liv (1888), 126, 131–135.

[73] Ibid., 130. See also F. P. Cobbe, *The Study of Physiology as a Branch of Education* (London, 1883); *Zoophilist* iii (1883–1884), 128–129.

[74] F. P. Cobbe, "The Scientific Spirit of the Age," *Contemporary Review* liv (1888), 130–133; F. P. Cobbe, *Letters of Head School-Masters* (London, 1886). See also S. D. Collingwood, *The Life and Letters of Lewis Carroll* (New York, 1891), pp. 167–171, 187–191.

[75] Cf. V. Lee [V. Paget], "Vivisection: An Evolutionist to Evolutionists," *Contemporary Review* xli (1882), 794.

franchise upon the formation of public opinion and the maintenance of personal faith, such men conducted a continual polemical battle against the ascendancy of such scientists as Tyndall and Huxley to positions as social and intellectual prophets.[76] Four men who numbered antivivisection among their many causes, Cardinal Manning, Lord Shaftesbury, R. H. Hutton, and F. W. Newman, sufficiently indicate the variety of intellectual and religious temperament that could substantially agree in uniting to combat the encroachments of modern science. This combat rose to its greatest intensity during the seventies, just as antivivisection emerged as a national political issue and an ethical dilemma.[77] One can find no better epitaph for the issues raised in this chapter than Manning's declaration,

> I believe that science consists in the knowledge of truth obtained by the processes which are in conformity with the nature of God, who, the Holy Scripture says, is the Lord of all sciences . . . I believe in science most profoundly within its own limits; but it has its own limits, and, when the word Science is applied to matter which is beyond those limits I don't believe in it.[78]

It was precisely this perception that science was seeking authority beyond its legitimate purview that triggered a disenchantment with science on the part of humble believers and sophisticated intellectuals alike. Disenchantment took various forms, and many of them were crystallized and focused by the practical issue of antivivisection. Thus the movement transcended its apparent simplicity of purpose and showed itself to be less sui generis than a faithful mirror of the society in which it flourished.

[76] See, e.g., A. W. Brown, *The Metaphysical Society. Victorian Minds in Crisis, 1869–1880* (New York, 1947), pp. 231–242.

[77] Ibid., pp. 185–186.

[78] V. A. McClelland, *Cardinal Manning. His Public Life and Influence, 1865–1892* (London, 1962), p. 210.

11. The Mind of Antivivisection: Animals

It is a commonplace that Britons are inordinately fond of animals, and especially of pets. This endearing national trait, the subject of many a quip and satire, was reflected by the country's precedence in legislating animal protection and its role in promoting similar measures overseas. No doubt the roots of the phenomenon are complex, dating from well before the chronological scope of this study. Little research has been done on the subject and certainly it is not possible to do justice to it here. Nevertheless, the antivivisection movement took place in the context of the Victorian cult of pets, which, however ill understood, cannot be ignored.

Victorian England was a rapidly urbanizing society in which land ownership and rural ways of life retained profound cultural significance. The industrial titan transformed into country squire by the purchase of an appropriate estate was a cliche of individual social progress by mid-century. Few urban dwellers were so fortunate. Eager to sample the amenities of the countryside, amenities redolent of social prestige, health, and the simple verities, the urban middle class took day trips by rail, cultivated gardens at home, collected the artifacts of natural history,—and kept pets. Animals were a link to the land, which was in turn the basis upon which British society was still structured. They were a reminder that England was not all mud and soot and stone, but a place where a multitude of living things grew and thrived. They were a symbol of the potent myth of uncomplicated rural existence that was so prominent a part of the conventional wisdom. The cult of pets flourished in an urban society starved for points of reference to the life-style of the landed.

Though the model was rural, the relationship between animal and master could not be translated into the urban setting. Domestic animals in the countryside were used by their masters for working roles: riding, milking, hunting, herding, and shearing, to name some. In city homes, however, pets were members of the household as such. Though dogs might protect the premises, and cats catch mice, pets were kept for essentially emotional rather than practical reasons. City dwellers' attitudes toward animals were, therefore, rather different from those of rural society. While the squires and farmers grew up in an environment with many animals and were familiar with the daily round of hunting, predation, crude agricultural surgery, and other vicissitudes of the animal creation, their urban counterparts tended to think in terms of the unique sufferings of a particular pet. Little wonder, then, that the power base of both the animal protection and the antivivisection movements was urban.

The intense and often bizarre emotion directed toward animals in much of antivivisectionist literature inspired unsympathetic observers to liken it to a modern day revival of totemism or to christen its followers "Bestiarians." One point is clear from the outset: although the literature of the movement speaks of "animals" in the generic sense, the animals foremost in the minds of almost all antivivisectionists were those usually taken as pets—dogs, cats, and horses, that trio significantly singled out for special protection in the original government bill. With the exception of a very few—notably the vegetarians—antivivisectionists were motivated by thoughts of their pets, or animals like their pets, being subjected to the supposed agonies of vivisection.[1] It is in this concern for pets, for dogs and cats,

[1] See, e.g., W. Ballantine, *Some Experiences of a Barrister's Life* (New York, 1882), p. 395. Edward Maitland, a vegetarian, protested vigorously that the special protection for dogs, cats, and horses in Carnarvon's bill was an "utterly unworthy" provision that demonstrated that people were selfishly sheltering their own feelings at the expense of any real inroads into the extent of animal suffering per se. E. Maitland, "The Doctors and the Vivisection Bill," *The Examiner* (1876), 682–684.

that we must look for the key to antivivisectionist fervor, which was undoubtedly shared in varying degrees by other segments of the animal-protection movement and by the general population. The discussion will suggest that the supposedly abstract philosophical and theological treatments of the rights of animals or of their prospects of immortality make more sense when seen as the outgrowth of attitudes toward pets, rather than as the preexistent and wholly rational intellectual substructure of the movement.

The particular nature and scope of antivivisectionist concern for animals was perfectly captured by John Robertson in an appraisal of the movement published in Annie Besant's *Our Corner:*

> Miss Cobbe is a typical anti-vivisectionist in that she does what we have said the majority [of mankind] do not do—carry her sympathy with animals to the extent of attributing to them—or some of them—a moral and emotional nature closely resembling our own. She lets it appear very plainly that she is moved by purely personal sentiment as distinguished from an impartial sense of justice. Her dog and cat are a great deal to her, and it is the idea of their suffering which excites her . . . . she is not defending a right inherent in sentient things as such; she is doing special pleading for some of them for which she has a special liking.[2]

Hand in hand with a special anxiety about dogs and cats, then, went an almost equally universal tendency to anthropomorphize these animals. This tendency was classically exemplified in antivivisectionist short stories, where animals often think and talk to one another, and in the numerous anecdotes about noble dogs and intelligent cats retailed by periodicals of the movement. Frances Power Cobbe, for one, constructed a "canine psychology" of "how a dog thinks and feels" upon a base of anecdote and the categories

[2] J. Robertson, "The Ethics of Vivisection," *Our Corner* vi (1885), 89, extracted in "Physiology and its Opponents," *Journal of Science* vii (1885), 602.

of a very naive psychology of human faculties.[3] Dogs not only had such fine qualities as nobility and loyalty, they possessed a conscience and were capable of exercising thought and judgment. Furthermore, they were widely different from one to another, a fact that certain people could not seem to understand:

> That every dog has his idiosyncracy no less than his master has his own; that his capacities, tempers, gifts, graces and propensities, vary through the whole gamut of intellect, will, and emotion; and that it would be as easy to find two human as two canine Sosias, are facts which the vulgar and dog-ignorant mind has never grasped.[4]

Given this view of canine moral capacity and individuality, one could indict some dogs with such human failings as gluttony, pride, avarice, superstition, even crime:

> I have used the word "faults" but I am not sure that we might not equally properly speak of the crimes of dogs, for the turpitude of some of their actions certainly surpasses mere failure in justice or benevolence. There are traitor dogs who have basely accepted bribes of raw meat and remained silent when it was their imperative duty as sentinels to challenge the intruder with the loudest of barks . . .[5]

As one of Cobbe's friends said to her, "I don't understand your feelings about animals at all. To me a *dog is a dog*. To you *it* seems to be something else!"[6]

Antivivisectionists saw pets in sentimental terms, endowing them with human virtues and frailties. Furthermore, such characteristics were often generalized from domestic or tame animals to include wild animals also. It is in this

[3] [F. P. Cobbe], "The Consciousness of Dogs," *Quarterly Review* cxxxiii (1872), 419–451.

[4] F. P. Cobbe, "Dogs Whom I Have Met," *Cornhill Magazine* xxvi (1872), 662.

[5] Ibid., 667.

[6] *Life of Frances Power Cobbe* (Boston, 1894), i. 163. Italics are in the original.

context of anthropomorphism that discussions of the prospects of a life to come for animals seem most logically connected to the broader controversy. Many antivivisectionists were prepared to entertain the possibility that higher animals might have an immortal soul,[7] and they had some impeccable authorities to call upon, notably Bishop Butler and John Wesley. The resulting literature consists largely of such quotation, tortuous biblical exegesis, and emotional exhortation based on the impossibility of consigning to darkness beings capable of such sagacity, loyalty, and affection. There was more to it than this, of course. God created animals distinctly from man, the Holy Spirit vivifies them as it vivifies man, yet they suffer and die after the Fall, which was the responsibility of man alone. Is it reasonable then that God's plan is to deny them the final restitution that the millennium will bring? Is the convenant with God entered into by the animal creation, no less than by man, at the time of Noah, to come thus to naught? Is the pain and extinction of animals over the ages, revealed by geology and paleontology, to be meaningless and unredeemed? As the most sophisticated statement appearing in the antivivisectionist literature summarized the case for animal immortality:

> We have seen that there are scriptures to show that, as they [animals] unwillingly fell with Adam under the curse, so they will share the blessings of that age when Christ shall have all things under him: and hence the reasonable inference that they will be interested also in the eternal period that will follow, when the last enemy

[7] Among them were Cobbe, Jesse, Shaftesbury, Morris, Kingsford, and Mary Somerville, the astronomer. For a cross-section of the discussion see "Have the Lower Animals a Future Life?," *Home Chronicler* (1877), 916–918, 932–934; "The Consciousness of Dogs," *Quarterly Review* cxxxiii (1872), 419–451; "The Future of the Lower Animals," *Zoophilist* vi (1886), 91–93; R. Eyton, *Our Brute Friends in the Life to Come* (London, 1891); F. O. Morris, *The Curse of Cruelty* (London, 1886); G. MacDonald, *The Hope of the Universe* (London, 1893); F. P. Cobbe, "The Ethics of Zoophily. A Reply," *Contemporary Review* lxviii (1895), 497–508.

> shall have been destroyed . . . It is further adducible in favour of the doctrine, that the purpose of their creation had at no time exclusive reference to man; that their life, like his, is derived from the Spirit of God; that, though not like him made after the image of God, still they, in many cases, manifest dispositions and powers that are worthy of admiration; and that analogy would lead us to think are destined rather to be increased, even above what they once were, than to cease by annihilation forever; that a compensation seems not unlikely to be made them for their present sufferings; that they have been from the days of Noah, in covenant relationship with the Lord, a relationship, which there is some ground for thinking, will be eternal; finally there seems to be no scripture that is clearly opposed to the view supported by the foregoing considerations.[8]

To some, if such considerations did not suffice to establish the doctrine of animal immortality, the conviction of similar arguments for a future life for man seemed inevitably in question. Furthermore, the knowledge of the sufferings of animals at the hands of science gave particular weight to a consideration of a possible future life:

> [Antivivisectionists] implore to be shown how they ought to think of it [vivisection] consistently with their reliance on the Judge of all the earth to do right, and their faith that in His universe there can be no final and remediless injustice. . . . there is no possible solution of this heart-wearing question save the bold assumption *that the existence of the vivisected animal* (and of course, as a consequence, of other creatures of the same rank in nature) *does not end at death.*

Any other conclusion would logically entail the blasphemous view that the Creator had given a sentient creature an existence of unmitigated and unredeemed suffering.[9]

[8] "Have the Lower Animals a Future Life?," *Home Chronicler* (1877), 933.

[9] F. P. C., "The Future of the Lower Animals," *Zoophilist* vi (1886), 92. Italics are in the original. For an excellent discussion of

Religious speculations about animals' prospects of heaven were highly important to certain antivivisectionists. So, too, were attempts by the movement to enunciate or to utilize doctrines of animals' rights.[10] One thinks, for example, of the editorial credo of the *Animals' Guardian,* which revolved around the "enslavement" of animals by man.[11] But this theological and ethical discussion was less a fundamental underpinning for the antivivisection movement as a whole than a by-product of psychological and intellectual forces manifested by the movement, for which animals became symbolic. The most important indicator to these forces is the anthropomorphization of animals.

The attribution to animals of human qualities and individuality was by no means solely an antivivisectionist predilection. The work of Darwin, Romanes, Lloyd Morgan, and others was only the best of an enormous popular literature on animal psychology that dated from the early part of the century, much of it as naive as the most childish antivivisectionist fable. As the success of genre painter Edwin Landseer (1802–1873) demonstrates, the tendency antedated the emergence of antivivisection as a national issue and was diffused widely through bourgeois and aristocratic society. Landseer was an animal painter of superb technical skill, who augmented his considerable popularity and fortune by turning out moralizing, anecdotal pictures of animals of quasihuman mien (see Figures 13, 14, and 15). His pictures achieved an enthusiastic public response and critical acclaim. Landseer became the most famous British painter of his day, a favorite and intimate of royalty, a man whose work was guaranteed international notice. For antivivisectionists, works like Landseer's "The

---

these issues, see L. G. Stevenson, "Religious Elements in the Background of the British Anti-Vivisection Movement," *Yale Journal of Biology and Medicine* xxix (1956), 126–127, 143–153.

[10] For two attempts to establish such rights, see E. B. Nicholson, *The Rights of an Animal: A New Essay in Ethics* (London, 1879), and H. S. Salt, *Animals' Rights Considered in Relation to Social Progress* (New York and London, 1894).

[11] *Animals' Guardian* (October 1890).

Figure 13. Edwin Landseer's "A Distinguished Member of the Humane Society" was adopted as part of the Victoria Street Society's logotype. *Courtesy of the Tate Gallery.*
Figure 14. Landseer's "Alexander and Diogenes," a classical example of anthropomorphism. *Courtesy of the Tate Gallery.*

Old Shepherd's Chief Mourner"—a collie grieving on its master's humble casket—were examples par excellence of the perceptive treatment of the animal nature.[12] The praise of antivivisectionist William Ballantine, a celebrated barrister, was typical:

> He [Landseer] has spiritualized animal life, and has given it an affinity to the human race, and yet has neither destroyed nor altered its natural characteristics. We see upon his canvas the nobility, of which in the higher animals so many examples have been proved to exist, and in dealing with those which fill a humbler space in nature he has created a poetry essentially his own. One wonders what those two squirrels are saying to each other.[13]

Illustrations of animals in antivivisectionist literature frequently aped Landseer's treatment; an engraving on one pamphlet portrayed a dog with a pipe in his mouth sitting contentedly beside a glass of beer and asked "Shall such creatures be allowed to be cut to pieces alive?"[14] Serious critics of art have been less charitable toward Landseer's willingness to accede to the popular tastes of his time by infusing animals with human traits.[15] Humphrey House captured the essence of the matter when he wrote,

> All the anthropomorphism and the sentimental treatment of animals . . . was a means of infusing into what was otherwise a purely materialistic "scientific" representation of the external world some kind of "spirit." It was, as

[12] See, e.g., *Zoophilist* (1881), 136, for one of many panegyrics to Landseer and his work.

[13] W. Ballantine, *Some Experiences of a Barrister's Life* (New York, 1882), pp. 350–351. Cf. Cobbe's response to the work of another animal painter, Rosa Bonheur: *Life of Frances Power Cobbe* (Boston, 1894), ii. 360–361.

[14] E. von Weber, *The Torture Chamber of Science* (6th ed., London, c. 1880), p. 14.

[15] See the issue on Edwin Landseer of *Masters in Art* liii, No. v (May 1904), published in Boston. I am grateful to Professor E. D. H. Johnson for advice about Landseer.

Figure 15. "High Life" and "Low Life," allegories of Victorian social divisions. *Courtesy of the Tate Gallery.*

> I have already suggested, an afterthought. The effect of what we call sentimentality—that is of flabby and exaggerated sentiment—comes about partly because it is an afterthought, an added embellishment, extraneous to the first conception, and partly because the social and moral circumstances of the time required in art as well as in life the overassertion of certain modes of feeling.[16]

For present purposes, two elements of House's analysis may be taken as keys to the symbolic importance of (quasi-human) animals to antivivisectionists. The first element, the attempt to infuse some kind of spirit into a materialistic, scientific world, captures in the antithesis of spirit and science an issue of crucial intellectual significance for the movement. Experimental medicine treated animals precisely as the antivivisectionist indictment of science as rampant materialism would predict. Physiologists saw animals as species, a group of interchangeable biological systems on which to carry out mechanistically conceived experiments, in the attempt to generate general conclusions about bodily function. In contrast, antivivisectionists saw animals as unique individuals, whose varying mental, moral, and spiritual qualities were infinitely more important than their physiological ones. These conflicting attitudes were derivative of the confrontation of two world-views whose manifold differences have been explored at length. For antivivisectionists, animals and their treatment came to symbolize the attack on the moral and spiritual bankruptcy of science.

Both sides in the vivisection controversy found evolutionary doctrines of the relationship between man and animals a weapon of debate to be used gingerly: such doctrines could be adduced both in support of the validity of extrapolating experimental results obtained with animals to human beings and in support of arguments on behalf of animals as "our weaker brethren" or "our dumb fellow creatures," frequently put forth by antivivisectionists. For

[16] H. House, "Man and Nature: Some Artists' Views," in *Ideas and Beliefs of the Victorians* (London, 1949), p. 229.

the latter, the conduct of experimental physiologists and biologists in the face of the obvious consequences of their beliefs about man's relationship to animals was a particularly bitter paradox:

> It is, indeed, the scientists themselves who have proved to us the close relationship existing between man and animals, and their probable development from the same origin. It is they who instruct us to cast aside the old theology which makes men differ from the beasts of the field, inasmuch as he was created in "the image of God," and yet would arbitrarily keep, for their own convenience, the line of division which such a belief marked out between man and animals.[17]

Evolutionary doctrines, which had probably been for many antivivisectionists a crucial instigator of suspicion toward science, seemed to have become, in the hands of those who originally promulgated them, a philosophy of "race selfishness" at the expense of "sinless fellow creatures" to be subjected to vivisection.[18] Nevertheless, those antivivisectionists who accepted evolution did so in terms that emphasized the *psychic* and *spiritual* continuities between animal and man while they attempted to ignore the physical links that were at the heart of the theory and of the rationale for experimental medicine.

A second element in House's analysis, "the social and moral circumstances of the time," directs attention to a

[17] A. Armitt, *Man and His Relatives: A Question of Morality* (London, 1885), pp. 6–7. See also "The True Teaching of Evolution as Regards Human Duties Towards Animals," *Anti-Vivisectionist* ix (1882), 93; A. Kingsford, *Unscientific Science* (Edinburgh, 1883), p. 32; J. H. Clarke, *Monkeys' Brains Once More. Schaeffer v. Ferrier* (London, 1888), p. 1; E. Berdoe, *An Address on the Attitude of the Christian Church Towards Vivisection, Delivered before the Ruri-Decanal Chapter of Rugeley, Staffordshire* (London, 1891), p. 3; H. J. Reid, *Experimental Physiology. What it is, and What it asks* (London, 1892), p. 3.

[18] F. P. Cobbe, *The Moral Aspects of Vivisection* (4th ed., London, 1882), pp. 17, 21. See also Cobbe's, *The Significance of Vivisection* (London, 1891), p. 2, and *Zoophilist* iv (1884–1885), 149.

somewhat more complex aspect of antivivisectionist attitudes toward animals. The preoccupation with the human or protohuman qualities of animals arose very much as a consequence of psychological pressures in a Victorian moral atmosphere of repression and prudery. For eighteenth and early nineteenth-century painting, animals served as powerful and passionate symbols of elemental forces such as lust and aggression. This depiction of untrammeled appetities and drives gave way, with Landseer and Victorian popular taste, to polite and sentimental homilies in which animals were no longer the symbols of projected brute instincts.[19] Instead, the animal world, especially the canine portion thereof, came to resemble a microcosm of human society in which fleshly indulgences were (ideally) absent or implicitly masked as failures of "conscience" or "moral judgment." As Pelles remarks, animals

> . . . were humanized as bearers of a philosophic, sentimental, or moral message. Landseer's dogs were English nineteenth century allegorical figures, evading the sensual female allegories which had long prevailed in France.[20]

Vestiges of the earlier attitudes toward animals remained through the Victorian period and beyond, however, in the attribution of human faults such as cruelty or sensuality to the "spirit of the Beast" or the "animalism" of man. The task of human life was to transcend corporeal appetites, or animalism, through spiritual purity.[21] Antivivisectionists had a particular horror of the corporeal and the bodily, in which medical science "animalistically" grovelled, and a particularly prudish brief for repression, seeing it as the objective for which art must be mobilized.

[19] See G. Pelles, *Art, Artists, and Society* (Englewood Cliffs, N.J., 1963), pp. 134–136.

[20] Ibid., p. 136.

[21] W. E. Houghton, *The Victorian Frame of Mind 1830–1870* (Paperback edition, New Haven and London, 1957), pp. 354–356, 368; P. T. Cominos, "Late Victorian Sexual Respectability and the Social System," *International Review of Social History* viii (1963), 23–25.

Frances Power Cobbe provides us with one especially apposite example of righteous prudery:

> Among the delicate and beautiful things which Science brushes away from life, I cannot omit to reckon a certain modesty which has hitherto prevailed among educated people. The decline of decency in England, apparent to everyone old enough to recall earlier manners and topics of conversation, is due in great measure, I think, to the scientific (medical) spirit. Who would have thought thirty years ago of seeing young men in public reading-rooms snatching at the *Lancet* and the *British Medical Journal* from layers of what ought to be more attractive literature, and poring over hideous diagrams and revolting details of disease and monstrosity? It is perfectly right, no doubt, for these professional journals to deal plainly with these horrors and with the thrice-abominable records of "gynaecology." But, being so, it follows that it is *not* proper that they should form the furniture of a reading-table at which young men and women sit for general—not medical—instruction. Nor is it only in the medical journals that disease-mongering now obtains.[22]

Antivivisectionist fear of the physical and of the dark and terrible "lower self" lurking within man, emerging with startling clarity in the practice of vivisection, is best illustrated by W. S. Lilly's essay "The New Naturalism," in which the author undertakes a lengthy examination of Zola's self-proclaimed association between experimental fiction and experimental medicine. Lilly was an enthusiastic antivivisectionist whose literary analysis was simultaneously and explicitly an attack on vivisection. To Lilly, the new, experimental naturalism was enslaved and obsessed by externals, by the senses, and hence totally devoid of, indeed antithetical to the moral and the spiritual, the sentimental, the ideal. Such materialistic, atheistic naturalism was recog-

[22] F. P. Cobbe, "The Scientific Spirit of the Age," *Contemporary Review* liv (1888), 137. Elsewhere, Cobbe's readers were informed that science kept "the material (or, as our forefathers would have put it, the *carnal*) fact" uppermost. See ibid., p. 130.

nizable both in science and in art, where preoccupation with bodily functions at the expense of inner moral and spiritual qualities stripped away Victorian defenses against the life of the emotions, revealing man as no more than another wild animal. According to Lilly,

> The New Naturalism . . . eliminates from man all but the ape and the tiger. It leaves of him nothing but the *bête humaine,* more subtle than any beast of the field, but cursed above all beasts of the field.[23]

In art and in science, as in life,

> There are moods of thought which do not yield in heinousness to the worst deeds—moods of madness, suicidal and polluting. To leave them in the dark is to help toward suppressing them. And this is a sacred duty.[24]

Given such an abhorrence for biological reality, such a horror of the corporeal, and such a belief in the need for repression of the "animal in man," it is not surprising to find antivivisectionists preoccupied with the anthropomorphization of pets. Mona Caird's dictum "Sentiment, in short, is the sole safeguard that the individual possesses against the crude and ferocious instincts of the human animal . . ."[25] leads us straight to Landseer's graphic homilies. Likewise, popular fear of "animalism" in man underlay a good deal of the hostile public reception of Darwinism and of compulsory vaccination. A scientific indication of man's blood relationship with animals could hardly be welcomed when "the great criterion of elevation in the order of existence is whether the higher or lower self . . . is dominent: the self of the appetites and passions, or the self of the reason and moral nature," and

[23] W. S. Lilly, "The New Naturalism," *Fortnightly Review* xxxviii (1885), 252.

[24] Ibid., 254.

[25] M. Caird, *A Sentimental View of Vivisection* (London, 1893), p. 20.

when the "lower self" was the "*bête humaine*."[26] Antivaccination literature consistently reflected popular fears in protesting the pollution of human blood or the "animalization" of human beings by the inoculation of animal fluids.[27] Significantly, English critical response to Zola and the French naturalists frequently expressed its disgust with details of "animal" functions by resort to medical terms, and dubbed the naturalist phenomenon "disease in fiction," although so far as I know Lilly was the only committed antivivisectionist among the critics.[28] Smith has noted in his study of the contagious diseases agitation that "Notions of obscenity took on a prurient clinical connotation which I do not find in obscenity allegations earlier in the century."[29] Clearly the whole problem of medical metaphors for what was regarded as obscene or disgusting in the later Victorian period would bear further investigation.

In sum, then, the "humanizing" of animals and the "deanimalizing" of man were coordinate aspects of a single psychic mechanism that attempted to deny and repress certain biological realities. Emphasis upon the quasihuman qualities of animals represents a means of defense, an evasion of the conflict inherent in explicit recognition of the drives projected onto animals. This recognition was

[26] W. S. Lilly, op. cit., 252. For an analysis coming to precisely the opposite conclusions, however, see D. Fleming, "Charles Darwin, the Anaesthetic Man," *Victorian Studies* iv (1961), 227–228.

[27] A recent psychiatric study suggests that opposition to health programs often comes from individuals who experience a morbid and excessive fear that the purity and integrity of their bodies will be violated. Such individuals extend similar concerns to animals, "by a simple process of extension or displacement." J. Marmor, V. W. Bernard, P. Ottenberg, "Psychodynamics of Group Opposition to Health Programs," *American Journal of Orthopsychiatry* xxx (1960), 339–340. I am indebted to Arthur Viseltear for this reference.

[28] C. R. Decker, *The Victorian Conscience* (New York, 1952), pp. 57–59, 83–84, 90–91, 96, 104, and cf. 148.

[29] F. B. Smith, "Ethics and Disease in the Later Nineteenth Century: The Contagious Diseases Acts," *Historical Studies* (Melbourne) xv (1971), 135. Cf. L. G. Stevenson, "Religious Elements in the Background of the British Anti-Vivisection Movement," *Yale Journal of Biology and Medicine* xxix (1956), 139.

conscious in earlier artistic treatments of animals, and remained so in the association of these drives in man with "animalism." By the mid-Victorian period, the "social and moral circumstances of the time" demanded a repression of the corporeal and the carnal such that these associations *even in animals,* became intolerable. Furthermore, it does not seem unwarranted to suggest that such social expectations would have pressed much more heavily upon women than upon men, and, hence, that women would have experienced the psychic consequences far more acutely than men.

There is additional evidence that indicates that psychological complications revolving around the relationship of humans to animals, especially to pets, occur more often in women than in men.[30] Modern clinical experience suggests that humans regularly use pets as substitute gratification for unfulfilled human relationships, including mother-child relationships.[31] I have found no concrete evidence that this was true for antivivisectionists, but the coincidence of the attribution of human characteristics to animals with the demands of the substitutive role is striking.

In any case, attribution of human qualities to animals was carried by some antivivisectionists to the lengths of

[30] "The relationship of patient to dog appears to have attracted the attention of analysts mostly among female patients," according to Heimann, who also speaks of the "preponderance of females or feminine attitudes in the establishment of such relationships." M. Heimann, "The Relationship Between Man and Dog," *Psychoanal. Quart.* xxv (1956), 568–585; quotes from p. 578.

[31] See ibid. and S. E. Jelliffe and L. Brink, "The Role of Animals in the Unconscious, with some Remarks on Theriomorphic Symbolism as Seen in Ovid," *Psychoanalytic Review* iv (1917), 253–271; K. A. Menninger, "Totemic Aspects of Contemporary Attitudes towards Animals," in G. B. Wilbur and W. Muensterberger, *Psychoanalysis and Culture* (New York, 1951), pp. 42–74; N. D. C. Lewis, "Some Theriomorphic Symbolisms and Mechanisms in Ancient Literature and Dreams," *Psychoanalytic Review* 1 (1963), 5–26; E. A. Rappaport, "Zoophily and Zooerasty," *Psychoanalytic Quarterly* xxxvii (1968), 565–587. The reader is referred to this literature for extensive discussion of the symbolic and substitutive role of animals, and of their function in the evasion of conflict between libidinal drives and social constraints, and in sublimation.

identification with them, producing a highly characteristic sense of martyrdom:

> It is not only a question of torturing horses and dogs and rabbits, it is a question of torturing men and women. *I* am tortured, and thousands of human beings are tortured with me everyday, by the knowledge that this infamous practice is being carried on in our midst with impunity. For my own part—and I know, but too well, that I express the feeling of a large number of my countrymen—it is literally true that the whole of my life is embittered by the existence of this awful wrong. Since I have known what vivisection is, and how it is practiced, I have moved and slept, eaten and studied, under the shadow of it, and its effluvium has poisoned for me the very air of heaven.[32]

Some antivivisectionists, such as Anna Kingsford, actually thought of themselves as choosing animals over human beings, confirming the jibes frequently expressed that members of the movement loved animals in proportion to their dislike for humans:

> I do not love men and women. I dislike them too much to care to do them any good. They seem to be my natural enemies. It is not for them that I am taking up medicine and science, not to cure their ailments; but for the animals and for knowledge generally. I want to rescue the animals from cruelty and injustice, which are for me the worst, if not the only sins. And I can't love both the animals and those who systematically mistreat them.[33]

The peculiar intensity of these sentiments produced bizarre results; Kingsford was with difficulty dissuaded from offering herself as a subject for vivisection (in Paris) on the

[32] *Anti-Vivisectionist* vi (1879), 486. Letters in this vein were extremely common in the correspondence columns of the *Home Chronicler* and the *Anti-Vivisectionist. See* e.g., *Home Chronicler* (1878), 109.

[33] E. Maitland, *Anna Kingsford. Her Life Letters Diary and Work* (London, 1896), i. 48.

condition that the practice thenceforward be abandoned. Her companion Edward Maitland explained,

> . . . so far from her being credited with sincerity in making the offer, it would inevitably be ascribed, if not to downright insanity, to an inordinate vanity and craving for notoriety since no one would believe that she expected it to be accepted. She at length yielded to my representations, but declared that, if she could not sacrifice herself for the animals in that way, she would in some other which, if less painful, would be far more protracted. How she knew it she could not say, but she did know it, and it was her destiny to perish in saving them.[34]

Anna Kingsford's almost frantic desire for martyrdom serves as a fitting epitaph for this discussion of the psychological background to the antivivisection movement. I have been trying to suggest that antivivisection provided the opportunity for a conspicuously public adherence to certain moral values, which were embodied in a characteristic set of attitudes toward animals. Kingsford's was only the extreme case of the antivivisectionist attempt to achieve catharsis of the conflicts that arose when medical science threatened the place of animals in the emotional life of the Victorians.

[34] Ibid. 1. 257–258.

12. Epilogue

The antivivisection movement did not stop experiments upon living animals; on the contrary, after 1882, experimental medicine enjoyed a spectacular growth, documented with great precision, ironically enough, in the annual returns submitted to Parliament by the Home Office and its inspectorate under the Act of 1876.[1] Thanks to the returns, this growth can be traced to the present day with an accuracy shared by no other field in no other country, for the records include annual figures on living animals utilized in experimental trials (including those used for teaching and medical technology) and on persons licensed and laboratories registered for performance of such experiments.

Experimental medicine in Britain[2] between 1880 and the present exhibits the now familiar pattern of exponential growth. If the number of experimental trials (vivisections)[3]

[1] The figures reported are taken from the returns published annually in *Parliamentary Papers*.

[2] The figures reported here do not include Ireland.

[3] The figures reported here are gross totals for all activity carried out under license, including research in experimental and veterinary medicine; teaching demonstrations (always under complete anesthesia); testing of water, food, blood samples, drugs, and commercial products such as cosmetics; preparation and testing of biological extracts, sera, and vaccines; and animal research activities in psychology. The dominant categories in a statistical sense are research, teaching demonstrations (significant for the nineteenth-century period only), and medical technology, such as preparation and testing (significant for the twentieth-century period). The overall profile presented is an excellent indicator of the growth of experimental medicine, but the absolute figures mask a variety of related activities that cannot be completely disaggregated. Antivivisectionists occasionally

be used as a parameter (Figure 16), the growth can be seen to take place in three main phases, each phase having a somewhat lesser rate of growth than its predecessor. The first phase, from 1880 until just before the outbreak of World War I, saw the number of annual experiments rise from 311 in 1880 to a peak of 95,731 in 1910. There followed a period of arrest, with the second phase beginning around 1922, when the figures again rose above the level of 1910. This second phase, including a brief check in the early 1930s during the depression, continued until the outbreak of World War II, reaching a peak of 958,761 experiments in 1938. After a brief respite during the early years of the war, growth began again in 1943 when the total experiments numbered more than one million for the first time. Growth has continued since the war, though at a lower rate than previously, to current levels on the order of five and one-half million experiments annually.

Roughly similar patterns of increase are observed if the size of experimental medicine is measured by numbers of research personnel licensed[4] or by numbers of laboratories registered[5] (Figures 17 and 18), although these data do not break down as clearly into separate phases as do those

---

claimed that researchers reported identical trials on a number of animals as one experimental trial or one vivisection. I have found no evidence that this was the case. It seems certain that the figures represent numbers of living animals utilized.

[4] The figures reported include all licensees. Since up to 1932 licenses were awarded on a yearly basis only, and since it was easier to renew an existing license than to apply for a new one, many licensees would renew even though they did not plan to be active in a given year. With the exception of the war years, between sixty-five and eighty per cent of the current licensees were active in any given year.

[5] The figures available are for "registered places," indicating locations where animals may legally be experimented upon or confined. A registered place may include several rooms and may be utilized by a number of licensees. There may be a great many registered places in a single institution or even in a single department of an institution. Nevertheless, the figures clearly indicate trends in growth and expansion of institutions. Data on registered places are not available for the years 1916 to 1918.

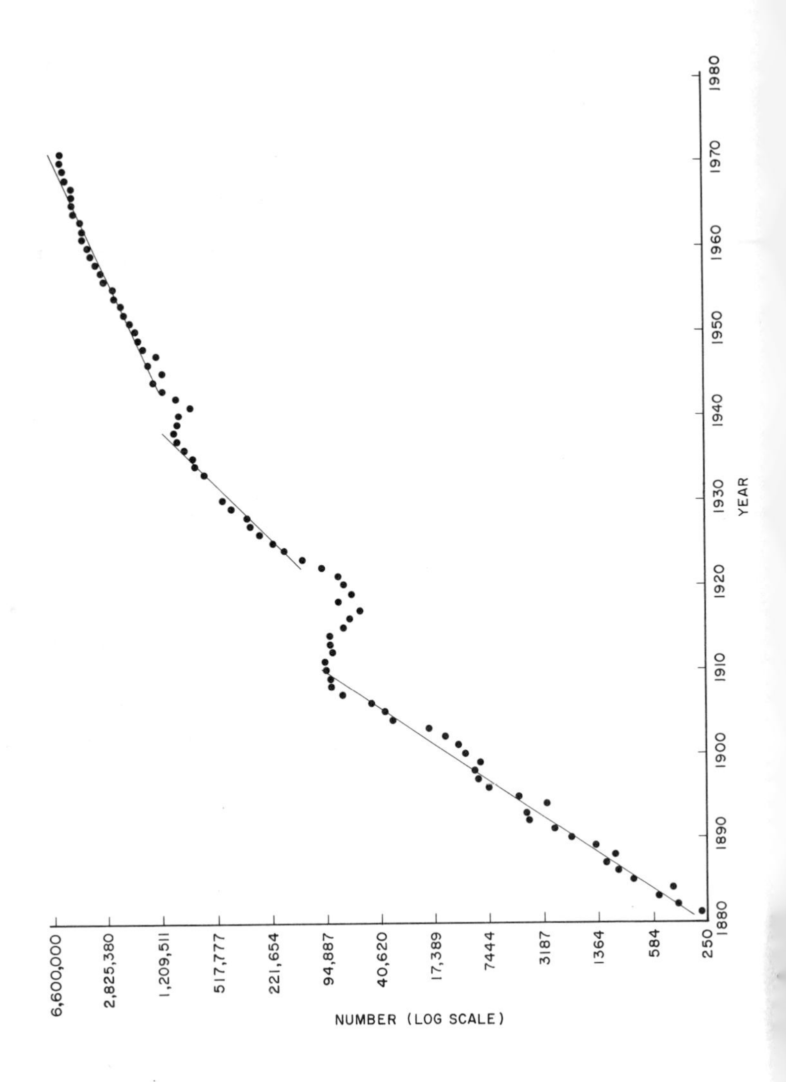

Figure 16. Number of vivisections in Britain between 1880 and 1971.

for numbers of experimental trials. Licensees increased in number from 33 in 1880 to 542 in 1910, to 2,152 in 1938, and to 15,638 in 1971. Registered places were reported as 14 in 1880, 85 in 1910, 404 in 1948, and 596 in 1971.

This kind of statistical analysis of the size of science was introduced by Price, whose studies of numbers of scientists publishing, numbers of publications, amount of research funding, and similar parameters indicate that the growth patterns for experimental medicine in Britain are not atypical. Price's work shows that general measures of the size of science usually show that it "doubles" in size every ten to fifteen years, doubling periods becoming as long as twenty years when parameters more selective for the quality of science are used.[6] Whereas in Europe, the typical doubling period turns out to be around fifteen years, in America it has been ten, in Russia around seven, and in China around five. Price interprets these data in terms of the increasing speed with which scientifically underdeveloped countries may adopt and promote science that has matured in other countries.[7] In the light of the lassitude of British experimental medicine during the period 1840 to 1870, when the subject was developing rapidly in France and Germany, it is interesting that during the first phase of growth in Britain for which precise evidence is available, 1880 to about 1910, the rate of growth was extraordinarily rapid. Depending upon the endpoints and parameters selected, doubling periods during the first phase of growth were between around 3.5 years and around 7.5 years. Obviously, the accomplishments of the likes of Pasteur, Bernard, Ludwig, and Koch provided a firm base from which British experimental medicine, once given appropriate resources, could "expand into a scientific vacuum."[8] Doubling periods for the second and third phases of growth in Britain were rather longer than those

[6] D. J. de Solla Price, *Little Science, Big Science* (New York, Columbia Paperback ed., 1965), p. 6.

[7] D. J. de Solla Price, *Science Since Babylon* (New Haven, Yale Paperback ed., 1962), pp. 110–112.

[8] The expression is Price's.

Figure 17. Number of research personnel licensed to perform vivisections between 1880 and 1971.

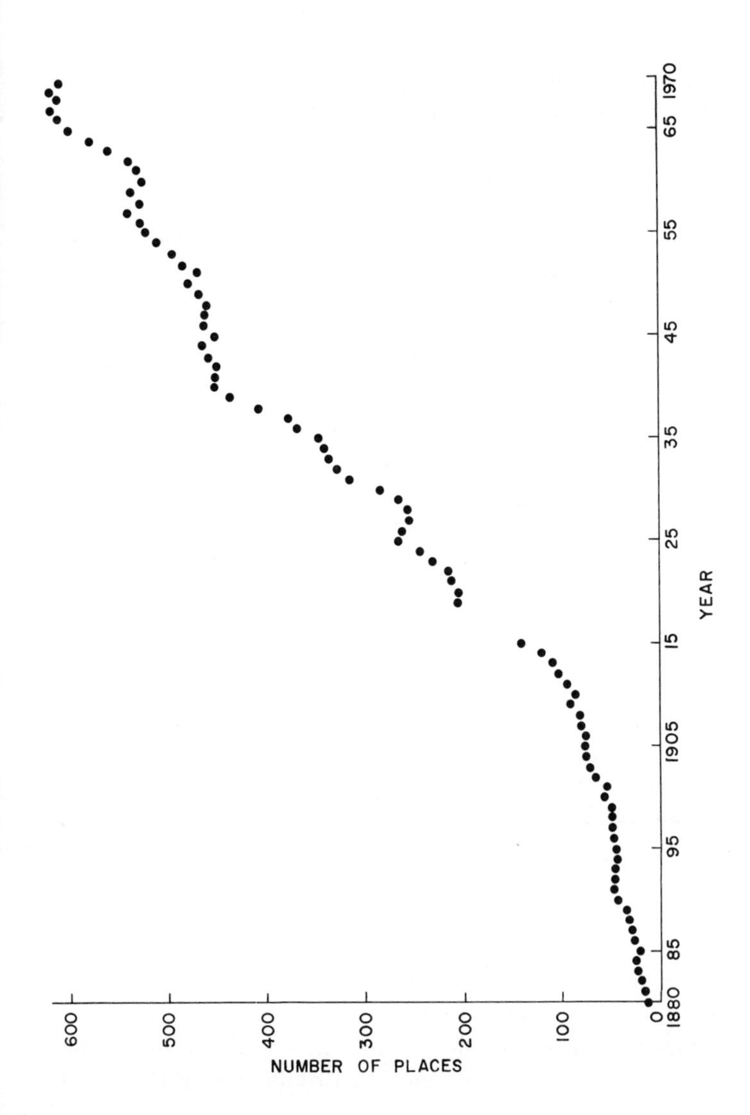

Figure 18. Number of places registered for vivisection between 1880 and 1970.

for the first phase, with post-World War II growth rates being quite typical of those found by Price for Western science in general.

Incidentally, it may be noted that these data confirm Price's conclusion that the effect of war upon science is to displace the growth curve sideways.[9] However, the displacement in this case does not coincide exactly with the beginnings and ends of the wars, nor does it fail to affect the relative rates of growth before and after displacement, in contrast to Price's findings. Falterings in the growth curve during the nineteenth century were probably the result of institutional and internal scientific factors, while the chronology and extent of twentieth century displacements indicate that other extrinsic factors besides war, notably, perhaps, economic conditions and policy, must be considered in accounting for the patterns of growth of a particular scientific field. It may make sense to see the three phases of growth in this case as three separate "life cycles" of a field, rather than as a single curve.[10]

The composition of the growth curve for British experimental medicine may be disaggregated, in order of increasing numerical significance, into physiology, pharmacology, and pathology (for "pathology," include bacteriology and immunology). H.O. returns break down gross figures on experimental trials into these three disciplines for the years 1887 to 1916. As Figures 19, 20, and 21 show, the three disciplines show fairly similar patterns of growth, with pharmacology and especially pathology dominant in a statistical sense. An attempt to make sense of the particular fluctuations shown in these graphs would have to explore in more detail than is possible here the institutional and intellectual factors relevant to the fields in Britain during the period in question.

In physiology, virtually all of the experimental trials in Figure 19 were for research and for teaching demonstra-

[9] D. J. de Solla Price, *Little Science, Big Science*, pp. 17–18; *Science Since Babylon*, pp. 102–104.

[10] Cf. D. J. de Solla Price, *Little Science, Big Science*, pp. 19–30; *Science Since Babylon*, p. 104.

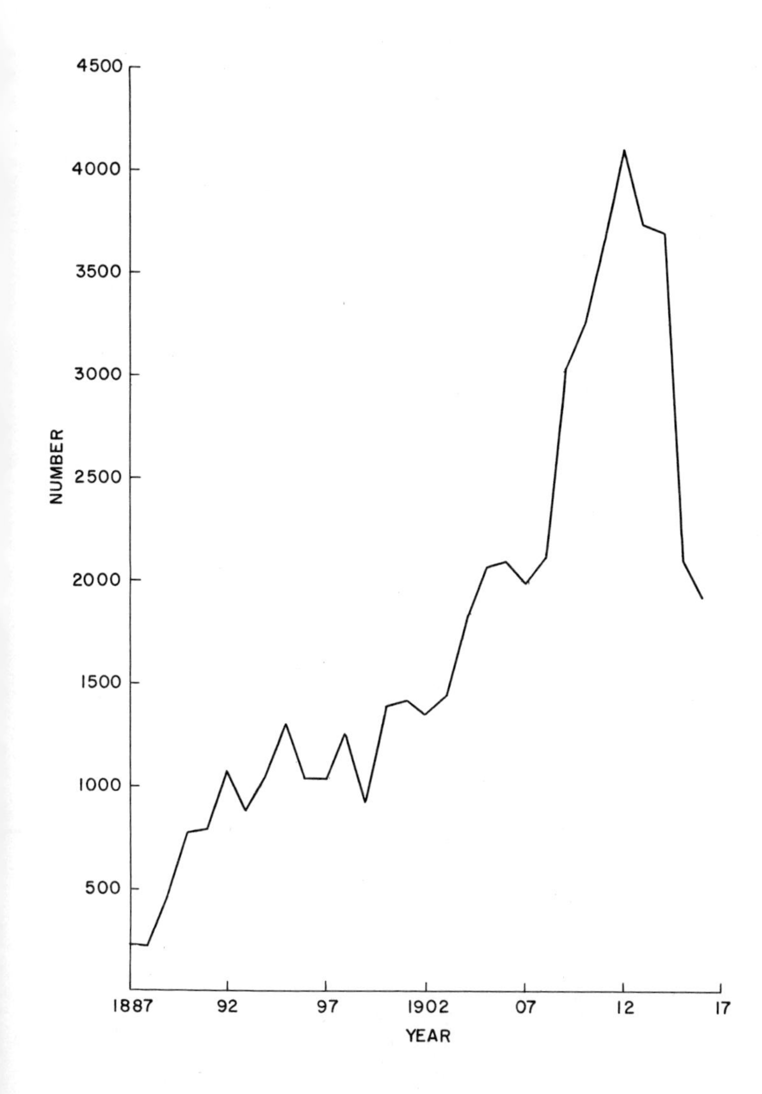

Figure 19. Number of vivisections in physiology between 1887 and 1916.

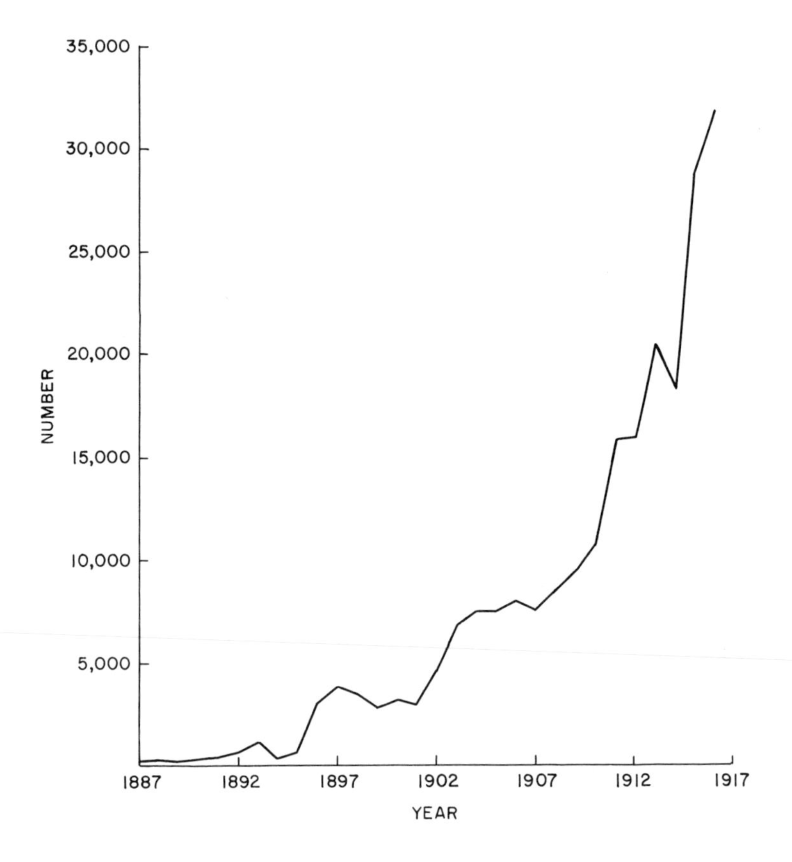

Figure 20. Number of vivisections in pharmacology between 1887 and 1916.

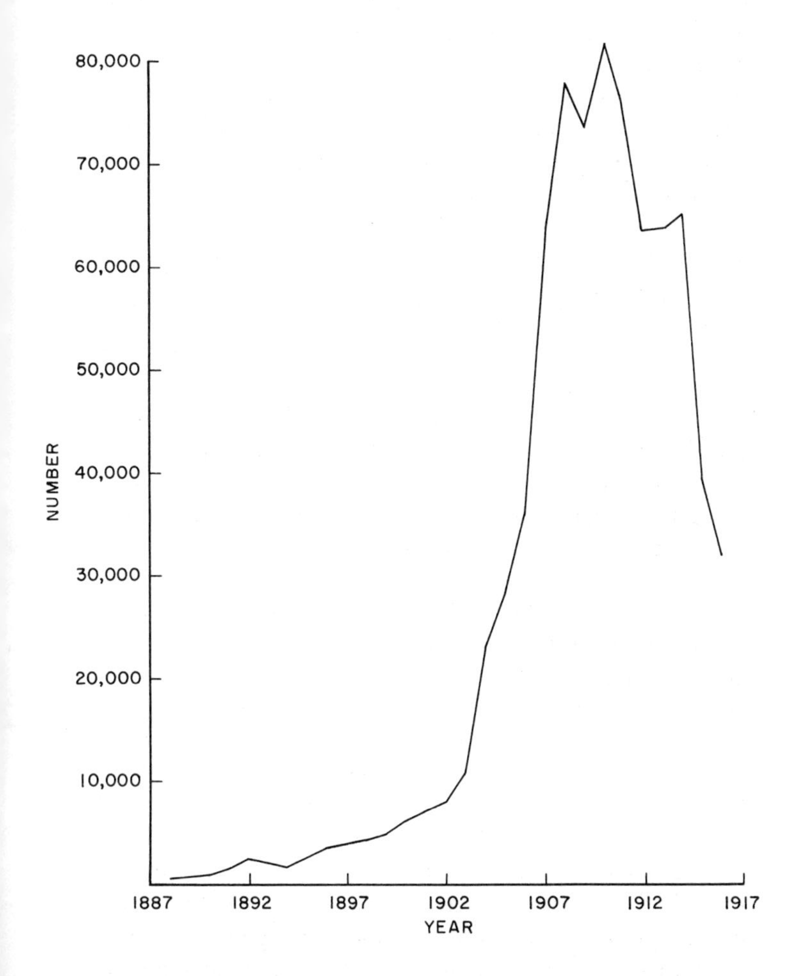

Figure 21. Number of vivisections in pathology between 1887 and 1916.

tions. The second generation of British physiologists, men like Charles Sherrington, E. H. Starling, and William Bayliss, brought their research to a point where, by 1900, their country was regarded as leading the world, and continental students were coming to work in British laboratories. British physiology has retained very significant influence, through the twentieth century, in such fields as the electrophysiology of nerve and muscle.

During the first phase of growth, Britain cannot be said to have been in the van of research in experimental pharmacology or experimental pathology. The contribution of Lister to surgical technique was vital, of course, but, although some other useful work was done by John Burdon Sanderson, Emanuel Klein, Thomas Lauder Brunton, and some of their colleagues, students, and successors, the most significant scientific advances were made on the continent. Here the validity of the germ theory of disease—which held that specific microscopic organisms were responsible for specific infectious diseases—was demonstrated during the sixties, seventies, and eighties by the work of Pasteur, Villemin, Koch, Cohn, and others. With the rival miasmatist theory laid to rest, bacteriologists rapidly began the systematic examination of pathogenic micro-organisms, mechanisms of infection, methods of treatment and of prevention. Silkworm disease, tuberculosis, anthrax, cholera, erysipelas, rabies, diphtheria, and typhoid were among the diseases that were identified with specified organisms and were investigated intensively. In the process, the scientific foundations for modern public health and preventive medicine were laid. Pasteur's research upon resistance to infection instigated the search for immunologic agents that could be used to prevent or abort the progress of infectious diseases. Ehrlich and others sought similar specifics, both biological and pharmacological. The increasing knowledge of the etiology of contagious diseases began to bear fruit in the last decade of the nineteenth century. At its beginning, von Behring and Kitasato announced their antidiphtheria serum. Pasteur began treatment of rabies. In pharmacology, Ehrlich and Hata initiated the era of chemotherapy with the discovery of arsphenamine as a specific for syphilis, introduced clinically in 1910. These accomplishments had enormous public impact. Biological products such as sera, vaccines, and antitoxins, and chemical ones such as arsphenamine could be administered to attack pathogenic micro-organisms selectively. Thus physicians no longer had to rely upon the old pharmacopial standbys that, if they had significant effect at all, tended to treat symptoms rather than diseases. The new prophylactic and therapeutic agents could be seen to prevent and cure disease.

If British scientists were not in the lead in these developments, nevertheless sufficient remained of the tradition of Jenner, Lister, and the sanitary administrators that Britain was able to move rapidly to take advantage of the continental advances, and this is reflected in Figures 19 and 20. The Lister Institute of Preventive Medicine was founded in 1891, and by 1894 it was producing diphtheria antitoxin.[11] During the mid-nineties, the first private research laboratories were being founded.[12] Steadily, an institutional infrastructure of university laboratories, research institutes, public and private funding bodies, hospitals, and pharmaceutical companies arose to support research, and to produce and deliver the new therapeutic agents. Experimental medicine came of age.

The twentieth century saw continued expansion in immunology and pharmacology, with Britain taking a major role. Large scale research and manufacture of chemical substances, presaged by the success of arsphenamine, began in earnest after Domagk's introduction, in 1932, of sulfanilamide against streptococcal infections. During the thirties, the concept of deficiency diseases was developed, and tremendous activity in the areas of vitamin supplements and hormone therapy followed. There were similar breakthroughs in the forties and fifties in antibiotic therapy. All these developments involved the use of enormous numbers

[11] A. Miles, "The Lister Institute of Preventive Medicine, 1891–1966," *Nature* ccxii (1966), 559–562.

[12] See, e.g., W. Dowson, *The Wellcome Physiological Research Laboratories, Founded 1894* (London, n.d.).

of animals. Compared to physiology (at least before nutritional research), pharmacological and pathological research required many more subjects, for frequently investigations turned upon statistical results for experimental versus control groups, rather than observation of specific functional phenomena. In addition, however, drug therapy, bacteriology, and immunology also demanded thousands of animals for producing biological extracts and for testing or standardizing such extracts, drugs, commercial products, clinical blood samples, water, and food. This medical technology began to be statistically significant around the turn of the century, and has mushroomed since that time. H.O. figures provide a flimsy basis for estimating the proportion of vivisections for which medical technology was responsible between 1900 and 1938;[13] it was in the area of twenty to thirty per cent before the lateral displacement of the growth curve about the time of World War I. During the first period of arrested growth it reached levels of up to eighty per cent of all activity utilizing animals. When growth resumed in the later twenties and thirties, around sixty per cent of the activity was medical technology, and it seems likely that present day proportions are at least of this magnitude, although after 1938 the data are unavailable or incomplete. Thus, the growth represented in Figure 16 represents both basic research and its technological spin-off.

The development of experimental medicine and its associated technology demanded institutional expansion, trained personnel, and allocation of resources on a major scale. It involved an intensification of the activity of individual

[13] From 1900 to 1938, one can guesstimate the medical technology component by totaling the figures for government research and testing with these for testing and preparation of sera, drugs, vaccines, etc. The figures are imprecise because (i) the government figures include an unknown proportion of research activity (including, it seems, the activity of the Medical Research Council after it was founded); (ii) after 1910, the figures for these categories were reported only to the nearest thousand. The percentages presented here may, therefore, be inaccurate by as much as several points, probably on the high side.

licensees, an increase in the number of licensees per registered laboratory, and increased capacity and usage of laboratory facilities. In 1880, the average licensee performed 12 vivisections, in 1969, it was 586. In 1880, there was an average of somewhat over 2 research workers per laboratory. The comparable figure for 1969 was 22.7. In 1880, the number of vivisections per laboratory was 22; in 1969, it was 8,789.

Antivivisection, then, did not end experiments on living animals. Indeed, the interesting question is why the movement flowered when such experiments were relatively few, and why it faded into insignificance while they grew enormously.

The highlights of that growth, which have been touched upon here, provide one kind of straightforward and perhaps somewhat circular response to the paradox. It was precisely the triumphs of bacteriology, immunology, and pharmacology, producing therapeutic tools of unprecedented power and discrimination, that finally provided convincing proof of the practical medical utility of vivisection as a research methodology. To a lay public increasingly preoccupied with bodily health, such a demonstration was extremely persuasive, as it was to most of those rank and file medical practitioners who had hitherto been indifferent or sceptical toward the claims of experimental medicine. On the simplest reading, the decline of antivivisection was in direct proportion to the success of the experimental approach, and it was symbolized by the tremendous resources placed at the disposal of medical research by a grateful public.

To stop here, however, is to miss the greater part of the point of this book, which is that science and medicine are ramified so diversely into the social fabric, influenced and utilized by such a variety of interests and sectors, that their history cannot be written on the assumption of a linear relationship between, say, something called "science" and something called "society." The key to the flowering of antivivisection during the last three decades of the nineteenth century lies not in the relative degree of conviction with which the therapeutic utility or scientific validity of vivisection could be argued at various times. Many antivivisectionists wanted to avoid such technical questions altogether, and those who did not frequently found that their tedious discussions of the subject went begging for a response from the scientific and medical interest. The resulting discussion was sterile, indeed aborted. Nor does an interpretation that sees the movement's heyday solely in terms of a response to the threat of cruelty to animals answer the case. Obviously, both of these factors were highly significant, but their manifest prominence should not obscure the number of other equally important factors that arose from the changing cultural milieu of later Victorian England.

The movement transcended the issue of vivisection. It arose from, and served as a palpable focus for a disparate array of social, cultural, and political forces. The ostensible purpose of the movement, preventing the infliction of pain upon animals by medical scientists, by no means comprehended all of the possible motives for an individual's participation. Antivivisection became a rallying point, a nodal issue that provided opportunity for a very public commitment to higher principle by people from a diversity of social, religious, and intellectual orientations. It was the strength of the movement that it could attract such a variety of people, while it was the heterogeneity of their motivations that confined the movement's literature and its institutional commitments solely to the straight and narrow of the attack on experimental medicine. It was the potential for fragmentation of the movement that induced the obsession with total abolition as an acid test for antivivisectionist reliability. The movement was caught. It could not develop a larger constituency without transcending its public image of fanatical monomania, but any attempt to do so threatened its preexistent power base. At the root of these tensions was the peculiar intensity of psychic investment in the movement by the faithful. In the leadership, one sees this intensity not only in the frantic rhetoric but also in the personal ambition exposed when a Frances Power Cobbe ruins the social propects of an Anna Kingsford in the interests of her own preeminence in the movement. For the rank and file, one sees the constant and rather pathetic affirmations of self-importance, moral superiority, and religious exemplitude reiterated again and again in the correspondence columns of the movement's periodicals. In a fundamental and important way the actual possibility of achieving a political solution, of making experiments on living animals illegal, was not more meaningful to participants in the movement than participation per se. To take a moral stand was an end in itself. Antivivisection was an exercise in "expressive politics," where "the main payoff . . . is that of a psychological or emotional kind—in satisfactions derived from expressing personal values in action."[14] As Frances Power Cobbe put it,

> We all believe in the saying (which I would almost take for our motto), that Mercy's battle, like that of Freedom, when—
>
> *"Once begun,*
> *Though often lost, is always won."*[15]

This book has been devoted in part to an exploration of some of the kinds of personal values expressed by antivivisectionist activity, and of some of the forces that helped to make the movement what it was. Each reader will find certain of these factors of more interest than others: the expanding role of women in public life or the traditions and tactics of popular agitations, the changing concepts and practice of state administration or Victorian psychosexual attitudes, etc. Inevitably, some factors, such as the discussion of intuitive versus utilitarian ethical philosophies, which was of such interest to Frances Power Cobbe, have been given short shrift. This book is too long as it stands, and rather than attempt an exhaustive study of every relevant issue, I have sought to highlight what I take to be

[14] F. Parkin, *Middle Class Radicalism. The Social Bases of the British Campaign for Nuclear Disarmament* (Manchester, 1968), p. 2. Cf. pp. 1–40.

[15] *Zoophilist* iv (1884), 72.

the principal ones, to emphasize the complexity of the problem, and to lay the basis for further exploration of particular issues should the reader become intrigued. My choice has been predicated upon the idea that the historical interest attaching to the movement involves not cruelty to animals so much as the tensions surrounding the roles of science and medicine in society.

Victorian England was profoundly shaken by the emergence of science as a major influence and a leading institution. The concern was multidimensional: what was the appropriate cultural role for science, what were its religious implications and its institutional perquisites? On what range of issues were its methods germane and its leaders legitimate spokesmen? What place should it hold in university and school curricula? The antivivisection movement provided for practical embodiment of some part of this largely free-floating but nevertheless genuine concern, much as has activism over the quality of the environment for some not dissimilar concerns in the present day. Thus, the rise of science to popular prestige, institutional strength, and broad intellectual pretensions generated deep suspicions in parts of the intelligentsia, some of whom found in antivivisection an instrument for a political counteroffensive against scientists. Similar sentiments today express themselves rather less directly, though they are equally manifest in declining enrolments in science courses and stagnating budgets for scientific research. In neither case will any real understanding emerge until there is an appreciation of the degree to which particular issues, such as vivisection, fail to comprehend the complex spectrum of motivations temporarily manifested in them.

The political threat to experimental medical science in England was a real one and it was vitally important in uniting the research community and stimulating the development of a political infrastructure within the community. Antivivisection brought the elite of the medical profession publicly into a virtually unanimous endorsement of experimental medicine. Medical politicians, Ernest Hart for example, aroused the rank and file of the profession, developing a power base from which a few leaders could negotiate in camera with the politicians. There followed a classic demonstration of the conciliatory, informal, insider politics that best permits the scientists to exercise the leverage of expertise and to exploit the politicians' usual deference toward them. The result was a measure ultimately administered to protect experimental medicine rather than restrict it, under which research upon living animals prospered as never before, and transformed the theory and practice of medicine.

This characteristic political style of science, which relies heavily upon the relations between the scientific and the political elites, has served the scientific community well from the later nineteenth century almost to the present day. Quite soon, experimental medicine outgrew its reliance on the political muscle of the medical profession, and traded on its own scientific successes. Other sciences were able to justify themselves in fairly similar terms: a confounding of "scientific freedom" with vague but insistent promise of a cornucopia of practical benefits. As the scientific community mushroomed in size, however, one aspect of its style emerged as a potentially significant flaw. Little if any reliance was placed upon the commitment to political action of individual, working scientists. As long as politicians retained their typical awe of science, this factor was relatively insignificant. There are unmistakable signs in a number of the Western democracies that that era of political deference is over, and that science must begin to develop the kind of bargaining leverage that depends upon the mobilization of individual members of the profession—tactics previously eschewed by science, or at least held at arm's length. As the rhetoric of practical utility and national prestige sounds increasingly hollow to the hard-pressed guardians of the public purse, newly impressed by general public scepticism and a marked hostility to science in some cultural quarters, the scientific community faces a crisis of political tactics and leadership. Although it shows little sign of recognizing it yet, the community must choose between clinging to its traditional political style or painfully developing a power base to support interest-group tactics similar to those already exercised by most other professions. The latter option is fraught with difficulties, but the choice of the former guarantees that science will be increasingly subject to the vagaries of popular opinion and at the mercy of the vast policy analysis resources commanded by modern governments. The scientific community is being pushed from its lengthy tenure of a privileged position near the centers of power. Its members see the results, but few of them understand the process, much less how to neutralize it. The successes of 1876, 1882, and afterward are significant not only as fascinating cases of a historically effective political style for science, but as counterpoint to the frustrations of a similar style in the present day.

Besides the question of science, its place in society, and its political resources, a second major theme running through antivivisectionist ideology was a sense of malaise signified by a sense of the decline of duty and morality in certain of the educated classes. In fact, with benefit of hindsight, one can see that the matter was more complex and more extensive than most members of the movement realized. For what they were in fact protesting were certain characteristic features of the modern profession, in this case the profession of medicine. The complex of legal and institutional changes that gave birth to a profession of medicine in Britain during the sixth and seventh decades of the nineteenth century involved the demarcation of bona fide members of the profession from would-be members, who were regarded by the majority as quacks and charlatans. To effect this demarcation, to police the new profession, very substantial power and influence were placed in the hands of the medical leadership, a leadership dominated by eminent London practitioners and their professorial colleagues in the ancient universities. One method of separating the sheep from the goats, common to all professions, was the development and guardianship of an esoteric body of knowledge and technique, with access confined to genuine initiates. A second and related aspect of the process involved the articulation of a set of professional ethics and values. The leaders of the Victorian profession saw in experimental medicine a rigorous and demanding intellectual underpinning for the profession, one guaranteed to expose the incompetent and to awe the laity by its "scientific" cachet and by a revolutionary new therapeutic technology. Thus the medical elite embraced the experimental disciplines and sanctified their promotion as one of the primary duties of the profession.

Antivivisectionists greeted the results of professionalization with dismay and indignation. Not for them the dogma of scientific medicine nor the uniform ranks of its newly trained proponents, with their hideous vivisectionally-based "science cures," such as vaccination. They yearned for the days before a collusive "trades unionism" isolated patients from their doctors, the days before the imperatives of technique distorted the humanity of the profession, the days when medical men were more variegated in approach, more personal in manner, and hence, somehow more humane. The antivivisectionist indictment was thus based upon much more than medicine's support for experiments upon living animals. It was in fact a prescient foreshadowing of current critiques of the impersonality, the ethical insularity, not to mention the hypertrophied scientism, of modern medicine. As such, it may occasion reflections similar to those inspired by the case of science: to wit, what is the future of the assumptions and practice of a huge basic medical research enterprise in the face of persistent public dissatisfaction and the increasing leverage of lay people upon the profession, through the health bureaucracy and even the courts?

Both science and medicine find themselves newly threatened by the combination of a resurgence of persistent attitudes of various and complex origins, and far-reaching economic, institutional, and political changes in their professional environments. Thus far their leadership has succeeded in largely ignoring the diverse veins of criticism that seem always to lie close to the surface of public opinion. Can they afford to continue?

A final word on antivivisection. The movement liked to see its cause as linked to fundamental questions about the future of society, and in the broadest sense, its spokesmen were not wrong. For the movement saw in science the agent that shattered an idealized past, the symbol and cause of the anomie generated by rapid change. In the last part of the last chapter of his memoirs, Stephen Coleridge held science responsible not only for "a gradual decay of the sense of duty in all classes" and "a steady disappearance of great men in all walks of life," but also the evils of mass production, Post-Impressionist art, the vulgarization of the English language, the Black Country, the degradation of craftsmanship, the indignity of modern warfare, the death of religion, and the lack of repose in modern life.[16] Coleridge's diatribe will strike some as silly, but it makes the point. Antivivisectionists foresaw the cold, barren, alienation of a future dominated by the imperatives of technique and expertise. It was not experiments on animals they were protesting, it was the shape of the century to come.

[16] S. Coleridge, *Memories* (London, 1913), pp. 228–239.

Appendix I

Report of the Committee appointed to consider the subject of Physiological Experimentation*

i. No experiment which can be performed under the influence of an anaesthetic ought to be done without it.

ii. No painful experiment is justifiable for the mere purpose of illustrating a law or fact already demonstrated; in other words, experimentation without the employment of anaesthetics is not a fitting exhibition for teaching purposes.

iii. Whenever, for the investigation of new truth, it is necessary to make a painful experiment, every effort should be made to ensure success, in order that the suffering inflicted may not be wasted. For this reason, no painful experiment ought to be performed by an unskilled person with insufficient instruments and assistance, or in places not suitable to the purpose, that is to say, anywhere except in physiological and pathological laboratories, under proper regulations.

iv. In the scientific preparation for veterinary practice, operations ought not to be performed upon living animals for the mere purpose of obtaining greater operative dexterity.

Signed by:— M. A. Lawson, Oxford.
G. M. Humphry, Cambridge.
John H. Balfour, Edinburgh.
Arthur Gamgee, Edinburgh.
William Flower, Royal College of Surgeons, London.
J. Burdon Sanderson, London.
George Rolleston, *Secretary,* Oxford.

(Members of the committee who did not sign the report were Michael Foster and Professors Redfern and Macalister. I have been unable to discover any reason for their omitting to do so.)

* *Report of the British Association for the Advancement of Science* (41st meeting, Edinburgh, 1871; London, 1872) p. 144.

Appendix II

Extract from Dr. George Hoggan's letter to the *Morning Post,* 2 February 1875:

I venture to record a little of my own experience in the matter, part of which was gained as an assistant in the laboratory of one of the greatest living experimental physiologists. In that laboratory we sacrificed daily from one to three dogs, besides rabbits and other animals, and after four months' experience I am of opinion that not one of those experiments on animals was justified or necessary. The idea of the good of humanity was simply out of the question, and would be laughed at, the great aim being to keep up with, or get ahead of, one's contemporaries in science, even at the price of an incalculable amount of torture needlessly and iniquitously inflicted on the poor animals. During three campaigns I have witnessed many harsh sights, but I think the saddest sight I ever witnessed was when the dogs were brought up from the cellar to the laboratory for sacrifice. Instead of appearing pleased with the change from darkness to light, they seemed seized with horror as soon as they smelt the air of the place, divining, apparently, their approaching fate. They would make friendly advances to each of the three or four persons present, and as far as eyes, ears, and tail could make a mute appeal for mercy eloquent, they tried it in vain.

Were the feelings of the experimental physiologists not blunted, they could not long continue the practice of vivisection. They are always ready to repudiate any implied want of tender feeling but I must say that they seldom show much pity; on the contrary, in practice they frequently show the reverse. Hundreds of times I have seen, when an animal writhed with pain and thereby deranged the tissues, during a delicate dissection, instead of being soothed it would receive a slap and an angry order to be quiet and behave itself. At other times, when an animal had endured great pain for hours without struggling or giving more than an occasional low whine, instead of letting the poor mangled wretch loose to crawl painfully about the place in reserve for another day's torture, it would receive pity so far that it would be said to have behaved well enough to merit death; and, as a reward, would be killed at once by breaking up the medulla with a needle, or "pithing," as this operation is called. I have often heard the professor say, when one side of an animal had been so mangled and the tissue so obscured by clotted blood that it was difficult to find the part searched for, "Why don't you begin on the other side?" or "Why don't you take another dog? What is the use of being so economical?" One of the most revolting features in the laboratory was the custom of giving an animal, on which the professor had completed his experiment, and which had still some life left, to the assistants to practice the finding of arteries, nerves, etc., in the living animal, or for performing what are called fundamental experiments upon it—in other words, repeating those which are recommended in the laboratory handbooks. I am inclined to look upon anaesthetics as the greatest curse to vivisectible animals. They alter too much the normal conditions of life to give accurate results, and they are therefore little depended upon. They, indeed, prove far more efficacious in lulling public feeling towards the vivisectors than pain in the vivisected.

Index

Library of Congress Cataloging in Publication Data

French, Richard D 1947–
Antivivisection and medical science in Victorian society.
1. Vivisection. 2. Medicine—Great Britain—History. 3. Medical research—Great Britain—History. I. Title.
[DNLM: 1. Vivisection—History—England. HV4943.G6F875a]
HV4943.G55F73 322.4'4 74-2966
ISBN 0-691-05226-3
ISBN 0-691-10027-6 pbk.